# INTRODUCTION TO VOLCANIC SEISMOLOGY

# INTRODUCTION TO VOLCANIC SEISMOLOGY

**V.M. Zobin**
Observatorio Vulcanologico, Colima, Mexico

ELSEVIER
Amsterdam – Boston – Heidelberg – London – New York – Oxford
Paris – San Diego – San Francisco – Singapore – Sydney – Tokyo

ELSEVIER SCIENCE B.V.
Sara Burgerhartstraat 25
P.O. Box 211, 1000 AE Amsterdam, The Netherlands



First edition 2003

Library of Congress Cataloging in Publication Data
A catalog record from the Library of Congress has been applied for.

British Library Cataloguing in Publication Data
Zobin, Vyacheslav M.
Introduction to volcanic seismology. - (Developments in volcanology ; 6)
1.Seismology 2.Volcanism 3.Volcanic activity prediction
4.Volcanism - Remote sensing
I. Title
551.2'1

ISBN: 0-444-51340-X

∞ The paper used in this publication meets the requirements of ANSI/NISO Z39.48-1992 (Permanence of Paper).
Printed in Hungary.

# Preface

I just returned from my first field seismological practice in Altai. It was October of 1964, and I was a 20-year old student of the Faculty of Geology and Geophysics at Novosibirsk University in Siberia, Russia. Our guys told me that one man from Kamchatka wants to see me. It was Dr. Pavel Tokarev, he had just become a chief of a new department for seismic monitoring of volcanic activity in the Insitutute of Volcanology and recruiting personnel for his department. He invited me to go to Kamchatka the following summer for my second field seismological practice at volcanoes and proposed the following theme for my year's essay: "Volcanic earthquakes and their relationship with volcanic eruptions".

I went to our University library and found a few papers concerning this problem. Amongst them was a famous paper by Takeshi Minakami (1960) about the prediction of volcanic eruptions, the papers by Georgy Gorshkov and by Pavel Tokarev about seismic study of Klyuchevskoy and Bezymianny volcanoes in Kamchatka, and that was all. No more. Thus I understood that this field of science was quite open for researchers, and I was lucky to enter to this "terra incognita".

I wrote this essay, presented it at the University student conference, received as a prize a beautiful book of "Elementary Seismology" by Charles Richter, and next summer went to my Kamchatka practice. I met a wonderful land. I climbed Avachinsky volcano, I had adventures at the flank cones of Klyuchevskoy volcano, and I decided to work there, in Kamchatka. Later, I wrote my M.S. thesis on the seismic wave attenuation beneath Klyuchevskoy volcano and was then accepted in the staff of the Institute of Volcanology in Petropavlovsk-Kamchatky. Two months later, in October 1966, I had my first eruption of the Piip flank cone of Klyuchevskoy volcano; my first publication (together with P. Tokarev and V. Shirokov) about seismic signals associated with this eruption appeared in 1968. I began my life in volcanic seismology.

I spent in Kamchatka about 30 years. In 1971, the General Assembly of the International Union of Geophysics and Geodesy took place in Moscow. It was a great opportunity to meet many outstanding volcanologists of the world. Takeshi Minakami and Izumi Yokoyama became my friends; I was happy to receive Christmas cards from them for many years. In 1975, the great Tolbachik fissure eruption began in Kamchatka. A study of seismic signals related to this eruption strongly improved our knowledge of the nature of volcanic earthquakes. I wrote a book about the source properties of volcanic earthquakes, it was published in Russian in Moscow in 1979. It was one of the first books devoted to volcanic seismology.

I have lived in México for the last nine years. I have prepared and delivered twice the courses in *Volcanic Seismology* for post-graduate students at the scientific center CICESE in Baja California, México and recently decided to write a book. You have it now.

Colima, México, March 5, 2003

This photo was taken in Moscow during the 1971 IUGG Assembly. The second in the first row (from the left) is Georgy Gorshkov, he was elected as a President of IAVCEI that year; the fourth is Takeshi Minakami. I am the fourth in the third row. The last in the fourth row is Izumi Yokoyama.

# Contents

# Chapter 1
# Introduction

> *The subject of* **Volcanic Seismology** *is not only the most beautiful and spectacular, but also the most difficult to study of all the subjects seismologists have encountered on Earth. This is because seismic sources in volcanoes involve dynamic motion of gas, fluid and solid, and propagation paths in volcanoes are usually extremely heterogeneous, anisotropic and absorptive, with irregular topographies and interfaces including cracks of all scales and orientations. Thus, volcanic seismology is the most challenging to seismologists requiring ingenuity in designing experiments and interpreting observations.*
>
> Keiiti Aki, 1982. From the Preface to the Collection of papers *"Volcanic Seismology"*, IAVCEI Proceedings in Volcanology, Vol. 3, Springer-Verlag, Berlin Heidelberg New York

*Volcanic Seismology* is a science about seismic signals originating from volcanoes and associated with volcanic activity. The study of the origin of these signals, their spatial-temporal distributions, their relationships with volcanic processes, and using them as an instrument to investigate the volcano deep structure and to predict a volcanic eruption together create the subject of *Volcanic Seismology.*

## *1.1. Terms and definitions*

The subject of Volcanic Seismology lies in the interaction between volcanic and seismo-tectonic processes. Therefore, it is necessary to define the main terms of volcanology, tectonics and seismology at the very beginning of the book.

### 1.1.1. *Volcanic terms*

A *volcano* is a vent or chimney, which connects a reservoir of molten matter known as *magma*, in the depths of the crust of the Earth, with the surface of the Earth. The material ejected through the vent frequently accumulates around the opening, building up a *volcanic edifice*. The material ejected consists of liquid lava and broken fragments of partially or completely solidified rock (pyroclastic debris), as well as great quantities of gases. The gases are the motivating force and the most important factor in volcanic action (Bullard, 1963).

Fig. 1.1 shows the main elements of a volcano. A *volcanic cone* is the result of the accumulation of ejected material around the vent. A *crater* is the surface connection of the *volcanic conduit* through which the ejected material reaches the surface. The crater may be situated on the summit of the cone (*summit crater*) or on a slope of the cone (*flank crater*). An enormous crater representing a large depression in the truncated summit of volcano is called *caldera*.

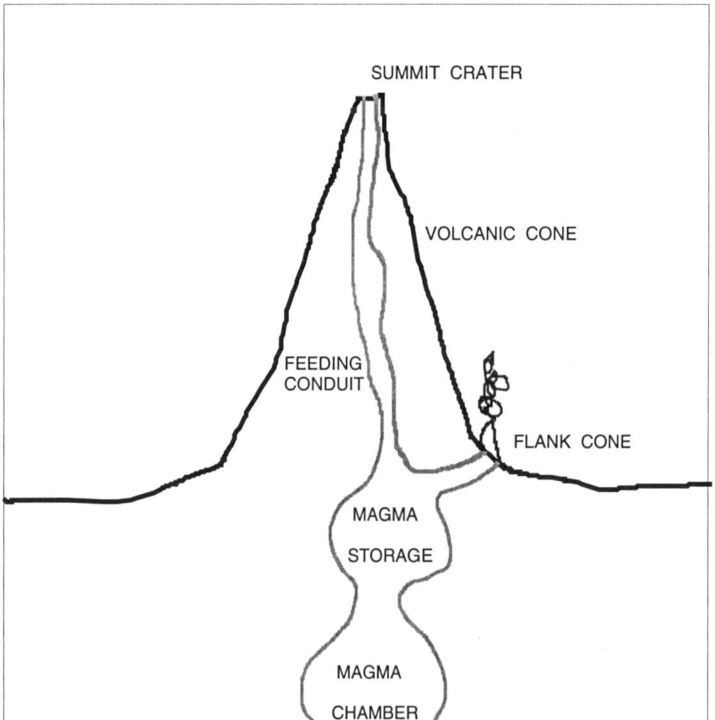


Fig. 1.1. The main elements of volcanic edifice.

Beneath the volcano, we have the magma feeding system of the volcano that consists of a deep *magma chamber* which is a reservoir of magma in the shallow part of lithosphere from which volcanic material is derived; an intermediate *magma storage*, a place between a magma chamber and the Earth's surface, where magma may be collected before an eruption; and the *feeding conduit*, connecting the magma collectors with the surface. The surface appearance of magma is called *lava* (Bullard, 1963). Volcanoes are widely distributed on the surface of the Earth and on the ocean floor (Fig. 1.2). The volcanic belts go along the *subduction zones* as well as along the *oceanic ridges*. Simkin and Siebert (1994) have noted more than 1300 active volcanoes with more than 5500 dated eruptions.

The *volcanic eruption* is a process of material transport from the Earth's depths to the surface. The magnitude of eruption is described by the *Volcanic Explosivity Index (VEI)* that combines total volume of explosive products, eruptive cloud height, descriptive terms, eruption type, duration, etc., and have the grades from 0 to 8 (Newhall and Self, 1982). Depending on the state of stress and the content of magma, the volcanic eruptions may be of three main types. The main characteristics of each type of eruption are the index of explosivity (IE), which is based on the ratio of fragmental material to the total quantity of ejecta (in %), and the index of $SiO_2$ content.

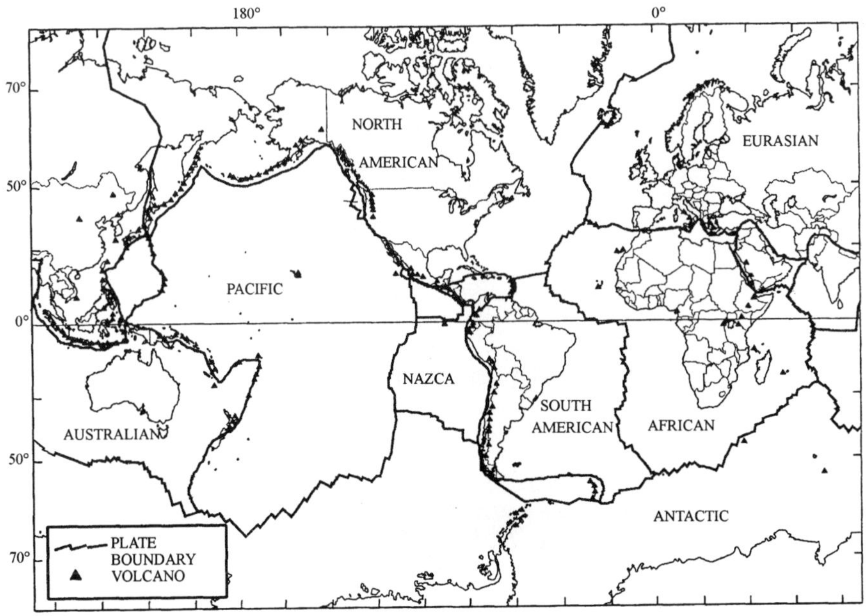


Fig. 1.2. Global distribution of volcanoes. Courtesy of U.S.G.S.

Fig. 1.3. Magmatic product of Pelean eruption: Dacitic 310-m-height spine that was formed in the crater of Mont Pelée volcano, Martinique during the 1902-1903 eruption. Courtesy of Institut de Physique du Globe de Paris.

Fig. 1.4. Magmatic product of Vulcanian eruption: A view of Bezymianny volcano, Kamchatka, showing the dome, and andesitic lava moving down the flank. Photograph by A.B. Belousov. Courtesy of Institute of Volcanology.

Fig. 1.5. Magmatic product of Hawaiian eruption: Basaltic lava fountain and spatter cone on Kilauea Volcano, Hawaii. Photograph by J.D. Griggs. Courtesy of U.S.G.S.

The first type, *Pelean*, is characterized by high IE and acidic, stiff and viscous dacitic and andesitic lavas, containing more than 66% $SiO_2$. The *Vulcanian* type of eruption is intermediate in IE and the content of $SiO_2$ (between 52 and 66%) in characteristic andesito-basaltic lavas. The last, *Hawaiian*, type is characterized by low IE (about 0) and the liquid basaltic lavas with less than 52% of $SiO_2$. Each of these types has some local subtypes. Figures 1.3-1.5 show the typical surface manifestation of each type of eruption. Gas and water are the moving agents of the eruptive process and their concentration within magma defines the type of explosions during the eruption. The volcanic eruptions may be of the *central* type with an eruption of the summit crater and of the *flank* type, with an eruption of the flank crater.

### 1.1.2. Tectonic terms

The movement of magma in the interior below the volcanic structure leads to the change in hydrostatic equilibrium and to the fracturing of the Earth material. Tectonic *faults* and *fractures* represent the breaks in rock material of different scales. We define *fractures* as the ruptures of length less than 100 m; the greater ruptures would be the *faults*. As a rule, the system of fractures forms around the general faults and reflects the local rupturing of different parts of the fault. A type of movement along the fault plane characterizes tectonic faults. There are three main types of tectonic movements: normal faulting, thrust and strike-slip. The *normal fault* and the *thrust* are characterized by a low angle of inclination with reference to a horizontal plane. With normal faulting, the hanging wall of the fault has been depressed, relative to the footwall; the thrust is a reverse fault. The *strike-slip* represents the horizontal movement along the vertical plane. Volcanoes usually lie along a tectonic fault or intersection of faults, usually within a *rift* structure where normal faulting is dominant.

### 1.1.3. Seismological terms

The rupturing along the fractures and faults as well as the eruptive activity produce the seismic waves, which could vibrate the Earth surface as *earthquakes*. The site of initial radiation of seismic waves in the interior is *earthquake focus, or hypocenter*; its projection to the surface is *epicenter*.

The seismic waves are divided into *body waves* and *surface waves*. The body waves *P* (compressional waves with volumetric changes) and *S* (shear waves without volume change) radiate from the earthquake focus, the surface *Rayleigh* and *Love* waves are formed as a result of the interaction of body waves with the surface of the Earth. The record of seismic waves is a *seismogram* (Fig. 1.6). The earthquake size is described by *magnitude M (or m)*, the conventional value proportional to earthquake energy, and by *seismic moment Mo*. The *focal mechanism* characterizes the stress system acting in earthquake source zone, while the *stress drop* give the value of stress release during the earthquake.

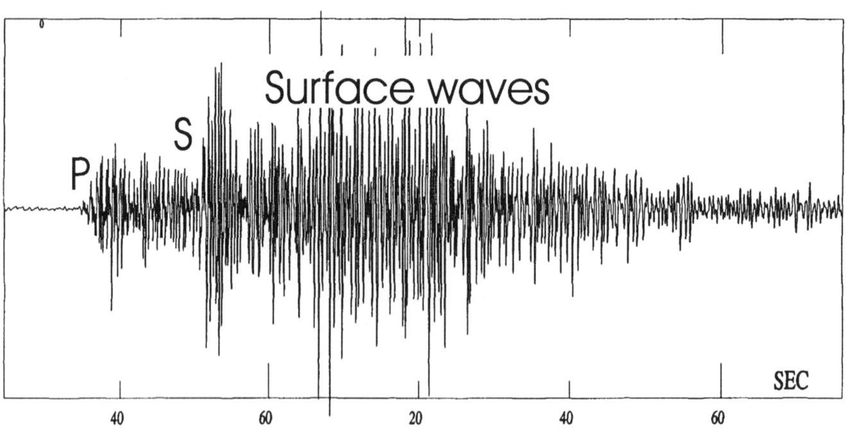


Fig. 1.6. Seismogram of shallow local earthquake. Vertical component of short-period instrument. Courtesy of Colima Volcano Observatory.

### *1.2. Subject of the book*

*Volcanic Seismology* has not yet any plain theory; it is more a science of observations and cumulative experience. This book is an introduction to *Volcanic Seismology.* It is devoted to the study of the relationships between volcanic earthquakes and eruptive process and of the nature of volcanic earthquakes.

The book consists of 15 chapters. Chapters 2 to 4 introduce us to the history and the topic of *Volcanic Seismology*; discuss the models that were developed for the study of the origin of volcanic earthquakes of both a volcano-tectonic and eruption nature.

The next three chapters give the case histories of seismic activity associated with the eruptions at basaltic, andesitic and dacitic volcanoes. For this book, I selected the 34 cases of volcanic eruptions (or episodes) of 17 volcanoes that occurred all over the world from 1910 to 1998. This selection was argued by different reasons. The most important of them was the possibility to have the catalog of seismic events for special analysis; among others, the quality of seismic observations, and the historical significance of the eruption.

Chapters 8 to 10 describe the general regularities of volcano-tectonic earthquakes, their participation in eruptive process, their source properties, and the hazard of strong volcano-tectonic earthquakes.

Then we have three chapters devoted to the description of eruption earthquakes: volcanic tremor, seismic noise of pyroclastic flows, and explosion earthquakes. I discuss their relationship to eruptive processes.

The final two chapters discuss the mitigation of volcanic hazard. I write shortly about the methodology of seismic monitoring of volcanic activity, about the experience in forecasting of volcanic eruptions by seismic methods; I discuss the unresolved problems of Volcanic Seismology.

### *1.3. Acknowledgements*

I thank my colleagues from Kamchatka, Russia and Colima, México for collaboration and useful discussions. My co-authors in the investigations in volcanic seismology Valentina Gorelchik, Pavel Tokarev, Valeria Levina, Tonatiuh Domínguez and Pavel Firstov helped me to understand better many aspects discussing now in this book. The seismic data obtained by the Kamchatkan regional seismic network and the Colima seismic network gave me an experience in the study of seismic signals recorded at volcanoes. The catalogs of volcanic earthquakes provided by Stefano Gresta (Etna volcano, Italy), Juan Manuel Espíndola (El Chichon volcano, México), Jennifer Nakata

(Kilauea volcano, Hawaii), Kodo Umakoshi (Unzen volcano, Japan), and Willy Aspinall (Soufriére Hills volcano, West Indies), as well as the Catalogs of earthquakes of Pinatubo volcano (Philippine) and Mt. St. Helens volcano (Cascades, U.S.A) prepared by the U.S.G.S., the Kamchatka Regional Earthquake Catalog, and the Catalog of earthquakes of Icelandic Meteorological Office (Grimsvøtn volcano) were very important in the preparation of this book. Izumi Yokoyama, Daisuke Shimozuru and Viktor Ivanov sent me the data and photos for biographical notes.

My work was benefited by the numerous comments made during the preparation of the papers that formed the basis for this book. Stefano Gresta, Kazuhiro Ishihara, Fred Klein, Jim Luhr, Steve McNutt, Seth Moran, Alejandro Nava, among others, helped me to better understand many aspects of volcanic seismology. Nick Varley read the book and made a gigantic work polishing my English.

I thank also all copyright holders, whom allowed me to reproduce the figures in my book: American Geophysical Union, Birkhäuser Publishers Ltd., Earthquake Research Institute of Tokyo University, Elsevier Science (Academic Press), Geological Society of London Publishing House, Istituto Nacionale di Geofisica, Seismological Society of America, Springer-Verlag, Tohoku University and University London College Press. U.S. Geological Survey allowed me to reproduce their published and Internet materials. J. Neuberg provided me the copy of seismogram. I thank M. Iguchi, M. Kasahara, S. Matsumura, H. Okada, K. Tanaka, H. Watanabe, H. Yamasato and I. Yokoyama for the kind permission to reproduce their published figures.

# Chapter 2
# Seismicity at Volcanoes

## *2.1. History of seismic monitoring of volcanic activity*

Earthquakes were associated with volcanic eruptions from ancient times. Pliny the Younger gave us the first scientific description of the eruption of Vesuvius in 79 A.D. He wrote about numerous earthquakes related to this eruption. For many years, the people considered volcanoes as the main origin of earthquakes in general. In the classification of earthquakes that was prepared in the beginning of the XX century, earthquakes of volcanic origin were in first place.

Vesuvius volcano occupies a special place in the history of the study of seismic signals associated with volcanic activity. It was the first volcano whose earthquakes were mentioned in scientific literature, it was the first volcano to have a volcano observatory in 1847, and it was the first to be monitored using seismological equipment. The Palmieri electromagnetic seismograph, built in 1862, was the first seismic instrument to record volcanic seismicity. The seismograms were written on a telegraph tape. Later, in 1903, the Bosch-Omori seismograph was installed at Mont Pelée, Martinique one year after the great eruption of 1902 (Ferrucci, 1995). Seismic observations at Kilauea, Hawaii, volcano had begun in 1912 by H. Wood just after the foundation of Hawaiian Volcano Observatory (Wood, 1913; 1915; Macdonald, 1972).

As a science *Volcanic Seismology* was born when the Japanese seismologist Fusakichi Omori had began his investigations of seismic signals related to the 1910 eruptions of Usu-san and Asama (Omori, 1911-1913; 1912) and the 1914 eruption of Sakurajima volcano (Omori, 1914; 1922). Omori defined the volcanic earthquake as "seismic disturbance, which is due to the direct action of the volcanic force, or one whose origin lies under, or in the immediate vicinity of, a volcano, whether active, dormant, or extinct" (Omori, 1912).

The 3-component seismic station, with a magnification of 100 and the period of the pendulum about 4 s, was installed by Omori in July-August of 1910 at a distance of about 1-2 km from the craterlets of Usu-san. This station allowed him to record the earthquakes originating from the volcano and the volcanic micro-tremors whose period was about 0.5 s (Fig. 2.1). The micro-tremors were observed only during volcanic activity. Omori demonstrated a good relation between the appearance of micro-tremors and explosions of the volcano.

In the same paper, Omori (1911-1913) noted also that many eruptions of Japanese volcanoes were preceded by numerous earthquakes. "In such a case, he wrote, seismograph observations made at the vicinity of the center of volcanic activity would give the people a warning of the approaching outburst". Analysis of volcanic earthquakes recorded in the course of volcanic activity by Omori had shown the great potential of seismic observations. They became the main instrument for monitoring volcanic activity. Among the most important stages in the development of *Volcanic Seismology* we can note the foundation of seismic observations at Asama Volcano Observatory on Honshu Island, Japan in 1934, and at Kamchatka Volcano Observatory in Klyuchi, Kamchatka, Soviet Union in 1946.

Asama Volcano Observatory was established by Takeshi Minakami, and from the very beginning it was equipped by continuously operated seismographs. The observations of volcanic and seismic activity at Asama volcano allowed Minakami to propose a classification of volcanic earthquakes into four types, which is the basis for the study of volcanic earthquakes up to the present day. He obtained a good correlation between the long-term variation in the number of volcanic earthquakes and volcanic eruptions (Fig. 2.2). The increase of volcanic B-type earthquakes before the volcanic outburst became the criteria for the development of the method of eruption prediction (Minakami, 1960).

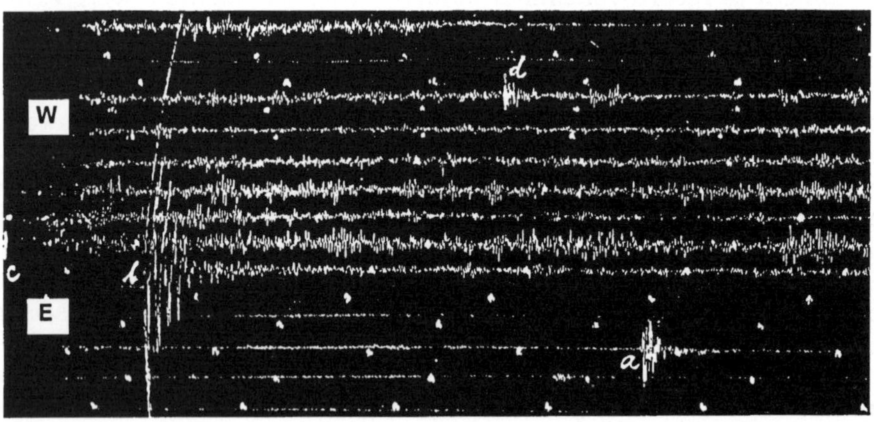


Fig. 2.1. Seismogram of volcanic earthquakes (a, b, d) and volcanic tremor (continuous signal) recorded on August 7-8, 1910 at Sobets seismic station near Usu volcano, Hokkaido, Japan, with a magnification of 100; component WE. From Omori (1911-1913).

*Prof.* ***Fusakichi Omori*** *(1868-1923) was a founder of Volcanic Seismology. He began his work in seismology with John Milne and became in 1896 the professor of Imperial (Tokyo) University and secretary of the Imperial Earthquake Investigation Committee. For decades, he was the natural leader in all Japanese investigations on volcanoes and earthquakes. Omori's instrumental observations of seismic events during the 1910 Usu, the 1910-1914 Asama and the 1914 Sakurajima eruptions showed the great significance of seismic monitoring of volcanic eruptions for the understanding of volcanic processes and forecasting eruptions. Prof. Omori left us the detailed descriptions of these seismic events; researchers still use his papers in volcanic seismology. His dearth was dramatic. D. Shimozuru wrote to me in 1999: "It is well known the dispute between Prof. Omori and Prof. Imamura about the possibility of an occurrence of a great earthquake in the Tokyo area in 1923. During his participation in the Congress at Australia, the great Kanto earthquake occurred on September 1, 1923. Prof. Omori was defeated. He was heartbroken during his return voyage from Australia, and soon after he passed away."*

The continuous seismological observations at Kamchatka Volcano Observatory began in 1946. They were headed by Georgy Gorshkov and later by Pavel Tokarev. Gorshkov was the first to discover that S-waves of distant earthquakes greatly attenuate the volcano; it allowed him to estimate the depth of upper mantle magma chamber of Klyuchevskoy volcano (Gorshkov, 1956). The great eruption of Bezymianny volcano in 1955-1956 gave numerous data for the detailed analysis of the interrelation of seismic and volcanic events. This was first complete history of a great eruption described in terms of seismic information (Gorshkov and Bogoyavlenskaya, 1965). The following analysis of the seismic activity related to the annual small eruptions of Bezymianny in 1957-1961 allowed Tokarev to develop a method of forecasting of volcanic eruptions based on a change in energy release of the preceding s

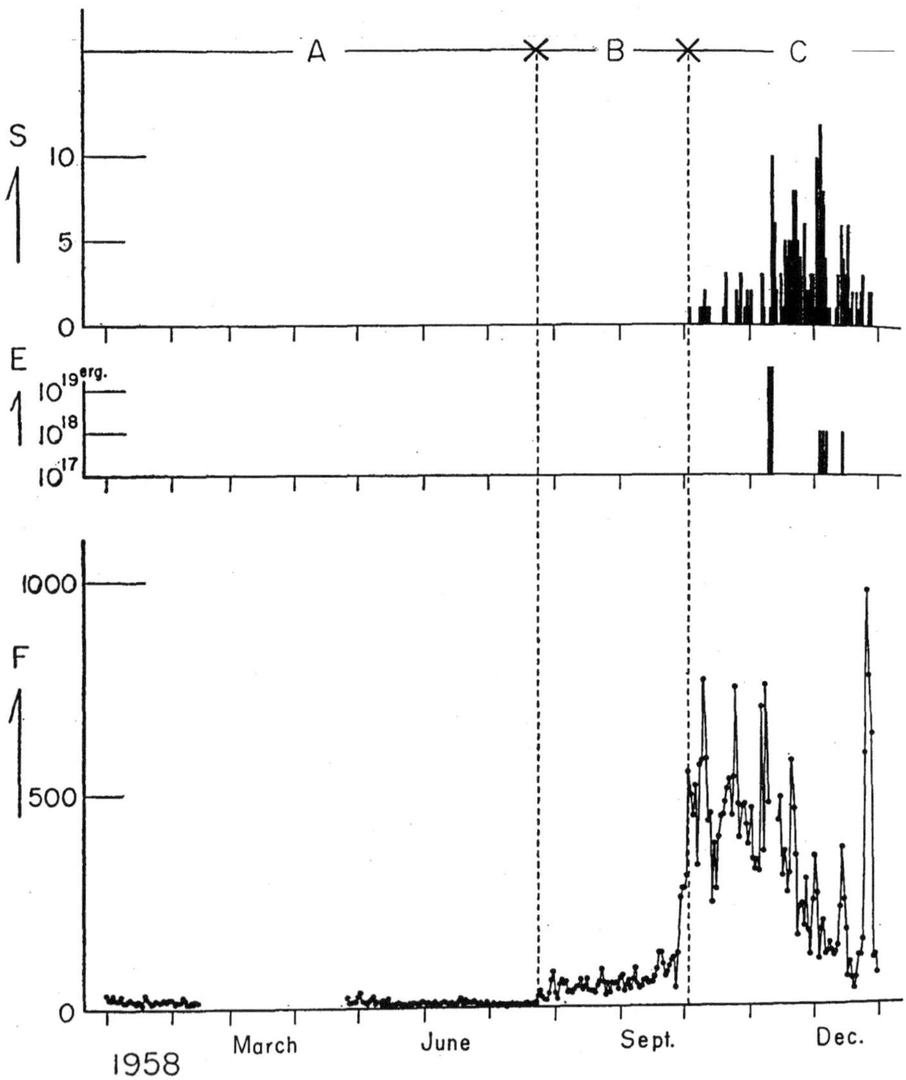


Fig.2.2. Relation between eruptions and shallow earthquakes (B-type) of Asama, Honshu, Japan. A, B, and C indicate the calm, pre-eruptive, and explosive eruption stages of the volcano. S is the daily frequency of explosions; E is the kinetic energy of strong explosions; and F is the daily frequency of shallow earthquakes. From Minakami (1960).

A network of high-gain seismometers at Hawaii Volcano Observatory has operated on Kilauea volcano since the 1950's, and up to 1985 it consisted of 50 seismic stations (Klein et al., 1987). The seismic studies allowed the construction of a fine model of the earth's crust near the Hawaiian volcanoes. The basic model of Kilauea was proposed (Eaton and Murata, 1960) with a magma reservoir at a depth of about 3-4 km beneath the caldera. The generalized histories of Kilauea earthquakes represent the detailed history of volcanic activity of Kilauea from 1970 (Klein et al., 1987). The regularities of inflation-deflation cycle of magma reservoir and the appearance of characteristic related seismic events combined with ground deformation monitoring allowed the development of the short-term forecasts of Hawaiian eruptions (Klein, 1984).

*Prof.* ***Takeshi Minakami*** *(1909-1985) graduated from Tokyo University and at the age of 25 became the first director of the first Japanese Volcano Observatory at Asama volcano. He initiated the seismological observations at Asama, and his researches of Asama volcanic earthquakes allowed him to elaborate their classification. He developed a methodology for the statistical forecastings of volcanic eruptions at Asama by seismic data. His 1960-year paper "Fundamental research for predicting volcanic eruptions" helped to turn a volcanic seismology into a science. Later T. Minakami studied seismic events associated with the eruptive activity of Usu, Miyake-sima, and Arenal volcanoes. As Professor of Earthquake Research Institute of Tokyo University he prepared a new generation of Japanese volcano-seismologists. D. Shimozuru wrote in 1971 that Prof. Minakami "obtained pioneering results, which led to great progress towards the better understanding of volcanic phenomena".*

*Dr.* ***Georgy Stepanovich Gorshkov*** *(1921-1975) entered volcanic seismology in the very beginning of his volcanological carrier. In 1948 he was appointed Director of Kamchatka Volcano Observatory in Klyuchi village. A seismic station was installed here in the end of 1946, and Gorshkov became the first who studied the seismic records of the Kamchatka volcanoes. Two of his seismological results were fundamental. He discovered the screening effect of magma on* S *waves of distant earthquakes and determined the depth of the magma chamber of Klyuchevskoy volcano in the upper mantle. He studied in detail the relationships between seismic energy release and volcanic activity associated with the great 1955-1956 eruption of Bezymianny volcano. Dr. Gorshkov worked in volcanic seismology only for about ten years. He was a petrologist, and later he worked mainly in this area of volcanology. His monographs* "Bezymianny Volcano and the Peculiarities of its Last Eruption, 1955-1963" *(1965; with G.E. Bogoyavlenskaya) and* "Volcanism and Upper Mantle" *(1969) became the important events in a volcanological science. At the end of his life, Georgy Gorshkov was elected as the President of the International Association of Volcanology and Chemistry of the Earth's Interior.*

These basic investigations had become the foundation for the construction of *Volcanic Seismology* as a science. During the last 30 years of the twentieth century, the detailed and complex studies of some eruptions (among them, New Tolbachik volcanoes, Kamchatka, 1975-1976; Mt. St. Helens, Cascades, 1980; Redoubt, Alaska, 1989-1990; Pinatubo, Philippines, 1991; Soufriére Hills, Montserrat, 1995-1996; Volcán de Colima, México, 1997-2000) strongly increased our understanding of the seismo-volcanic process. It was shown that the source of the majority of A-type volcanic earthquakes is similar to the source of tectonic earthquakes (Zobin, 1970; 1972); it was shown that the records of explosive (and B-type) volcanic earthquakes can be interpreted as Lamb pulses excited by a nearly vertical single force that represents the counter force of the eruption (Kanamori and Given, 1983; Nishimura, 1998); it was shown that shallow earthquakes in volcanic areas often have mechanisms with isotropic components, indicating volume changes of either explosive or implosive polarity (Miller et al.,1998).

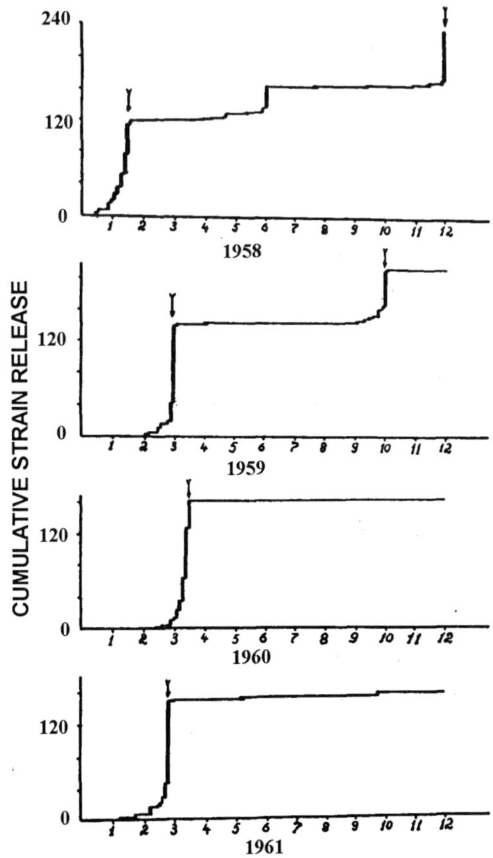


Fig. 2.3. Variations of the cumulative strain release [in $erg^{1/2}$] recorded at Bezymianny volcano, Kamchatka in 1958-1961. Arrows show the eruptions. The curves indicate an exponential increase in seismic energy release before an eruption. From Tokarev (1963).

Many models have been proposed for explaining the nature of volcanic tremor (Steinberg and Steinberg, 1975; Aki et al., 1977; Schick, 1981; Ripepe and Gordeev, 1999; etc) and long period volcanic earthquakes (Chouet et al., 1994; etc). And, of course, the development of a methodology for volcanic eruption forecasting (Tokarev, 1978; Malone et al, 1981; Harlow et al., 1996; Reyes-Dávila and De la Cruz-Reyna, 2002; etc). The installing of new broadband digital equipment at volcanoes gives a good perspective for better understanding the nature of seismic signals originating with volcanic activity (Chouet, 1996a).

*2.2. Classification of volcanic earthquakes*

The first observations of volcanic earthquakes showed that the seismic signals originating from volcanic activity had different shapes. Studying the seismic signals during the 1910 Usu eruption, Omori (1911-1913) wrote about “occurrence besides the proper volcanic earthquakes, of well defined small quick unfelt vibrations, which may be termed micro-tremors” (See Fig. 2.1). The first and still very popular classification of volcanic earthquakes was proposed by Minakami (1960; 1974a). He divided them into four types (Fig. 2.4) according to location of their foci, their relationship to eruptions and the nature of the earthquake motion.

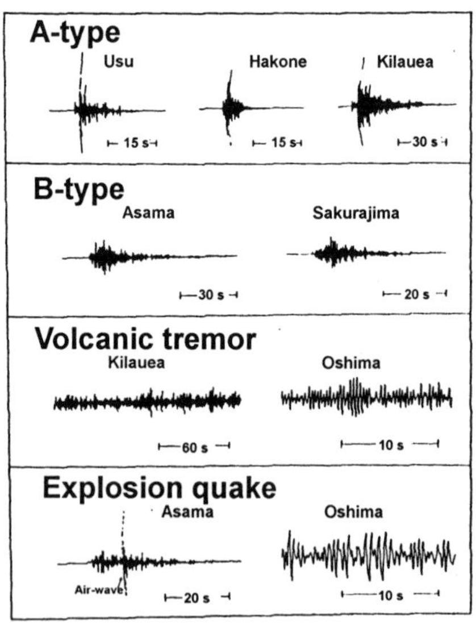


Fig.2.4. An illustration of Minakami's classification of volcanic earthquakes. From Minakami (1974a).

*A-type.* These are earthquakes originate from the bases of volcanoes or from a depth of about 1 to 20 km. These earthquakes take place previous to and in the first stage of eruptive activity and occur in swarms. They are generally less than 6 in magnitude. The nature of earthquake-motions cannot be distinguished from those of general shallow earthquakes (tectonic earthquakes). The P-phase and S-phase of seismic waves also are clearly defined.

*B-type.* The hypocenters are limited to an area of about 1 km in radius around the active crater. The hypocenters of these earthquakes are shallower than those of the A-type volcanic earthquakes. These earthquakes take place in swarms in the shallow part ranging from the earth's surface to depths of several hundred meters. Their magnitudes are generally extremely small. The surface waves are predominating and the S-phase is not clear.

*Explosion earthquakes.* These are earthquakes, which accompany individual explosive eruptions. The amplitude of these earthquakes is related to the magnitude of the explosive eruptions. The earthquake motions show an exceeding predominance of longer wavelength as comparable with those of the A-type volcanic and tectonic earthquakes; the initial motion is "push" in every direction. On the seismograms of explosion earthquakes, we often find disturbances caused by air shocks. The hypocenters are situated beneath the active crater floor.

*Volcanic pulsation or continuous volcanic microtremors.* The main part of these vibrations consists of the surface waves. Volcanic tremor has a form of an irregular sinusoid of rather long duration compared with earthquakes of the same amplitude.

Volcanic earthquakes of all four types can participate in eruptive activity. Fig. 2.5 illustrates the seismic history of the 1996 flank eruption of Klyuchevskoy volcano, Kamchatka (Tokarev et al., 1968). Prior to the eruption, volcanic earthquakes of A and B types were observed (bars 5). An eruptive fissure opened after the ceasing of the earthquake sequence. With the beginning of the eruption, volcanic tremor (curve 3) started simultaneously with explosion earthquakes (curve 2). Volcanic tremor stopped just before the second swarm of B-type earthquakes (curve 4) but explosion earthquakes continued up to the end of the eruption.

Minakami's classification stays basic and popular up to today. Of course, for every volcano, the seismologists use their own classification with more detailed description of every subtype of earthquakes (Tokarev, 1966; Latter, 1979; Malone, 1983; Lahr et al., 1994; Harlow et al., 1996; Ratdomopurbo and Poupinet, 2000). McNutt (1996; 2000a) presented his generalization of modern volcanic earthquake classification in the term of frequency (Fig. 2.6). A-type earthquakes can be of high-frequency and mix-frequency events (Fig. 2.6A, B); and B-type earthquakes can be of mix-frequencies and low-frequency events (Fig. 2.6B, C and F).

In this book, I consider two groups of volcanic earthquakes that include all these types. The first group includes the earthquakes generated by stresses, which arise during magma migration in the Earth's crust. They may be caused by shear failure, tensile failure or fluid pressurization processes and occur during magma movement from depth to the Earth's surface through conduits, dikes, etc, producing the rupture at the tip of magma body and within adjacent fault systems. I call them *volcano-tectonic earthquakes*; they correspond to high-, mixed and low-frequency earthquakes of Fig. 2.6 or A- and B-types of Fig. 2.4. Their study is a study of the physics of magma movement to the surface. These earthquakes are very important for forecasting of volcanic eruptions.

The second group includes earthquakes originated during the eruption process; I call them *eruption earthquakes*. They include explosion earthquakes, various types of volcanic tremor, and earthquakes related to the formation of pyroclastic flows and rockfalls.

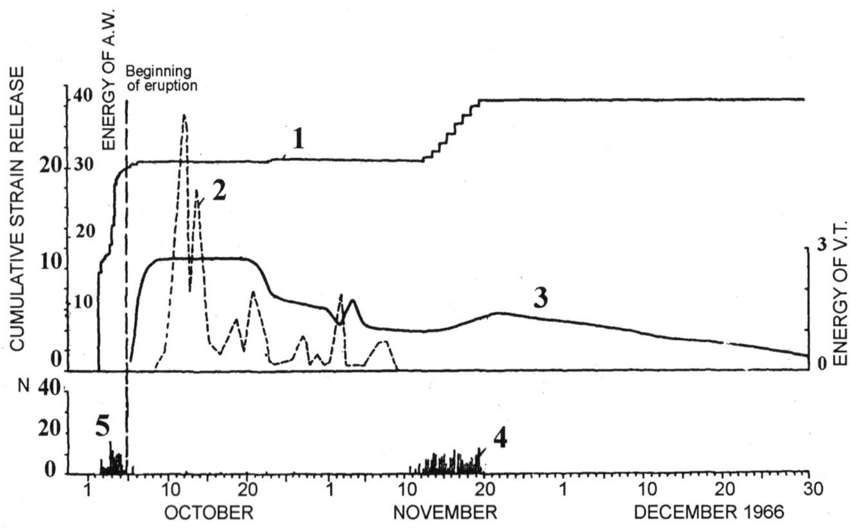


Fig. 2.5. Seismic history of the 1966 flank eruption of Klyuchevskoy volcano, Kamchatka. The curves 1-5 show the variations: (1) in the cumulative strain release measured in [$10^{-8}$] $erg^{1/2}$ for A- and B-type earthquakes; (2) in the energy of air waves (A.W.) produced by explosions and measured in [$10^{-15}$] erg/day; (3) in the energy of volcanic tremor (V.T.) measured in [$10^{-14}$] erg/day; (4) in the number of B-type earthquakes of the second swarm; (5) in the number of A- and B-type earthquakes of magnitude M > 0. N is the number of earthquakes. From Tokarev et al. (1968).

*Dr.* ***Pavel Ivanovich Tokarev*** *(1923-1993) graduated from Moscow University, Russia and began his scientific activity in the Laboratory of Volcanology in Moscow in the middle of the 50s. His first papers were published in 1958-1959 and were devoted to the study of the structure of the Kurile-Kamchatka subduction zone and the relationship between seismicity and volcanism in this zone. In 1958, he moved to Kamchatka and began to work at Kamchatka Volcano Observatory in the village of Klyuchi. Tokarev installed the first seismic network for the monitoring of Klyuchevskaya Group of volcanoes and began to develop the method of seismic forecasting of volcanic eruptions for Bezymianny volcano. His first publications in this field appeared in 1963-1971 and became classic. Later Tokarev headed the Department of Geophysical Monitoring of Volcanic Activity in the Institute of Volcanology in Petropavlovsk-Kamchatsky and studied the seismic activity of Klyuchevskoy, Bezymianny, Karymsky and New Tolbachik volcanoes. His forecasting of the 1975 fissure eruption of New Tolbachik volcanoes was published in regional newspaper a few days before the beginning of the eruption and allowed a group of volcanologists to travel to the site of forthcoming eruption one week before the event begun. Pavel Ivanovich founded a system of seismic monitoring for volcanic activity in Kamchatka and prepared a team of seismo-volcanologists who continue now his efforts in seismic studies of volcanoes.*

Their study is a study of the physics of volcanic eruptions.

### *2.3. Sequences of volcanic earthquakes*

Majority of volcanic earthquakes occur in sequences. Mogi (1963) described three main types of earthquake sequences (Fig. 2.7) as the main shock -aftershocks sequence (Type 1), the foreshocks – main shock – aftershocks sequence (Type 2), and the swarm (Type 3).

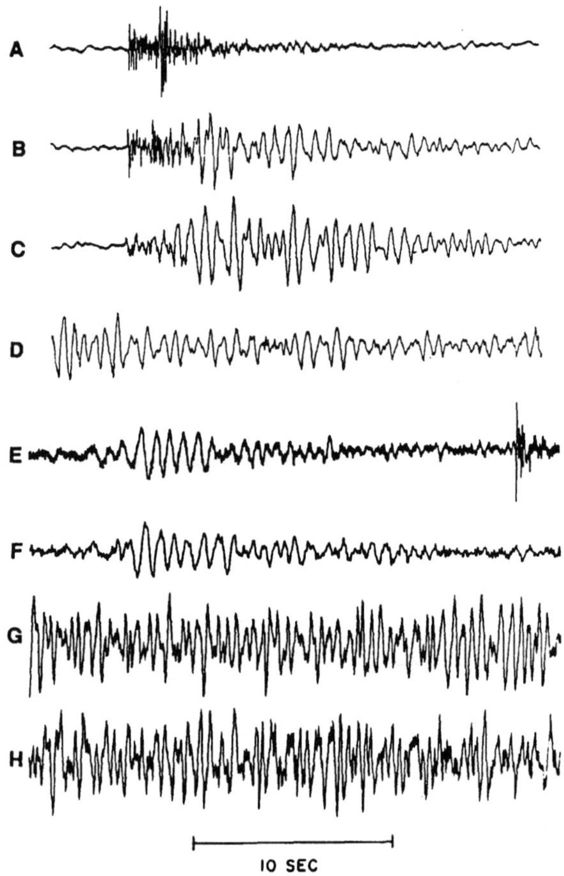


Fig. 2.6. An illustration of McNutt's classification of volcanic earthquakes. A, high-frequency or volcano-tectonic earthquake; B, hybrid or mix-frequency event; C, low-frequency or long-period earthquake; D, Volcanic tremor; E, Explosion quake; F, Low-frequency or B-type earthquake; G, Volcanic tremor before eruption; and H, volcanic tremor during eruption. Earthquakes A-D were recorded at Redoubt volcano, 8 km from the vent; earthquakes E-F, at Pavlof volcano, 8.5 km from the vent; and G-H, at Mt. Spurr volcano, 4.8 km from the vent. From McNutt (1996).

Earthquake swarms are generally defined as a sequence of events closely clustered in time and space without a single outstanding shock. In practice, the difference in magnitude between the largest and second largest event of a swarm is 0.5 magnitude units or less (McNutt, 2000a).

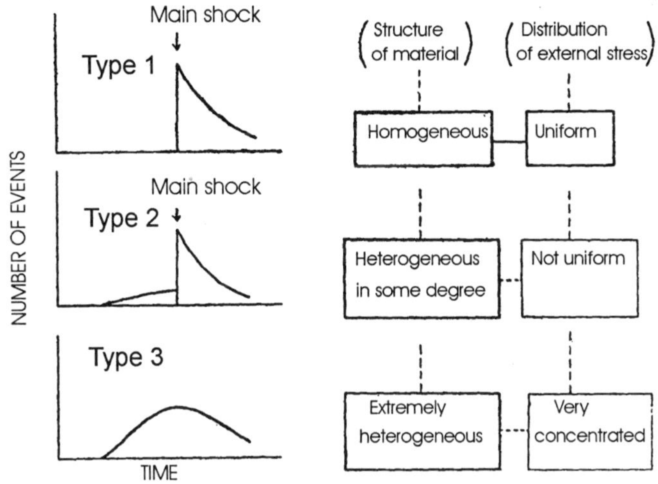


Fig. 2.7. Three main types of successive occurrences of elastic shocks accompanying fractures that were obtained from experimental destruction of material and their relations to their structures and the applied stresses. From Mogi (1963).

If this difference is greater, we have a main shock – aftershock sequence.

### 2.3.1. Sequences of volcano-tectonic earthquakes

Volcano-tectonic earthquakes may occur in all three types of sequences. The two first types of sequences are rare, and they occur prior to flank or fissure eruptions. The swarm sequences are the most common for volcanic earthquakes.

The type 1 sequence was observed prior to the September 30, 1996 flank eruption of subglacial volcano Grimsvøtn in Iceland (Zobin, 1999; Konstantinou et al., 2000). The main shock of magnitude mb 5.2 was the first in the earthquake sequence (Fig. 2.8A); the second largest earthquake occurred after about 3 hours and was of magnitude mb 4.3 (ISC, 1996).

Seismic activity developed during three stages (Zobin, 1999). The first stage (first 19 hrs) included the main shock of magnitude mb 5.2 (Ms 5.4) and its aftershocks along the northern slope of the Bárdarbunga volcano, situated about 20 km to NW from Grimsvøtn (Fig. 2.8B). Then during the next 17 hrs the seismic foci marked a line connecting the two volcanoes, Bárdarbunga and Grimsvøtn. This stage culminated in the opening of a fissure and the beginning of the eruption. The third stage was observed during the eruption (midnight of September 30 - October 7), and included continuous seismic activity along the northern slope of the Bárdarbunga volcano as well as more western distributed epicenters, which together had clustered along the ring-fault epicentral zone.

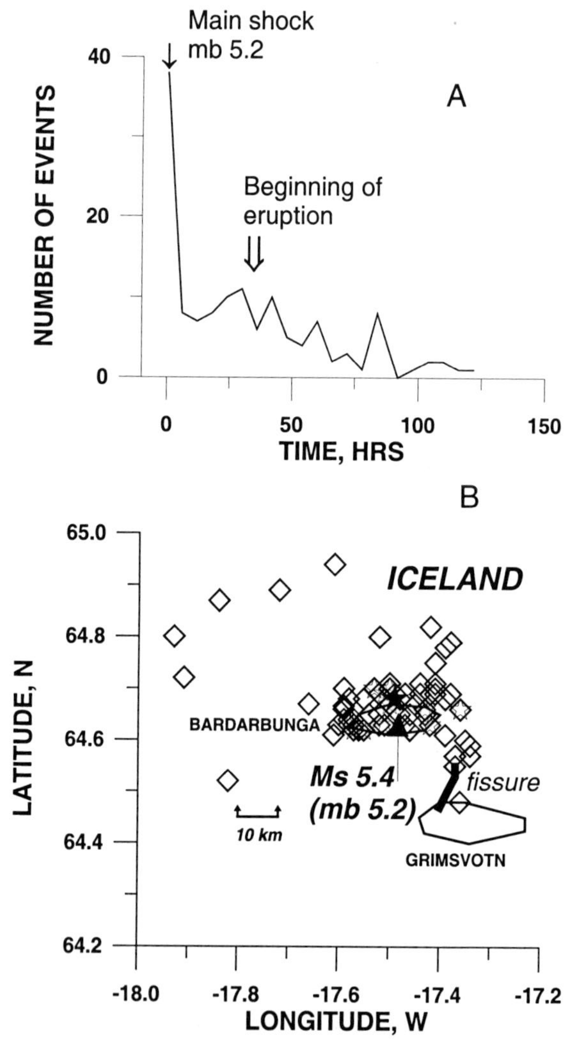


Fig. 2.8. Variations in 6-hour numbers (A) and epicenter distribution (B) of the earthquakes associated with the September 30, 1996 subglacial flank eruption of Grimsvøtn volcano, Iceland. The calderas of volcanoes Bárdarbunga and Grimsvøtn are shown. The diamonds are epicenters; the star shows the epicenter of the main shock. From Zobin (1999).

The type 2 sequence was observed before the January 1996 subaqueous phreatomagmatic eruption in Akademia Nauk caldera (Karymsky Lake), Kamchatka. This eruption occurred after 28,000 years of dormancy and resulted from the injection of juvenile basaltic magmas through a fissure (Fedotov, 1998). The preliminary seismic activity (Zobin and Levina, 1998) was represented by a foreshock – main shock – aftershock sequence with a main shock of magnitude Ms 6.7 (Fig. 2.9). The seismic foci of foreshocks migrated gradually from the forthcoming site of eruption to the south-west up to the Ms 6.6 earthquake epicenter. The aftershocks spread to the south-west and north-east from the main shock, reaching the site of eruption; the total area occupied by aftershocks was of about 50 x 20 $km^2$. The seismic foci were distributed at the depth range 0 to 60 km. The eruption in Karymsky Lake began with a sharp decrease in the number of aftershocks.

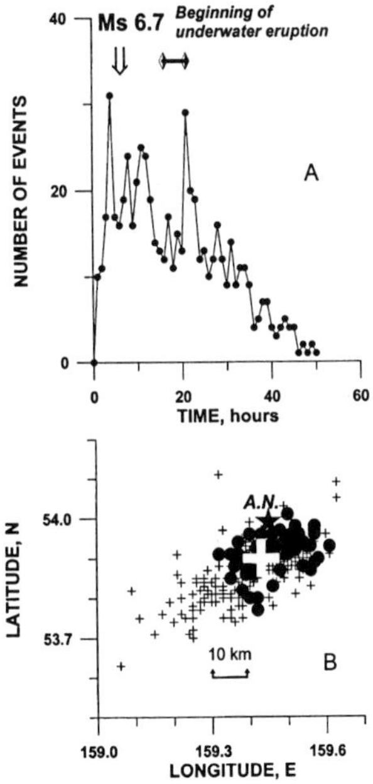


Fig. 2.9. Variation in hourly numbers (A) and epicenter distribution (B) of the earthquakes associated with the January 2, 1996 subaqueous phreatomagmatic eruption in Akademia Nauk caldera (Karymsky Lake), Kamchatka. The star shows the site of eruption in Akademia Nauk (A.N.) caldera; dots are the epicenters of foreshocks and crosses are the epicenters of aftershocks; large white cross shows the epicenter of the main Ms 6.7 shock. The proposed time interval of the beginning of the eruption is shown. Data from Kamchatka Regional Earthquake Catalog.

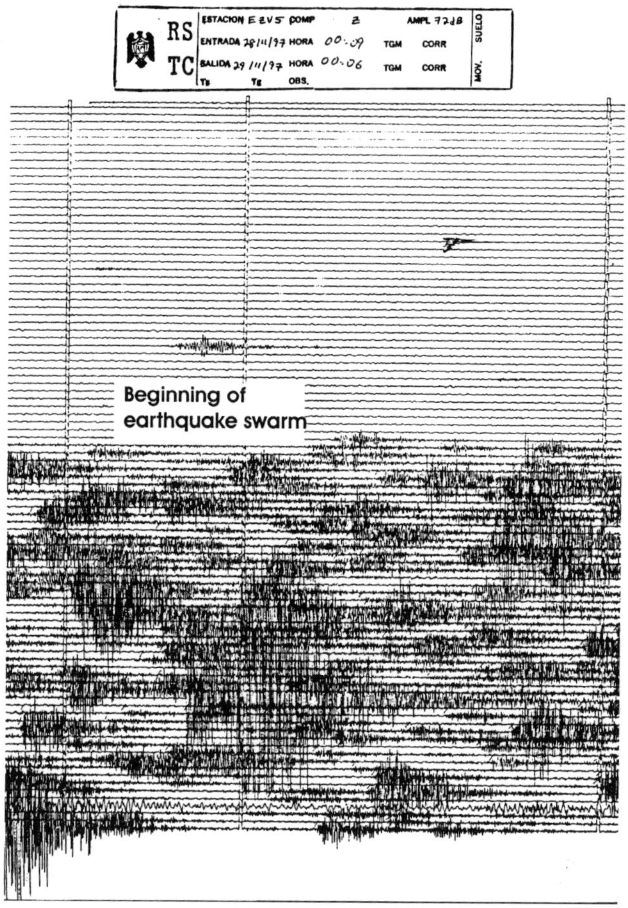


Fig. 2.10. Earthquake swarm at Volcán de Colima, México. November 1997. The seismogram was recorded at a distance of 4 km from the crater. Minute marks are shown. Courtesy of Colima Volcano Observatory.

The third type of earthquake sequence, and the most frequent type, is a swarm sequence. It occurs roughly with numerous earthquakes (Fig. 2.10). Fig. 2.11 shows the spatial-temporal characteristics of earthquake swarm associated with the September 1977 deflation of Krafla volcano, Iceland (Brandsdottir and Einarsson, 1979). A small basaltic eruption broke out on a 0.9-km-long fissure the northern rim of caldera and continued for about 5 hours. The earthquake activity and volcanic tremor began about two hours before the eruption and accompanied the eruption.

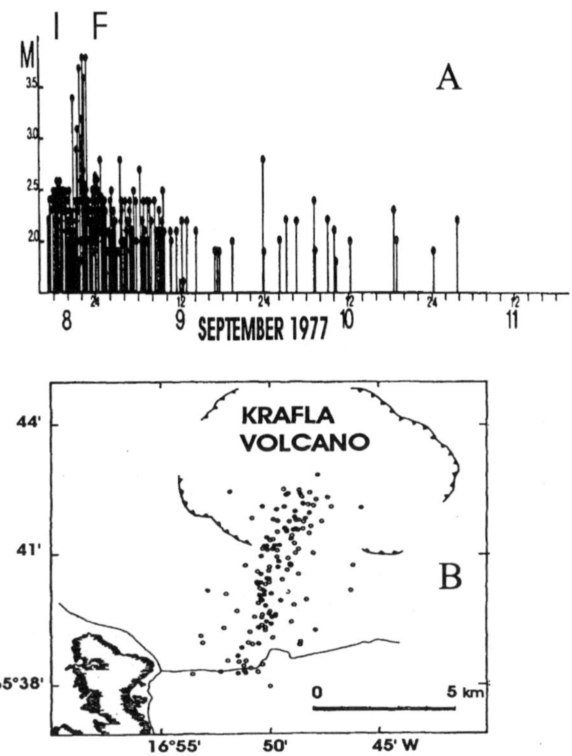


Fig. 2.11. The temporal variations of earthquake magnitudes in the swarm that occurred during the September 1977 deflation of Krafla volcano, Iceland (A) and the earthquake swarm epicenters (B). Symbol I marks the initial part of eruption; F marks the fin of eruption. It is clearly seen a group of equal-magnitude earthquakes in A. Dots are epicenters located with standard error less than 0.5 km; circles mark the epicenters of worse location. From Brandsdottir and Einarsson (1979).

The first very small earthquakes were located in the central part of the caldera; later the earthquake epicenters migrated southward with a rate of about 0.5 m/s. The seismic activity culminated just after ceasing of lava eruption with 8 earthquakes of magnitude 3 and greater (Fig. 2.11). It is interesting to note that the 1138-m-depth producing steam well situated on the way of earthquake migration throw out the glowing pumice at the time of earthquake migration. It indicates that the earthquake foci migration was directly associated with the magma lateral migration.

The swarms may occur before an eruption, during an eruption, and after the eruption of volcano. Very frequently the appearance of a swarm near a volcano may indicate a foregoing eruption of the volcano. Two main groups of volcano-tectonic earthquakes swarms can be distinguished according to their character of occurrence through time, their spatial position and their relationship with volcanic activity (Gorelchik and Zobin, 1971).

*Group 1.* These swarms precede generally the central eruptions of andesitic and dacitic volcanoes. The swarms consist of shallow volcanic earthquakes, the epicenters of which lie at a distance of 3-5 km from the crater and the foci of which are located 0-5 km above sea level within the volcanic edifice. Depending on the deep structure of the feeding zones at andesitic volcanoes, earthquake having focal depths of 10-15 km are feasible. The maximum magnitudes of earthquakes may reach 5-6. The swarm duration can be from 7-10 days to 3-5 weeks. During these swarms, the total energy, frequency, and maximum energy of earthquakes increase continuously and evenly, reaching their maximum values at the moment of eruption.

*Group 2.* These earthquake swarms precede generally flank eruptions at basaltic and andesitic volcanoes. Very shallow earthquakes predominate in the swarms. The epicenters of the earthquakes lie 3-10 km from the place of formation of a new crater, and the foci are located at a depth of 0-5 km. Prior to flank eruptions at stratovolcanoes, the swarms can last for 1-10 days, maximum earthquake magnitudes reaching 5-7. In these swarms the number of events and their maximum energy at first sharply increase, and on reaching their maximum, start to decrease gradually; and in a few hours before the flank fissure opening, the earthquakes cease.

The characteristic feature of volcano-tectonic earthquake swarms is the appearance of an earthquake family (a group of earthquakes having a similar wave form at fixed seismic station). Kaufman and Burdick (1980) have noted that during a swarm of volcano-tectonic earthquakes accompanied the collapse of Fernandina caldera, Galapagos Islands in June 1968, the teleseismic records of large volcanic earthquake of magnitude Ms 5.0 to 5.5 were spectacularly consistent from event to event (Fig. 2.12). The major seismic phases like P, PP, vertical S, and the Rayleigh waves correlate closely. Even the fine details of the coda between the major phases match.

The families of small earthquakes, recorded at a distance of about 2 km, were founded during the dacitic dome forming at Usu volcano, Hokkaido (Okada et al., 1981; Okada, 1983). The waveforms in each family (Fig. 2.12) resemble each other in details from the onset to the coda, even in high frequencies (≥ 10 Hz). Hypocenters of one family cluster in a very small region. Fig. 2.12 shows that the families OU1 and OU2 have completely different waveforms, although they are located less than 200 m apart.

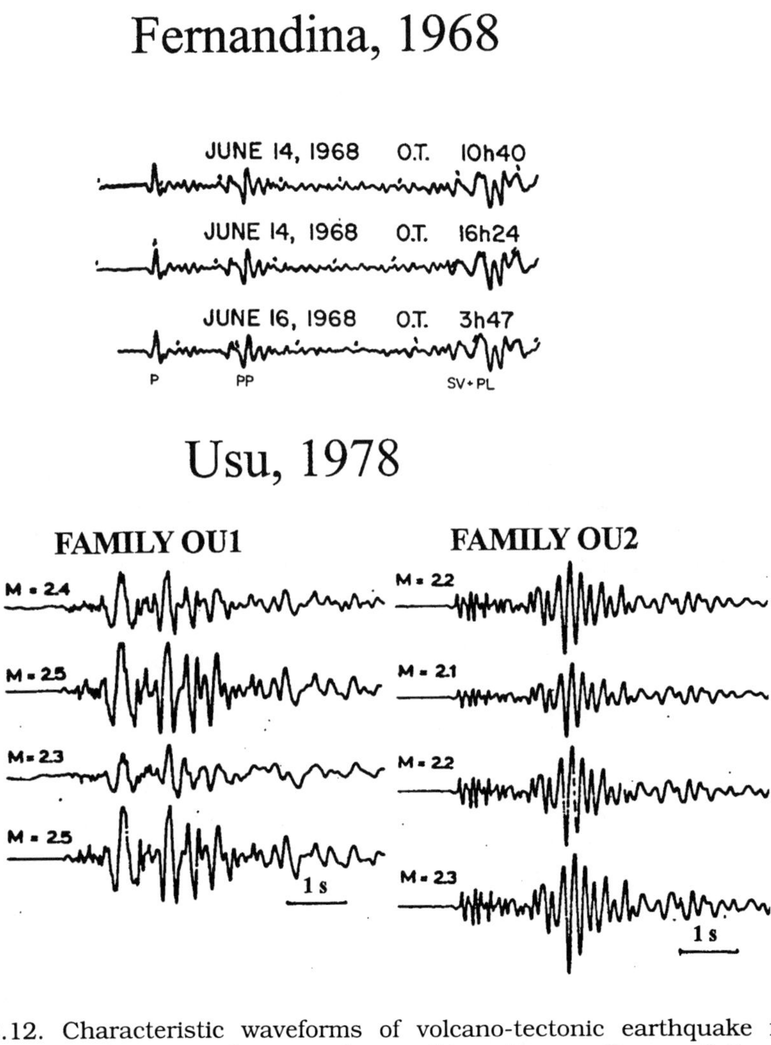


Fig. 2.12. Characteristic waveforms of volcano-tectonic earthquake families recorded at teleseismic distance during the caldera collapse of Fernandina volcano, Galapagos in 1968 (long-period records), and at near distance (about 2 km) during the dome growth at Usu volcano, Hokkaido in 1978 (short-period records). For Fernandina records, the minute marks are shown. From Kaufman and Burdick (1980) and Okada et al. (1981).

Okada (1983) noted that about 80% of 2300 located at Usu volcanic earthquakes (M>3.5) occurred only in four families. Earthquakes of any particular family tend to exist over a narrow range of their size distribution. The waveforms for family earthquakes vary only in the amplitude and not in the spectral shape.

### 2.3.2. Sequences of eruption earthquakes

Eruption earthquakes also may occur in sequences. Their clustering depends on the style of volcanic activity. The destruction of lava dome may cause a long-term sequence of seismic signals produced by the collapse of the lava body pieces and their movement along the volcanic flanks (Fig. 2.13). The permanent process of degassing can produce a sequence of explosion earthquakes (Fig. 2.14). The magmatic processes beneath the crater floor generate continuous volcanic tremor (Fig. 2.15). The nature of these sequences of volcanic earthquakes differs from the nature of volcano-tectonic earthquake swarms and related to the dynamics of surface manifestations of magmatic activity.

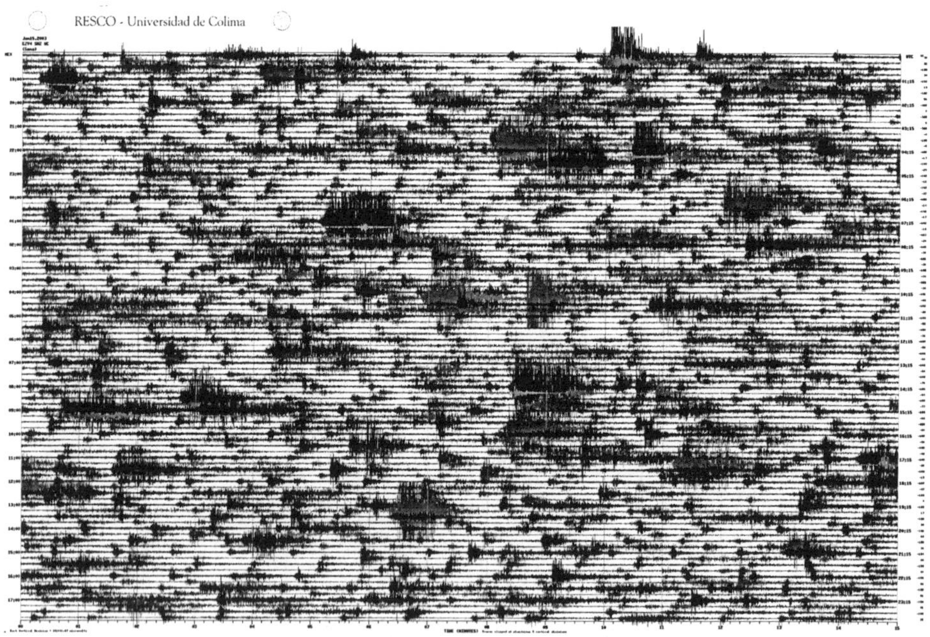


Fig. 2.13. Sequence of the seismic signals produced by rockfalls during the 2002 block-lava eruption at Volcán de Colima, México. One-day short-period records at a distance of 1.7 km from the crater are shown. Courtesy of Colima Volcano Observatory.

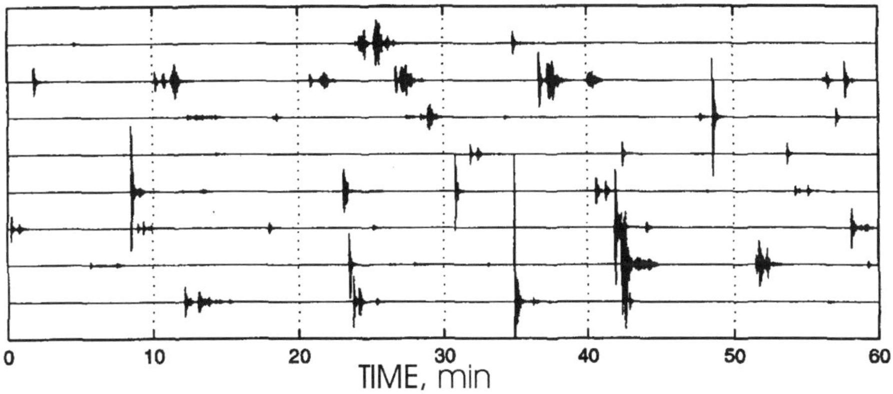


Fig. 2.14. Sequence of explosion earthquakes recorded at Karymsky volcano, Kamchatka during the 1997 eruption. The eight-hour broadband records at a distance of 1.5 km from the crater are shown. From Johnson and Lees (2000).

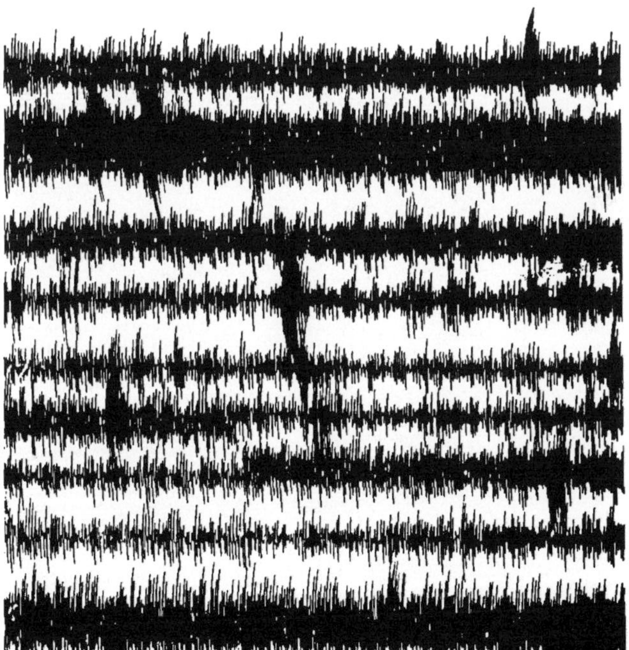

Fig. 2.15. Sequence of continuous volcanic tremors recorded by short-period instrument during sustained lava eruption of Kilauea volcano, Hawaii. From Koyanagi et al. (1987).

# Chapter 3
# Origin of Volcano-tectonic Earthquakes

Volcanic earthquakes occur within and around the volcanic edifice and reflect the interaction of two general geological processes: the magma migration to the Earth´s surface and the crustal tectonic activity. Fig. 3.1 gives a general scheme of magma melting in the mantle and its rising through the ruptured crust to the surface at zones of ocean spreading and subduction.

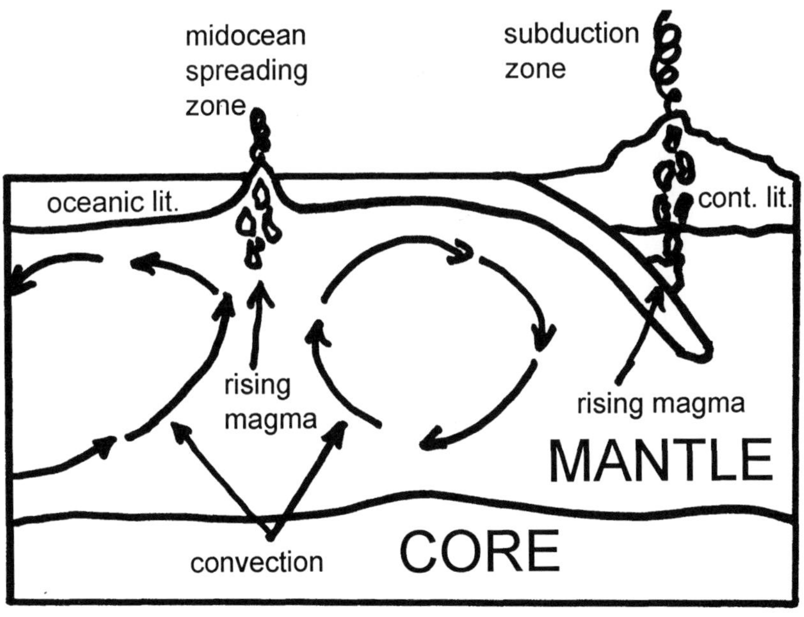


Fig. 3.1. An illustration of the process of magma melting in the mantle and it's rising at the midocean spreading centers and subduction zones to the surface. Volcanoes overlie the zones where slowly moving oceanic lithosphere is pushed beneath the thicker lithosphere of the continent. The engines that drive this movement are rising heat currents in the mantle. The molten magma rises buoyantly towards the Earth's surface. Based on Fisher et al. (1998).

### *3.1. Magma and its physical and chemical properties*

Magma is generated by a partial melting of rocks at depth. Migration of the melt then occurs in the Earth's interior by porous flow along grain-size channels, and by flow through larger conduits (Daines, 2000).

Magma is formed in the mantle in two principal ways: the first way involves simple decompression of hot, solid mantle material; whereas the second involves lowering the melting temperature of the mantle material by the addition of volatiles (Perfit and Davidson, 2000). Plate tectonics show (Fig. 3.1) that magma is generated mainly below divergent plate boundaries (oceanic spreading ridges) and below convergent plate boundaries (subduction zones). Intraplate volcanism is supported by magma that is generated within so called *hot spots* producing *mantle plumes*. The location of the magma generation constrains its physical properties and constitution.

Partial melting of the mantle generates basaltic magmas; the silicic magmas are generated by partial melting of crustal rocks (Rogers and Hawkesworth, 2000). Common magmas vary nearly continuously in composition from rhyolitic (75% $SiO_2$) through to basaltic (50% $SiO_2$). Generally three types of magma are distinguished: acid (dacitic-andesitic) containing more than 66% of $SiO_2$, intermediate (andesito-basaltic) containing between 52% and 66% of $SiO_2$, and basic basaltic, containing less than 52% of $SiO_2$.

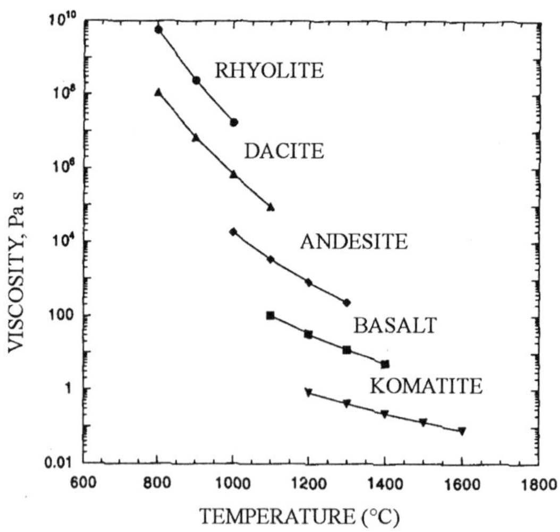


Fig. 3.2. Viscosity as a function of temperature at a pressure of 1 bar for natural melts spanning the compositional range rhyolite to komatite. All compositions are volatile free. From Spera (2000).

After magma generation, it commences its journey to the surface. Magma is not a pure liquid but has rheological properties and its ability to flow depends strongly on its viscosity. Fig. 3.2 (Spera, 2000) shows that the viscosity of different melted rocks at high temperature (about 1000 $^0$C) can change by 10 orders from komatite (less than 1 Pa s) to rhyolite ($10^{10}$ Pa s). This difference in viscosity determines the ability of different magma types to migrate through the cold rocks of the interior of the Earth. At mantle depths, the magma rises very slowly within convection currents without the fracturing of surrounding material. Within the shallow crust there is storage of the magma and differentiation begins to occur.

Seismic studies allow the location of the zones of magma storage since they represent a zone of low velocity. Fig. 3.3 and 3.4 show the locations of low velocity zones beneath the volcanoes Pinatubo and Mount St. Helens. A large zone representing a relative decrease of P-wave velocity (5 to 10%) beneath Pinatubo volcano (Fig. 3.3) was located between depths of 6 and 11 km and is estimated to have a volume of 40 to 90 cubic km (Mori et al., 1996a). For Mount St. Helens (Fig. 3.4), a P-wave low-velocity region began at 1.5 km depth, directly below the crater. This localized low-velocity zone extends to a depth of 3.5 km where it then appears to broaden (4-5 km diameter) and continues to a depth of 7.5 km (Lees, 1992). Therefore, the regions of magma storage are situated at very shallow depths.

Two types of igneous rocks are distinguished: intrusive igneous rocks and extrusive igneous rocks. Intrusive igneous rocks crystallize when molten rock intrudes into unmelted rock masses deep in the Earth's crust while extrusive igneous rocks are formed when magma erupts at the surface. The intrusive and extrusive rocks can have the same content but different textures. The main types of intrusions are *sills and dykes*. Sills are concordant, tabular bodies that are emplaced essentially parallel to the bedding of the country rock and occur in relatively unfolded regions at shallow crustal level. A low viscosity is required to produce this sheetlike form. Dykes are thin, tabular, discordant intrusive bodies that cut across the foliation or bedding of the country rock. They range in thickness from less than a meter to several hundred meters, and a few have been traced along strike for tens of kilometers. Typically emplaced into already existing fracture systems, they may occur singly or in swarms. In some areas, dikes occur as radiating swarms centered on an intrusion or on the flanks of a volcano. In rare cases, vertical- or outward-dipping ring dikes or inward-dipping cone sheets are distributed in oval or circular patterns around the intrusion (Blatt and Tracy, 1996).

Extrusive rocks form different types of lavas. The term *lava* is used both for magma that has been erupted onto the Earth's surface and for

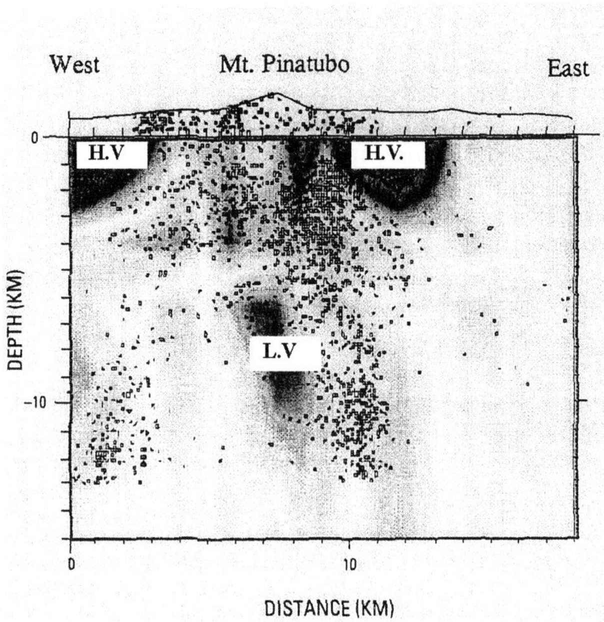


Fig. 3.3. Cross-section of the P-wave velocity perturbations from the average velocity at a given depth beneath Pinatubo volcano, Philippines. The dots show the hypocenters of post-eruption seismicity from June 29 to August 16, 1991. From Mori et al. (1996a).

the rocks that has solidified from it (Blatt and Tracy, 1996).

This intrusive-extrusive activity of magma indicates the seismic potential of magmatism within the upper crust with well-developed tectonics. When magma rises from its storage zone, it begins to move along the system of tectonic fractures and the volcanic conduit to the surface as a dike or sill. The difference in magma viscosity becomes decisive in the choice of magmatic route at shallow depths. Low-viscosity basaltic magma prefers to go along the stratified structures (or faults) in accordance with their strike and does not cut them. The intermediate-viscosity andesite-basaltic magma may choose both ways. High-viscosity dacitic magma slowly cut the stratified structures and forms obelisks. The average ascent rate varies from 0.001 m/s to 0.015 m/s for extrusive magmas. During explosive eruptions, the assessment of magma ascent rate is characterized by much greater variations and leaves the top of conduit at essentially sonic velocities for truly explosive eruptions (Rutherford and Gardner, 2000). So the main seismo-productive magma at shallow depths is the andesite-basaltic magma of intermediate viscosity that is transported in sills and dikes from the deep reservoir to the surface.

The three main types of volcanic eruptions, Pelean, Vulcanian and Hawaiian, are characterized the difference in content of erupted magma that causes the different surface manifestations of these eruptions (Table 3.1).

### *3.2. Volcanism and tectonics*

Volcanoes normally occur within rift (or graben) type structures along long-lived faults. A graben is often formed above intruding dikes in volcanic rift zones because a dike creates two zones of maximum tensional strain at the surface, parallel to the dike axis (Perfit and Davidson, 2000). The regional tectonic faults where the volcano is situated are accompanied commonly by numerous local faults.

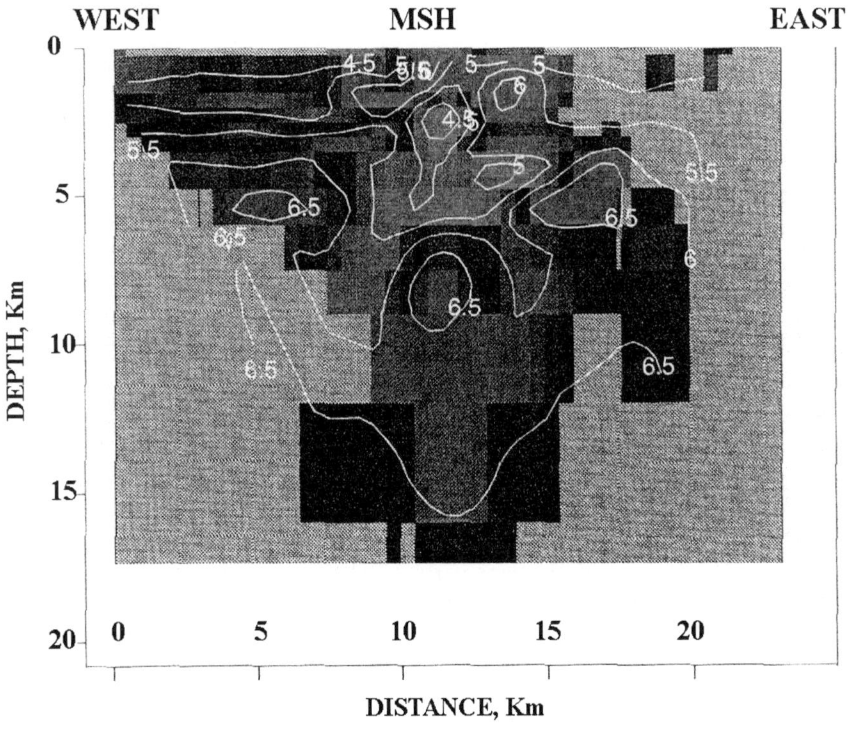


Fig. 3.4. Cross-section of P-wave velocity perturbations beneath Mount St. Helens (MSH) volcano, Cascades. Contours of the absolute P-wave velocities (background plus perturbation) are provided for reference. From Lees (1992).

**Table 3.1**
**Characteristic features of three types of volcanic eruptions**

| Type of eruption | Characteristic feature | Type of lava |
|---|---|---|
| Pelean | Central eruption, pyroclastic flows, lava dome | Dacitic and andesitic |
| Vulcanian (or Strombolian) | Central and flank eruptions, strong explosions without pyroclastic flows, but with a great cauliflower-shaped dark eruption cloud (white for Strombolian type), lava flow, somma structure | Andesito-basaltic |
| Hawaiian | Central, fissure and flank eruptions, lava flow and fountains, lava lake | Basaltic |

These structures could be mid-oceanic ridges or rift valleys (depressions) along the convergent boundary (subduction zone) or above a hot spot (See Fig. 3.1). They could be the basis for a long volcanic chains or a small volcanic group. Basaltic volcanoes have a well-defined framework of local faults where the flank cones are formed. Fig. 3.5 shows the structural map of Mt. Etna volcano, Sicily (Azarro, 1999). This volcano has grown at the intersection of two regional faults: the Malta Escarpment and the Messina-Fiumefreddo, striking respectively NNW and NE, and at the front of the Apennine-Maghrebian thrust belt, the crustal scale fault.

The tectonic setting of Etna's flanks results from the interaction of regional and local tectonics. The Timple system at the volcano's eastern flank represents the northernmost prolongation of the Malta Escarpment and forms a NNW-SSE-trending system of parallel east-facing step-faults of considerable length (8-10 km). To the north this system is interrupted by the E-W Pernicana fault, that cuts a large part of the volcanic edifice and by the NE-SW Ripe della Naca system. Two isolated structures, the Trecastagni and Tremestrieri faults are situated at the southern flank of the volcano.

The ground breakage that was observed in the volcanic region generally follows pre-existing active fault traces. Azarro (1999) shows also evidence of the movement along the hidden and semi-hidden local faults. The flank volcanic eruptions occur normally along the local tectonic faults, mainly on the eastern flank of the volcano (Gresta et al., 1990). Among them was the 1991-1993 eruption that occurred along the

NNW-SSE Timple system of faults (Ferrucci and Patané, 1993). The numerous cinder cones illustrate the process of magma intrusion along the local faults.

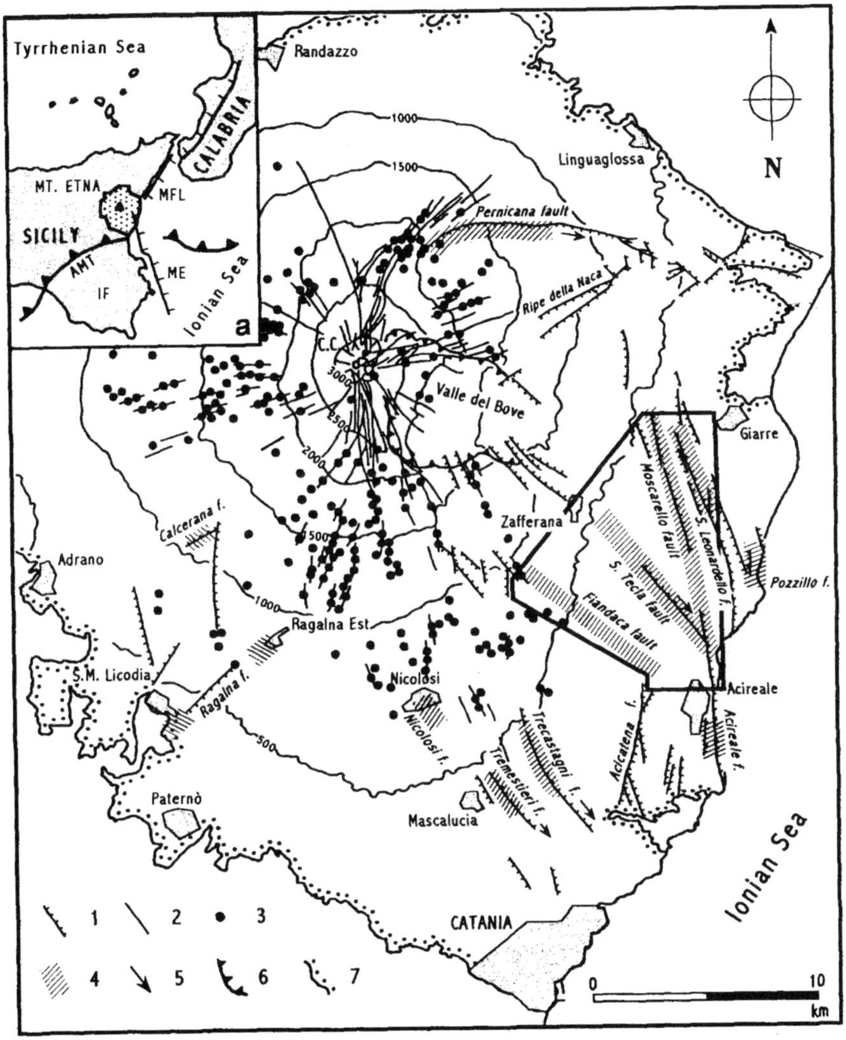


Fig. 3.5. Simplified structural map of Mt. Etna. 1, faults with downthrown side; 2, eruptive fissures; 3, cinder cones; 4, coseismic surface faulting zones; 5, strike-slip components; 6, caldera rims; 7, limit of Etna volcanics; C.C., central craters. Box indicates area of the Timple fault system. Insert map (a) shows the regional geological setting: AMF, front of the Apennine-Maghrebian thrust belt; IF, Iblean Foreland; ME, Malta Escarpment; MFL, Messina-Fiumefreddo line. From Azarro (1999).

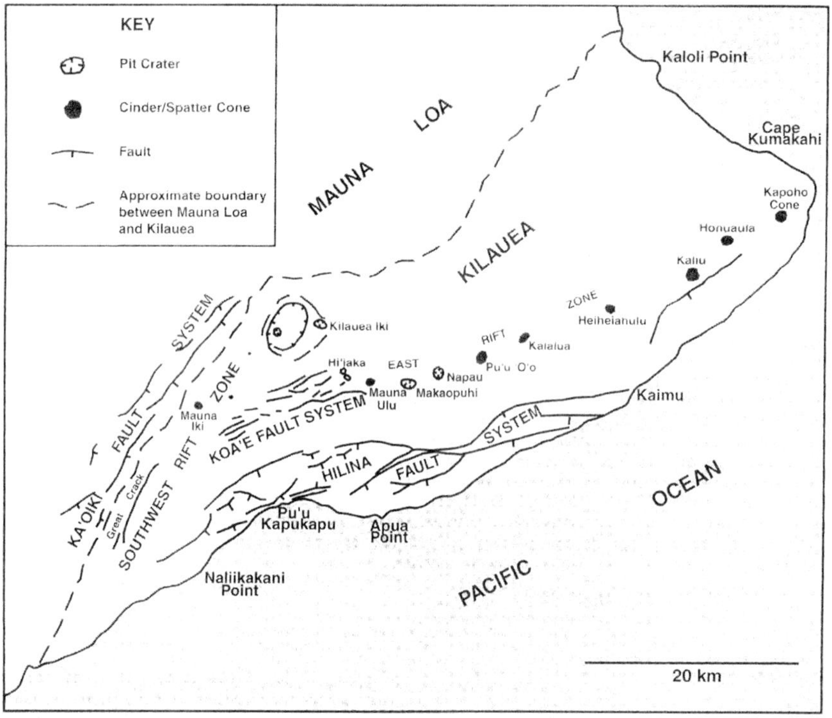


Fig. 3.6. Kilauea volcano, Hawaii, and its rifts and fault systems. From Parfitt and Peacock (2001).

Another basaltic volcano, Kilauea, Hawaii (Fig. 3.6) is situated within a complex tectonic framework formed by the Ka'Oki, Koa'E, and Hilina fault systems and the Southwest and East Rift zones. The normal fault systems at the south flank of Kilauea do not participate directly in magmatic activity but reflect longer-term deformation of the volcanic edifice. The fault systems consist of *en echelon* fault segments that have significant intersection and linkage (Parfitt and Peacock, 2001).

The rift zones, which intersect the flanks of the volcano, are filled by dikes that supply magma and form cinder cones. The dikes propagate downrift at narrow bladelike intrusions from shallow magma reservoirs beneath the summit caldera. Rift eruptions and many summit eruptions begin when the magma issues from long extension cracks that fracture the surface immediately prior to the eruptions. These eruptive fissures are oriented parallel to the regional trend of the rift zones and mark the intersection of propagating dikes with the Earth's surface (Dieterich, 1988).

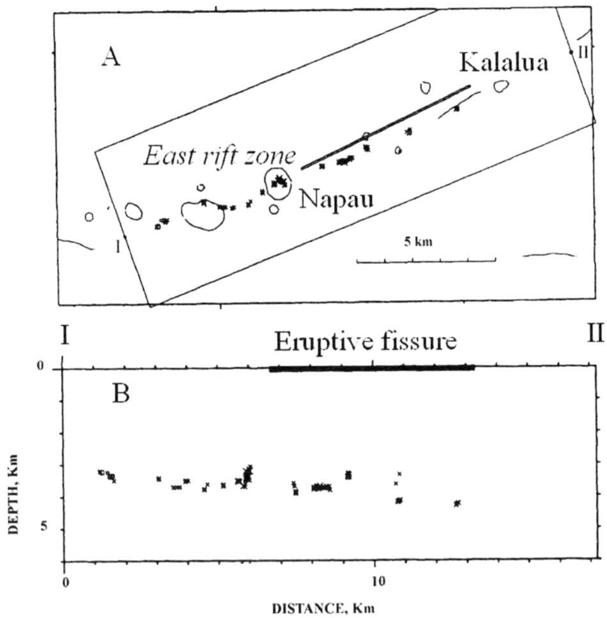


Fig. 3.7. Clustering of volcano-tectonic earthquakes (crosses) along the eruptive fissure (thick line) during the 1983 dike intrusion in East rift zone of Kilauea. A, epicenters of earthquakes; B, cross section along I-II line of Fig. 3.7A. The cinder cones are shown by ovals. From Rubin et al. (1998).

The precise location of small (magnitude 1 to 2) volcano-tectonic earthquakes that were observed during the 1983 dike intrusion fissure eruption (Rubin et al., 1998) shows a pronounced tightening of seismicity along the dike trend (Fig. 3.7A). The foci of earthquakes are located about 0.5 km south of the eruptive fissure, but a dike that dipped only 7° from vertical would pass through them. The hypocenters of the earthquakes lie between 3 and 4 km depth (Fig. 3.7B) that indicate the origin of seismicity from a very narrow depth interval, despite the fact that the dike breached the surface.

The andesitic stratovolcano Volcán de Colima is located within the N-S-trending Colima Rift Zone, in the western part of the Mexican Volcanic Belt, and together with the Pleistocene volcano Nevado de Colima forms the Colima Volcanic Complex (Fig. 3.8). The Colima Volcanic Complex is situated above the subduction zone where the Rivera plate goes beneath the North American plate. At the same time, the Colima Rift represents the eastern boundary of the Jalisco block that is moving towards ocean from the continent. The Colima volcano marks also the intersection of

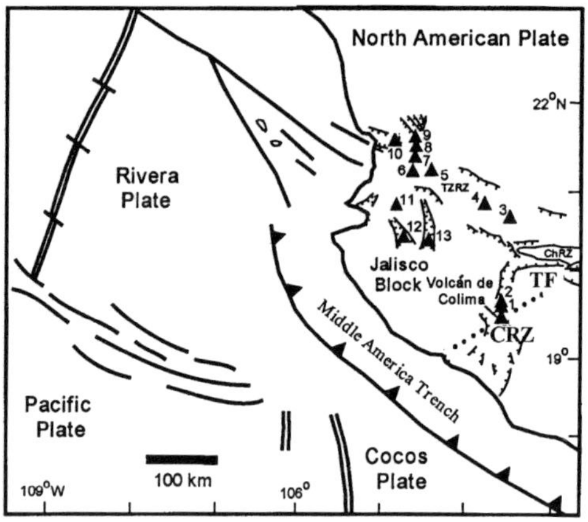


Fig. 3.8. A structural position of Volcán de Colima in southwestern México. Three major rift zones are shown: CRZ, Colima Rift Zone; ChRZ, Chapala Rift Zone; and TZRZ, Tepic-Zacoalco Rift Zone. The dotted line running through Volcán de Colima is the proposed Tamazula Fault zone (Garduño-Monroy et al., 1998). Other major volcanic centers from the western Mexican Volcanic Belt are shown as triangles: 1, Nevado de Colima; 2, Volcán Cántaro; 3, Sierra la Primavera; 4, Volcán Tequila; 5, Volcán Ceboruco; 6, Volcán San Pedro; 7, Volcán Tepetiltic; 8, Volcán Sangangüey; 9, Volcán las Navajas; 10, Volcán San Juan; 11, San Sebastian Volcanic Field; 12, Mascota Volcanic Field; 13, Los Volcanes Volcanic Field. From Zobin et al. (2002c).

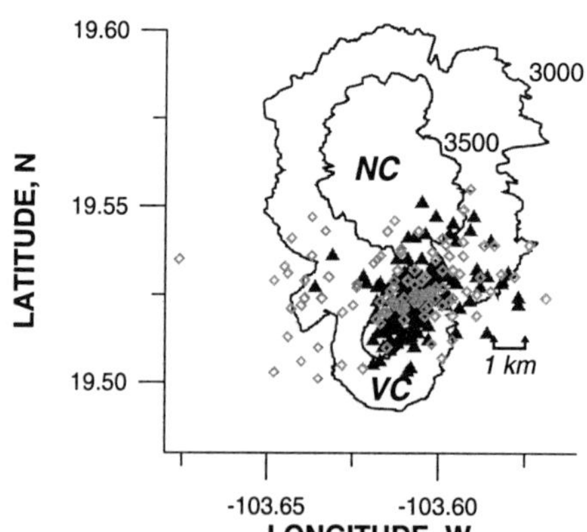


Fig. 3.9. Epicenters of earthquakes recorded at Volcán de Colima, México prior to the November 1998 block-lava extrusion in June-July (circles) and October-November, 1998 (stars). VC and NC are Volcán de Colima and Nevado de Colima, respectively. The contours of 3000m and 3500 m are shown. From Zobin et al. (2002b).

two large local tectonic structures, the N-S-trending Colima Rift and NE-SW-trending Tamazula fault (Garduño et al., 1998).

The November 1998 block-lava extrusion at Volcán de Colima was preceded by a 12-month period of seismic activity that included five earthquake swarms (Zobin et al., 2002b). The epicenters of the earthquakes that were recorded during June-July 1998 swarm clustered across the N-S-trending Colima Rift, while the epicenters of the October-November 1998 earthquake sequence were located along the rift structure (Fig. 3.9). This earthquake clustering may reflect the magma movement along the two local faults that were active at the different stages of magma uplift to the surface.

The dacitic volcano Usu, Hokkaido, is situated on the southern rim of the nearly circular Toya caldera that is of about 10 km in diameter, surrounded by a steep wall and partially filled now by Lake Toya (Katsui et al., 1981). Usu volcano has a characteristic distribution of its central and flank cones (Fig. 3.10) that outlines a system of circular faults in the volcanic edifice. The craterlets of the 1977 eruptions were formed within the summit caldera; whereas a chain of dacitic cones was formed in 1910 and 1943-1944 along the 200-m contour line of volcanic edifice at

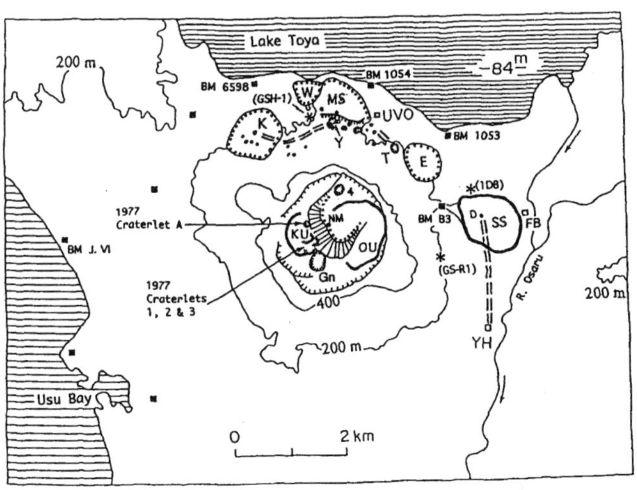


Fig. 3.10. A system of cones that formed at Usu volcano, Hokkaido, during the 1910, 1943-1944 and 1977 eruptions. Contours with dots show approximate outlines of cryptodomes. MS is the 1910 upheaval; K and E have no dates. Thick contours show lava domes. SS indicates the 1943 lava dome; Gn and NM are the 1977 Gin-numa craterlet and one of the 1977 peaks, respectively. The stars indicate the wells. Double broken lines show the lateral extent of volcanic activity. From Yokoyama and Seino (2000).

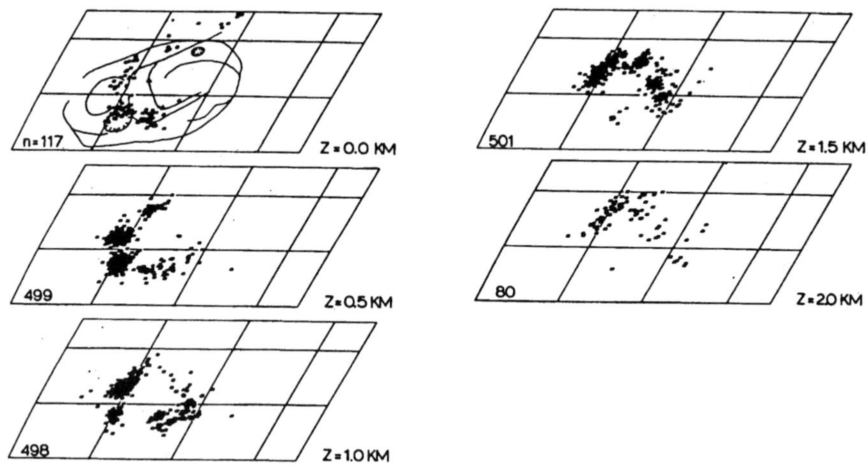


Fig. 3.11. The deep slices of earthquake foci located during the 1977-1978 eruption of Usu volcano for depths from 0 km (sea level) to 2.0 km. Numbers in the left corner show the number of located events for each depth slice. From Okada et al. (1981).

a distance of about 2-3 km from the summit (Yokoyama and Seino, 2000). These surface manifestations of volcanic activity reflect the movement of dacitic magma along the circular faults.

The earthquakes located during the 1977-1978 summit eruption of Usu (Fig. 3.11) show the circular distribution of earthquake clusters around its caldera (Okada et al., 1981). The earthquakes surround an event-free zone beneath the central part of the summit and indicate the deep circular fault distribution. Examining the slice at the 0.5 km depth, the epicenters for an arc in the SW part of volcano are seen, while with the 1.0 km and 1.5 km depth slices, the NE arc is outlined.

### *3. 3. Models of volcano-tectonic earthquake sequences*

It has been shown that volcano-tectonic earthquakes occur due to magmatic activity and tectonic fracturing. The movement of magma to the surface occurswithin a dike-type body. Rubin and Gillard (1998) proposed a model of dike-induced seismicity based on dike intrusion observations at Kilauea Volcano, Hawaii (Rubin and Gillard, 1998; Rubin et al., 1998). They study the inelastic outcrop-scale deformation produced within several meters of the dike tip when the internal pressure variations due to viscous flow of the magma must be considered.

The schematic diagram of blade-like dike is shown in Fig. 3.12. In this model, the dike moves within the host rocks that are subject to the action of ambient stress. The region near the tip of dike (tip cavity) where the magma cannot penetrate may be occupied by magmatic volatiles or host rock pore fluids. The region influenced by the increase in stress is comparable in size to the length of the tip cavity. The possibility of the magnitude 1 to 2 earthquakes with the source dimension of about 100 m in the vicinity of the tip cavity is discussed for three cases (Fig. 3.12):

1. Fault slip away from the tip cavity.
2. Fault slip near the tip cavity.
3. Shear failure of intact rock.

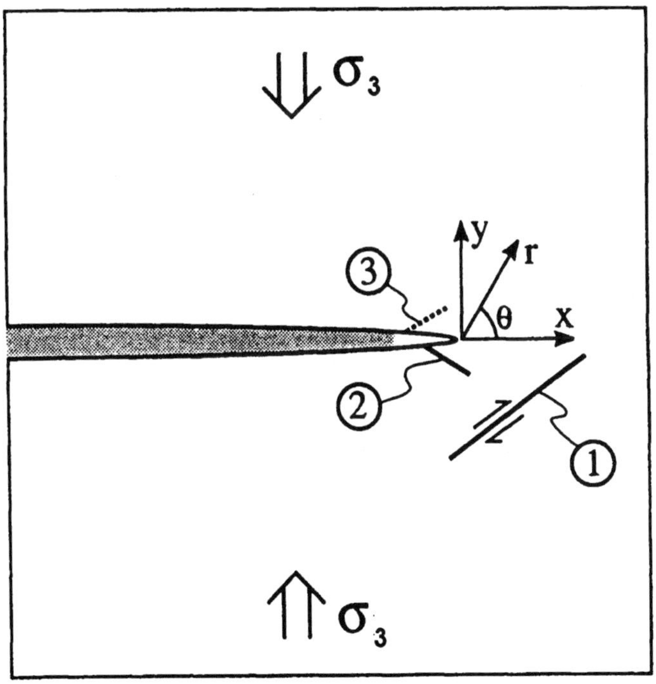


Fig. 3.12. Schematic diagram of blade-like dike (vertical section) and the three types of deformations that could be related to the moving dike: (1) slip on existing faults away from the tip cavity; (2) slip on existing faults adjacent to the tip cavity; and (3) shear failure of intact rock adjacent to the tip cavity. The least principal stress $\sigma_3$ is perpendicular to the dike. The tip cavity of dike is shown by empty space. From Rubin and Gillard (1998).

The examination of elastic stress fields surrounding propagating fluid-filled cracks showed that the stress state is favorable to failure occurring near the tip cavity when the cavity pressure is maintained by the influx of host rock pore fluids, rather than by exsolution of magmatic volatiles. Even in this case, however, shear fracture of previously intact rock seems unlikely. According to these results, most dike-induced seismicity should result from slip suitably aligned along existing fractures (Rubin and Gillard, 1998).

Therefore, Rubin and Gillard suggest that the distribution of earthquake foci of dike seismicity reflects the distribution of ambient stresses that are near to failure, and does not necessarily reflect the extent of the dike. They reject fracturing as the trajectory of the dike movement. At the same time, the dike can trigger the fracturing in the vicinity of its tip along existing fractures. We could consider these fractures as magmatic fractures. Rubin and Gillard (1998) consider that slips could occur at the distances over 70-100 m from the dike if a favorably oriented fault connected two en echelon dike segments.

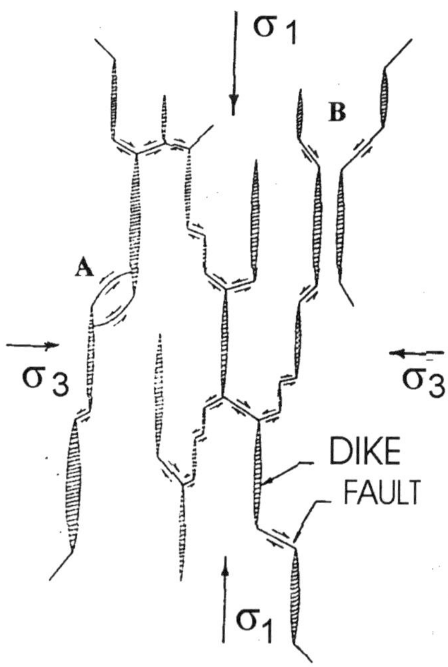


Fig. 3.13. Schematic representation of dikes and conjugate fault planes with respect to greatest and least principal stresses $\sigma_1$ and $\sigma_3$. A and B illustrate typical patterns of crack interactions near adjacent dike tips in homogeneous media. From Hill (1977).

The model that describes this situation was proposed by Hill (1977). He suggested that the clusters of magma-filled dikes exist within brittle volumes of the crust in regions of active magmatic intrusions, or volcanism, and the dikes within a cluster are systematically oriented in a regional deviatoric stress field. Their long dimension is parallel to the greatest principal stress $\sigma_1$. Hill (1977) considered shear failures occurring along oblique fault planes connecting adjacent tips of offset dikes when a critical combination of the fluid pressure *P* in the dikes and the regional stress difference $\sigma_1 - \sigma_3$ is attained.

Fig. 3.13 shows the schematic representation of the model. The dikes are forced open by a gradual increase in the tectonic stress difference, magma is passively drawn into the dikes. A swarm develops when the stress field within the volume reaches the critical state, and a series of shear failures occur along cracks between the dikes. Hill (1977) suggests that the distribution of earthquake magnitudes will be related to the size and spacing of dikes in the swarm volume in addition to such factors as the "roughness" in the variation of frictional properties along individual fractures.

These two models allow a description of the relationship between magmatic and tectonic fracturing. For basaltic volcanoes, where the liquid low-viscosity magma prefers to migrate along the stratified structures (or faults) in accordance with their strike and does not cut them, the Rubin-Gillard model works. For andesitic and dacitic volcanoes, the intermediate- and high viscosity magmas may choose both ways; therefore, the application of both Rubin-Gillard's and Hill's models could be more realistic.

# Chapter 4
# Origin of Eruption Earthquakes

Eruption earthquakes are related to the surface or subsurface manifestations of volcanic activity. They may be produced by phreatic, phreato-magmatic and magmatic processes in the crater of a volcano as well as by the movement of lava and pyroclastic flows on its slopes and associated falls of incandescent rocks. Therefore, the eruption earthquakes reflect the processes that occur during the outlet of eruption products to the surface and their destruction.

## *4.1. Processes of the outlet of eruption products from a volcanic crater*

### 4.1.1. Fragmentation of magma

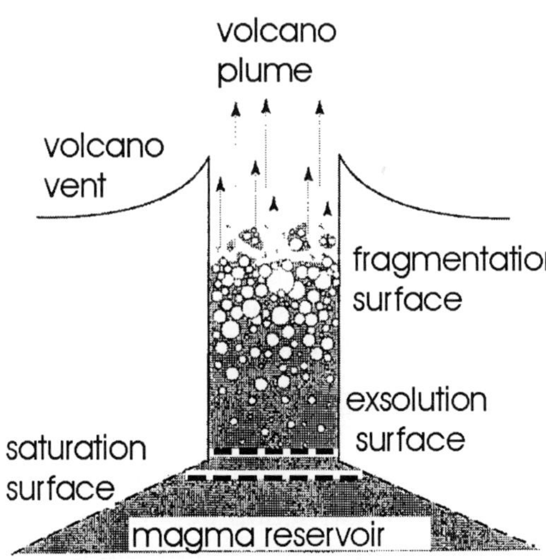


Fig. 4.1. Schematic illustration of processes that occur in volcanic conduits. From Cashman et al. (2000).

When magma passes from a magmatic reservoir up the conduit of volcano, it begins to fragmentize. The *magmatic fragmentation* is the transition from a continuous melt with a dispersed gas phase to disconnected parcels of bubbly melt with a continuous gas phase (Cashman et al., 2000).

Fig. 4.1 illustrates the processes of magma fragmentation. Magma is saturated in volatiles at the top of the reservoir (the saturation surface). Bubbles nucleate at a slightly higher level (the exsolution surface), as magma rises and experiences a decrease in pressure. Between the exsolution and fragmentation surfaces, the magma is a two-phase mixture of silicate melts and gas bubbles. Fragmentation occurs when gas occupies 70-80% of the available volume. At this point, the magma transforms from a liquid with suspended bubbles to a gas with suspended clots of liquid. The abrupt density decrease (expansion) that accompanies this transformation results in acceleration of the gas mixture out of the vent, in the form of a volcanic plume.

The actual process of fragmentation is probably different in magmas of different rheologies. The likely fragmentation mechanism in low viscosity magmas, such as Hawaiian fire fountains and strombolian bubble bursts, is the fluid instability. The fragmentation of an already-vesiculated magma occurs by sudden decompression and bubble rupture. This mechanism may be applied to the failure of volcanic domes, and may also describe explosive disruption of viscous melts during vulcanian eruptions. The relative rate of bubble growth, magma transport, and gas loss will control the style of eruptive behavior (explosive versus effusive) (Cashman et al., 2000).

The interaction of magma (or lava) with groundwater or surface water can lead to *phreato-magmatic fragmentation*. The depth of fragmentation is likely to be determined by the depth of the aquifer. Phreato-magmatic fragmentation results in a spectrum of eruptive activity ranging from passive quenching magma/lava to explosive ejection of ash (Morrissey et al., 2000).

The fragmentation processes generate volcanic tremor and low-frequency and explosion seismic signals.

### 4.1.2. Volcanic "fall-and-flow" processes

When magma merges from the conduit, it forms a lava dome and (or) lava flow. The type of lava construction depends on the type of feeding magma. Basaltic volcanoes form rather liquid low-viscosity lava flows. Andesitic and dacitic volcanoes form lava domes in the crater of the volcano and (or) high-viscosity lava flows. The gravitational collapse of a growing lava dome (Merapi type) or the explosive disruption of growing lava dome (Peléean type) and the collapse from an eruption column

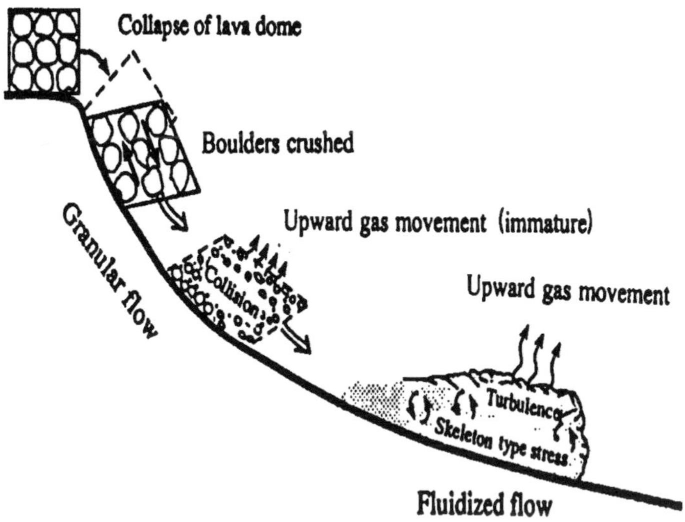


Fig. 4.2. The model of pyroclastic flow process. Collapse of the lava dome induces a granular flow, which changes to a fluidized flow. From Takahashi and Tsujimoto (2000).

Fig. 4.3. Pyroclastic flow that was observed at Volcán de Colima, México during the November 1998 andesitic block-lava flow emission (Zobin et al., 2002c). The resultant block-and-ash flow deposits extended out to 4.5 km from the summit. Photo by A. Cortés. Courtesy of Colima Volcano Observatory.

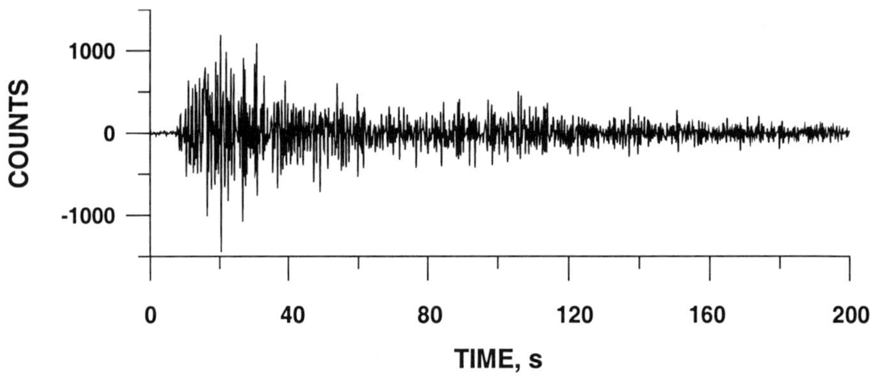


Fig. 4.4. Seismic record associated with the pyroclastic flow shown in Fig. 4.3. Seismic station was situated at a distance of 6 km from the lava dome, on the opposite side to the movement of the flow. Courtesy of Colima Volcano Observatory.

(Soufriére type) can generate pyroclastic flows and rockfalls (Francis, 1998). Pyroclastic flows are distinguished from rockfalls by their larger size, longer runout (≥ 0.5 km), greater production of fine ash and appreciable buoyant hot ash clouds (Calder et al., 2002).

Minor amounts of erupted magma can destabilize the central lava dome, causing its partial collapse. Collapse of a lava dome induces a granular flow, which later changes to a fluidized flow (Takahashi and Tsujimoto, 2000) and leads to the formation of pyroclastic flows of incandescent rocks and to numerous rockfalls down the slopes of the volcanic edifice (Fig. 4.2 and 4.3). Rockfalls and pyroclastic flows are associated with similar emergent, cigar-shaped waxing and waning seismic signals (Fig. 4.4) that had frequencies of 1-10 Hz and differed only in their durations and/or amplitudes (Calder et al., 2002).

### *4.2. Seismic modeling of eruption earthquakes*

Seismic modeling of eruption earthquakes may be done for seismic signals produced by an eruption as a single act or by a sequence of eruptive events. Section 4.1.1 shows that the main processes of the outlet of volcanic products that are able to generate seismic signals may be related to magma movement (for effusive eruptions) and the degassing of magma (for explosive eruptions); section 4.1.2 shows that the processes of gravitational and explosive destruction of a lava body are able to produce seismic signals too. Here I shortly discuss some models of seismic vibrations generated by an eruption and by the components of eruptive processes.

4.2.1. Model of seismic signals produced by an eruption

This model (Kanamori and Given, 1982; 1983; Kanamori et al., 1984) is based on the interpretation that seismic signals are produced by a volcanic eruption as Lamb pulses, excited by a nearly vertical single force that represents the counter force of the eruption. Lamb (1904) computed the transient response of a homogeneous elastic half space to a single force with various geometries. He showed that a pronounced

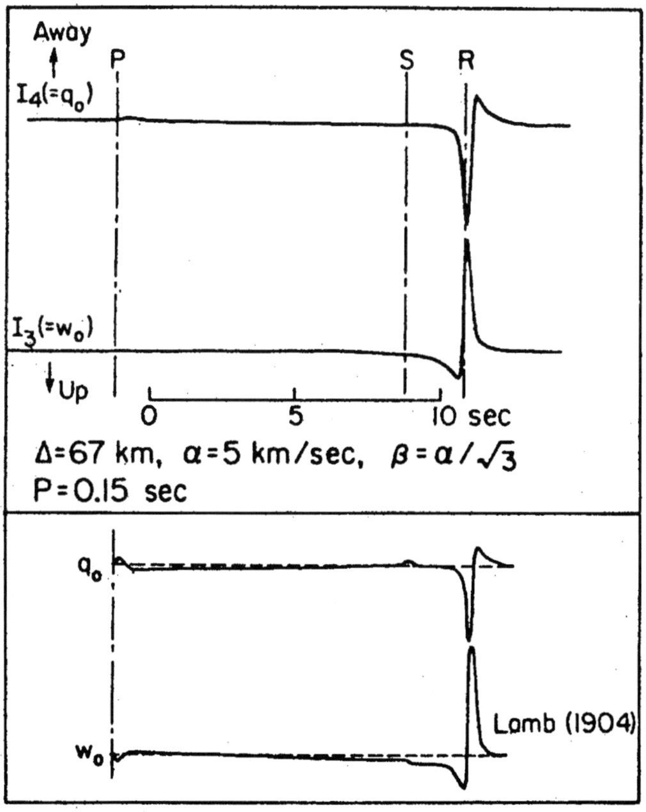


Fig. 4.5. Lamb pulses computed for a downward vertical single force applied at the free surface of a homogeneous half-space at a distance of 67 km. Here $\alpha$ and $\beta$ are P- and S-wave velocities, $I_3$ and $I_4$ correspond to vertical and radial components of the displacement, respectively. In this Fig., $I_3$ and $I_4$ are convolved with the time derivative of the source time function of the form $s(t) = p/(p^2 + t^2)$ that was used by Lamb (1904). Corresponding traces given by Lamb (1904) are shown below for comparison. $I_3$ and $I_4$ correspond to $w_0$ and $q_0$. From Kanamori and Given (1983).

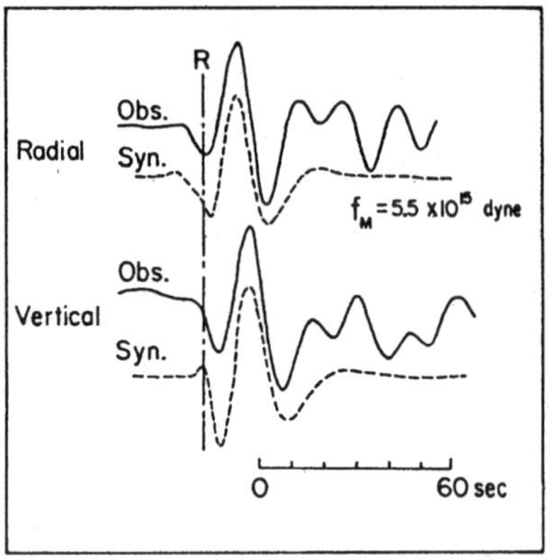


Fig. 4.6. Comparison of the synthetic for the vertical force with the observed long-period record of the June 13, 1980 Mt. St. Helens explosion at a distance of 67 km from the volcano. The synthetics are computed using a distance of 67 km, $\mu$ = $1.9\text{x}10^{11}$ dyne/cm$^2$, and $f_m$ (peak value of the force) = $5.6\text{x}10^{15}$ dyne. From Kanamori and Given (1983).

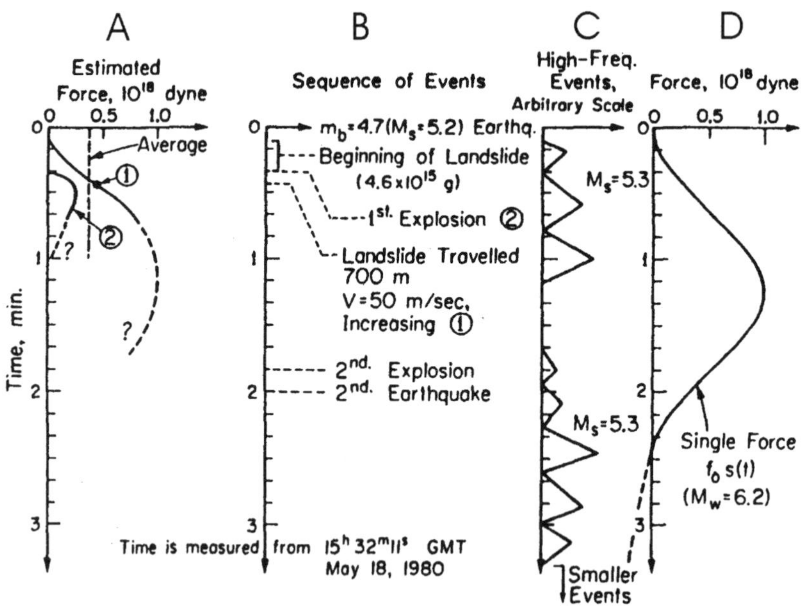


Fig. 4.7. Schematic diagrams showing the sequence of various events associated with the Mount St. Helens eruption of May 18, 1980. Time is taken positive downward. The time history of the single force determined from surface waves is shown on the right, and the magnitude and the time history estimated from the reported sequence are shown on the left. From Kanamori and Given (1982).

pulse propagates along the free surface with the Raleigh-wave velocity. This pulse is preceded by minor pulses, which propagate with P- and S-wave velocities (Fig. 4.5).

The comparison of observed seismograms for small Mt. St. Helens eruptions with synthetics constructed as Lamb's pulses excited by vertical force (Fig. 4.6) shows a good agreement between the observed and synthetic seismograms for the Rayleigh-wave signal. It allows the possibility of estimating the counter force of the eruption by comparing the amplitude of the synthetic with the observed.

Modeling of more complicated eruptions was allowed by supposing that a volcanic eruption generally may be modeled by a combination of a single force representing the trust of the eruption, and an isotropic source corresponding to the sudden pressure release in the magma chamber. For a simple model of eruption, the amplitude ratio of seismic waves excited by the single force to those excited by the isotropic source is approximately equal to the ratio of the seismic wave velocity to the particle velocity of the fluid or gas in the magma chamber. Since this ratio is about 10, a single force can adequately model the source.

An example of the modeling of a complex seismic signal produced by the great May 18, 1980 Mt. St. Helens volcano eruption is shown in Fig. 4.7. The eruption consisted of a few separate events, and all these events produced the parts of the total seismic signal. A sequence of geological events is shown in Fig. 4.7B. The sequence began with the magnitude Ms 5.2 earthquake. In 7 to 20 seconds after the earthquake, a 2.3-$km^3$ landslide began (event 1). During the landslide removal, an explosion (the lateral blast) occurred (event 2). The second explosion occurred just after the termination of landslide movement. Two minutes after the first earthquake, the second earthquake occurred. The total process continued for about 3 minutes.

The time history of the single force inferred from the long-period waves (Fig. 4.7D) characterizes the total process of the eruption sequence as a single event that could be equivalent to an earthquake of magnitude Mw 6.2. The value of this force reached its maximum when the landslide had a maximum velocity. The source time function inferred from the P-wave deconvolved record (Fig. 4.7C) allows the short-period characteristics of the source to be seen. There are two separate events of the same magnitude of 5.3 that might be associated with the first earthquake + first explosion and the second earthquake + second explosion (Fig. 4.7B).

Fig. 4.7A shows the comparative modeling of time history and magnitude of the forces caused by the landslide (1) and the lateral blast (2). According to these reconstructions, the most energetical was the process of landsliding; the resulting force reached a peak value of about

$10^{18}$ dyne. The effect of lateral blast was lower; its resulting force reached a maximum value of $3.4 \times 10^{17}$ dyne.

### 4.2.2. Models based on the vibration of magma body

The models have been developed in two main directions: (1) those based on the resonance of fluid-filled cavities with various geometries (sphere, pipe, or crack) and (2) those based on the oscillation of a number of vibrating conduits. A seismic signal is generated by pressure fluctuations in magmatic system. The choice of the model is very subjective and depends on the knowledge of specific characteristics of the volcano, as well as on the experience of scientist. Here I present the characteristic models of these two groups.

*The fluid-driven crack model (Chouet, 1986; 1996a).* This model may be used for modeling of volcanic tremor and low-frequency earthquakes. The model represents a rectangular fluid-filled crack contained within an infinite, linear elastic medium. The thickness of the crack is much smaller than the seismic wavelengths of interest. The fluid-driven crack is a composite of a trigger element (the energy source) and a resonator (the fluid-filled crack). At time $t = 0$, a pressure transient is applied over a small area of the crack wall triggering the acoustic resonance of the crack. An excess of pressure in the fluid would generate vibrations. Magma, moving through sudden openings of tiny cracks, produces tremor.

The resonance characteristics of the fluid-filled crack depend on the crack geometry, the physical properties of the fluid and solid, the impedance contrast between fluid and solid, the spatio-temporal characteristics of the pressure fluctuations driving the crack, and the boundary conditions in effect at the crack perimeter. The impedance contrast affects both the frequency and duration of the signal radiated by the crack.

The ground response to a buried source may be synthesized after calculating the source space-time function. Variations in the model characteristics enable the optimization of the synthetics modeling. Fig. 4.8 shows the synthetics calculated for different crack excitations and crack depths. The synthetics shape strongly depends on the model depth (compare the synthetics calculated for the depths of 0 m and 105 m). Fig. 4.8 demonstrates that the single excitation synthetics give a long-period event waveform, while the synthetics obtained by delaying and superposing multiple copies of the same signal, give the tremor record.

Fig. 4.9 illustrates a good fitting of the synthetics and observed record of long-period event recorded at Galeras volcano, Colombia.

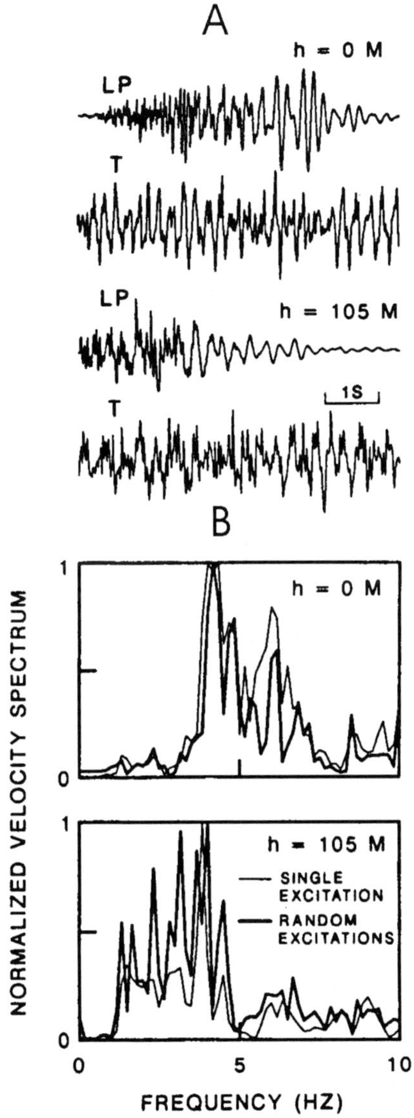


Fig. 4.8. Dependence of synthetics on crack depth and crack excitation. All synthetics are for a 100-m-wide, 200-m-long, vertical crack excited by a pressure transient applied at the center of the crack. Each synthetic represents a vertical component of ground velocity calculated at a distance of 1 km and azimuth of 25°. A, the synthetics of long-period (LP) event and volcanic tremor (T) calculated for the crack depths of 0 m and 105 m. B, the Fourier spectra of the same records. From Chouet (1996a).

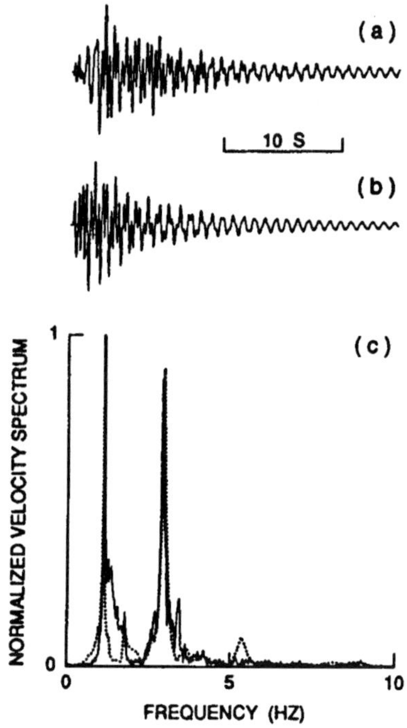


Fig. 4.9. Comparison between the vertical ground velocity recorded at a distance of 1 km from the crater (a) and the synthetics calculated at the same location for the excitation of a vertical fluid-filled crack buried at a depth of 1 km beneath the vent of a volcano (b) for the long-period event that preceded the January 14, 1993 eruption of Galeras volcano, Colombia. The normalized spectra for data (thin line) and synthetics (dotted line) are shown in Fig. c. From Chouet (1996a).

*The conduit-vibration model (Seidl et al., 1981; Schick et al., 1982; Gresta et al., 1991).* This model may be applied to volcanic tremor. It is supposed that the transient flows in magma in the conduits of the volcano can generate the tremor of some dominant frequencies, depending on the conduit geometries. Vapor and gas escaping through the open conduits are considered as responsible for turbulent motions in the magma; consequently the magma-filled pipes in the upper structures of the volcano behave as oscillators. Fig. 4.10 shows the model of conduit oscillators for Etna volcano, Sicily. It consists of a set of vibrators with characteristic frequencies from 0.9 to 4 Hz. These characteristic frequencies are comparable with the peak frequencies that were observed for volcanic tremor at Etna (Fig. 4.11).

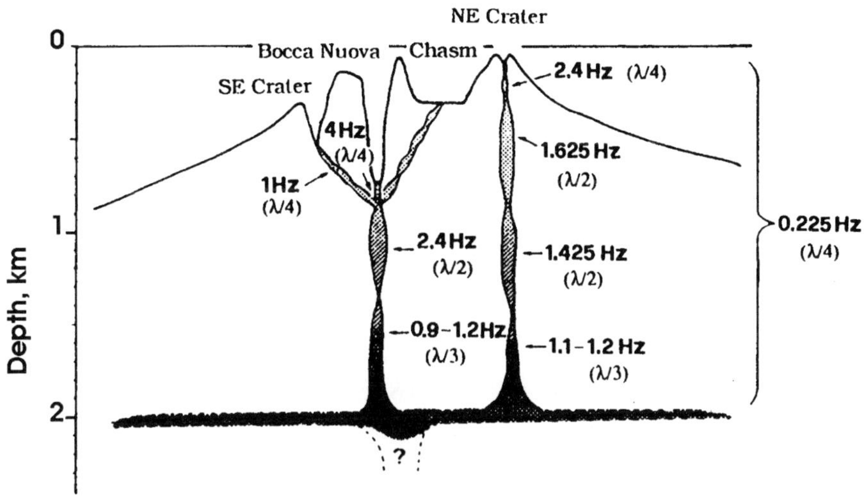


Fig. 4.10. Schematic model of main magma conduits at Etna. Bocca Nuova and Chasm are active craters of Etna. From Cosentino et al. (1984).

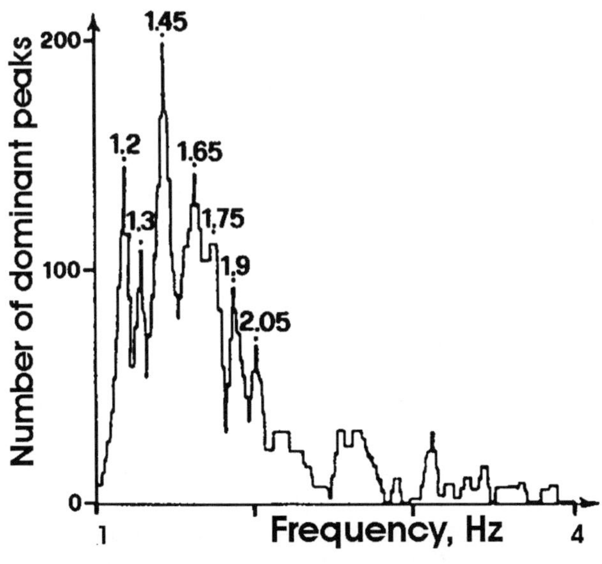


Fig. 4.11. Histogram of the first three dominant peaks of tremor recorded from January 1973 to January 1985 at Etna. From Gresta et al. (1991).

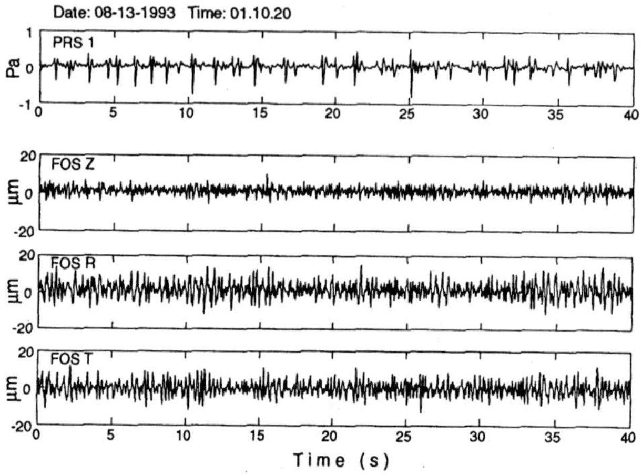


Fig. 4.12. Infrasonic pulses (upper line) and seismic signals (three lines below) recorded by the pressure sensor PRS 1 and the three-component (vertical Z, radial R, and transverse T) station FOS located 150 m from the active vent of Stromboli. From Ripepe and Gordeev (1999).

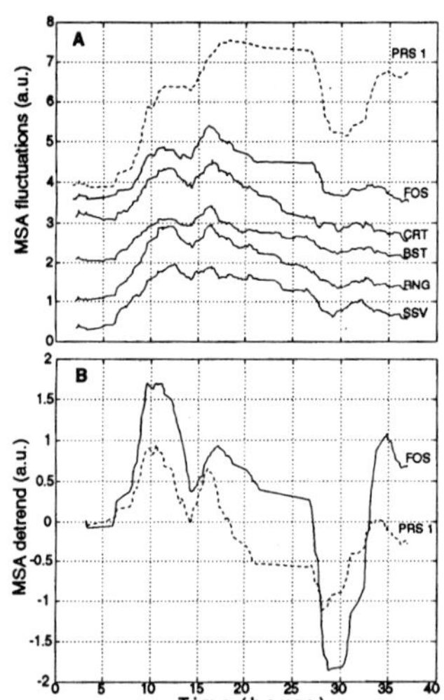


Fig. 4.13. Comparison of the mean-square amplitude (MSA) fluctuations of the infrasonic (PRS 1) and seismic signals. The seismic signals were recorded by 5 seismic stations at a different distance from the craters. The MSA curves of the both types signals are similar (A). In B, the infrasonic and seismic MSA functions have been detrended in order to filter out the long-term variations. Short-term fluctuations are similar in phase (B), indicating that the two signals are influenced by the same dynamical process. From Ripepe and Gordeev (1999).

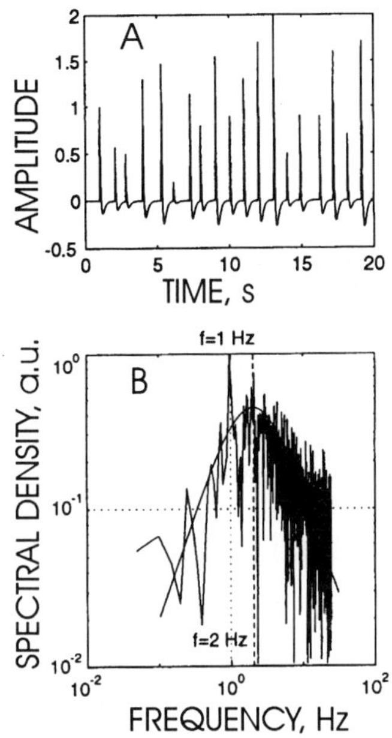


Fig. 4.14. Source time history of the gas bubble dynamic model simulated as series of pressure drops with normally distributed delay times around the mean value of 1 s (A) and the spectrum of this signal (B). The asymptote for the spectrum is shown that indicates the peak frequency of 2 Hz. Distribution of peaks in the spectrum reflects delay times between source events. It can be clearly seen in B that the strong peak at 1 Hz comes from the ~1-s delay time in the source distribution. A.U. is the arbitrary unit. From Ripepe and Gordeev (1999).

### 4.2.3. Models based on degassing process of magma

These models are derived from two approaches: to explain the generation of volcanic tremor and to explain the generation of explosion earthquakes.

*Gas bubble dynamics model* (Ripepe et al., 1996; Gordeev and Ripepe, 1998; Ripepe and Gordeev, 1999). This model is based on the observations of infrasonic and seismic signals. It was observed during simultaneous infrasonic and seismic records that volcanic tremor episodes at Stromboli volcano (Aeolian Islands, Italy) coincide with the small infrasonic transients, which occur rhythmically in time in a range between 0.8 and 1.2 s (Fig. 4.12).

Short-term energy release shows that infrasonic and seismic signals are linked to the same dynamical process (Fig. 4.13), however, when viewed over a longer time scale, it is evident that the two signals are controlled by two distinctive mechanisms. It was proposed that the possible physical model acts in two steps: first, gas bubble coalescence from a layer of small bubbles that produces a seismic signal, and then, gas bubble bursting when it reaches the magma surface. The second step is linked to the infrasonic signal.

Gas bubbles could form by free coalescence in magma or could be forced to coalesce by a structural barrier (such as a dyke). The forced coalescence is believed to be more realistic.

A pressure change can be induced in magma, two orders of magnitude higher than for free coalescence, and it best explains the tremor recorded at Stromboli. In this dynamic process of gases, the delay time of 1-2 s between infrasonic pulses could reflect the gas nucleation interval of basaltic magma. The predicted asymptotic decay of the frequency contents fits the spectral behavior of the volcanic tremor recorded at Stromboli (Fig. 4.14). The distribution of peaks in the spectrum reflects the delay times between source events.

*Vertical expansion source model* (Iguchi, 1994). It was developed for volcanic earthquakes of the andesitic Sakurajima volcano in Japan and may be applied to the modeling of low-frequency and explosion earthquakes. This model may be associated with the expansion process of a gas pocket in the conduit. Fig. 4.15 shows a scheme of the state of

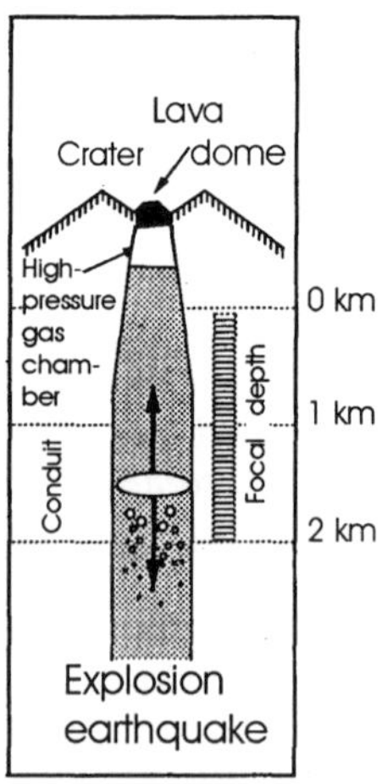


Fig. 4.15. State of the magma conduit prior to the occurrence of an explosion earthquake. Arrows indicate that gas pockets expand vertically along the conduit. Hexagons and ellipses in the conduit represent solidified lavas and gas pockets, respectively. From Iguchi (1994).

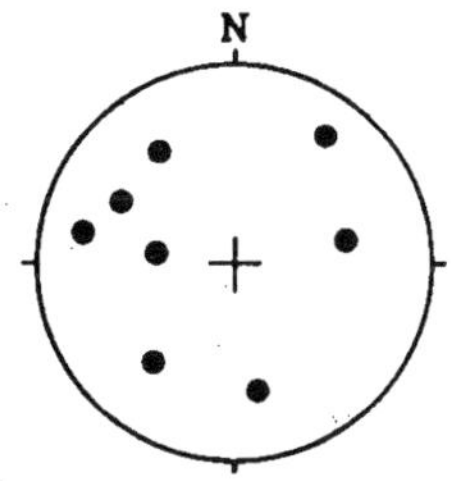


Fig. 4.16. Polarity distribution of the explosion earthquake of April 15, 1988 at Sakurajima volcano, Japan. Filled circles show compressional arrivals. From Iguchi (1994).

The evidence for the vertical expansion model was obtained from the analysis of seismic signals of explosion earthquakes. The polarities of first motions for these signals indicate compressional arrivals for all directions that is characteristic for an expansion source (Fig. 4.16).

Application of the polarization analysis shows that significant SV-waves arrive at the expected arrival time of S-waves, and SH-waves components appear a few seconds later (Fig. 4.17). This indicates the dominance of SV-waves. The moment acceleration tensor calculated for the signal shows that the vertical vector dipole corresponding to the tensor zz-component is dominant (Fig. 4.18).

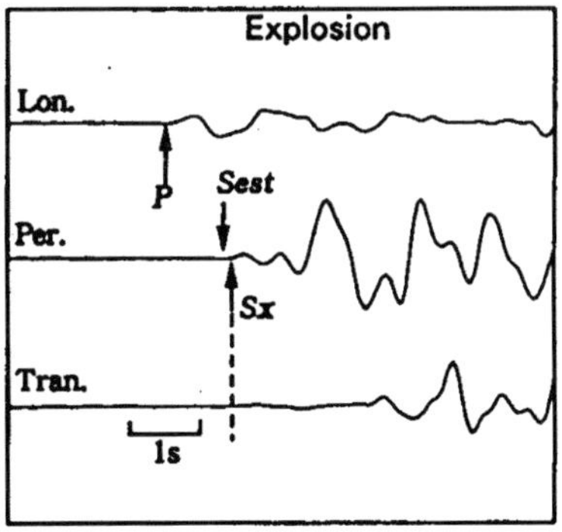


Fig. 4.17. Seismograms of explosion earthquakes of June 4, 1988 at Sakurajima volcano, Japan by applying a polarization filter. The traces Lon., Per. and Tran. correspond to the motions of P-wave, SV-wave and SH-wave, respectively. The symbols Sest and Sx indicate arrival times of S-waves estimated from focal distance and from the beginning times of the SV-wave component, respectively. From Iguchi (1994).

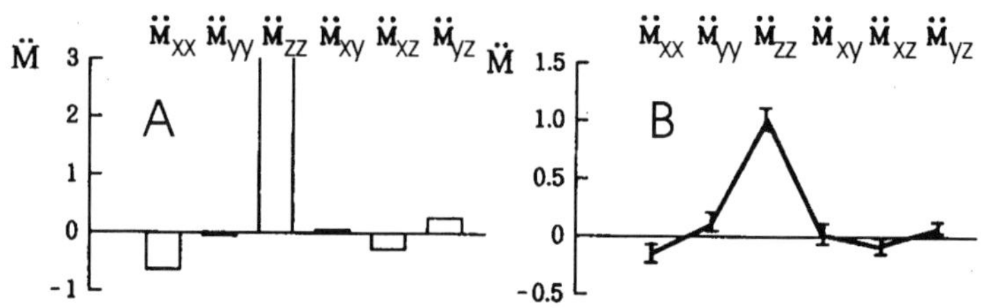


Fig. 4.18. Components of moment acceleration tensor (in units of $10^{14}$ N-m/s$^2$) calculated for the explosion earthquake of April 15, 1988 at Sakurajima volcano, Japan. A, the values of six tensor components; B, the average of all components normalized by the vertical vector dipole. Bars denote errors. The dominance of vertical zz-component is obvious. From Iguchi (1994).

All these seismic evidences allow the vertical expansion of a gas pocket to be considered as a source for explosion earthquakes.

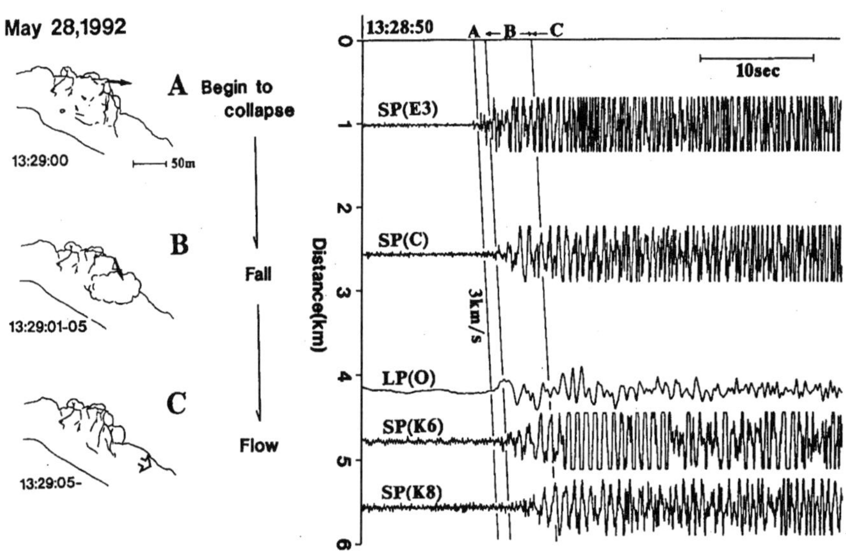


Fig. 4.19. Illustrations of the dome collapse sequence at Unzen volcano, Japan (left) and the corresponding vertical components of seismic signals (right). A sketch of the dome from the seismic station, situated about 3 km to S from the dome, is shown based on a video record taken by a portable 8 mm video recorder. SP indicates the short-period records; LP, the long-period records. The records are arranged according to the horizontal distance from the source. E3, C, O, K6 and K8 are seismic stations. From Yamasato (1997).

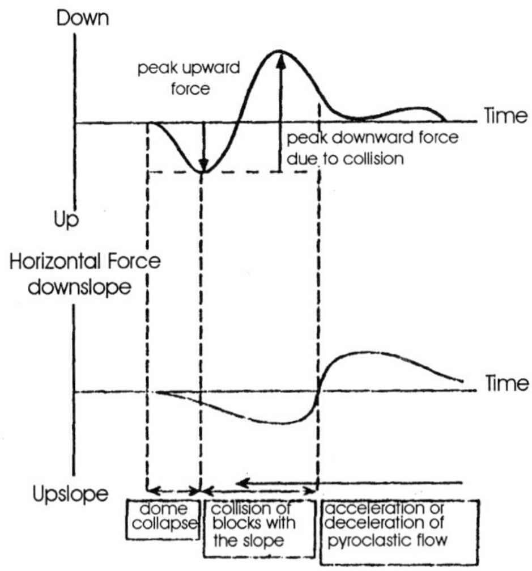


Fig. 4.20. Schematically illustrated source time functions for vertical and horizontal components of a single force exciting pyroclastic flow according to the dome collapse sequence shown in Fig. 4.19. From Uhira et al. (1994).

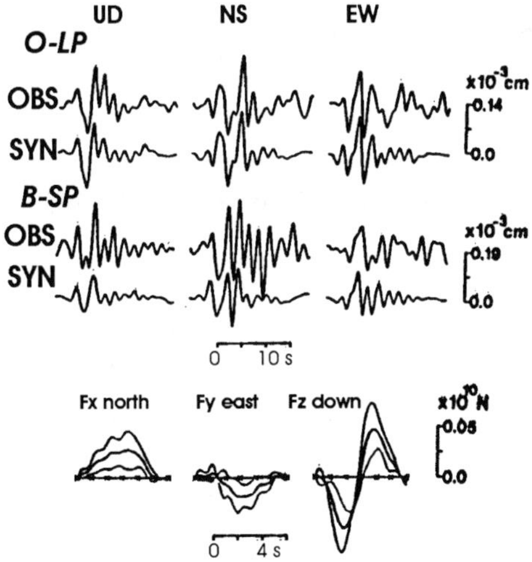


Fig. 4.21. Comparison of observed (OBS) and synthetic (SYN) waveforms of a seismic signal produced by the June 22, 1991 pyroclastic flow at Unzen volcano, Japan. The inversion was made corresponding to the Fig. 4.20 shape of the source time function. The single force components Fx, Fy and Fz that were used for inversion are shown below. The inversion was realized for the short-period record of station B and the long-period record of station O. From Uhira et al. (1994).

### 4.2.4. Modeling of seismic signals generated by pyroclastic flows and rockfalls

*Three-stages dome collapse model (Yamasato, 1997; Uhira et al., 1994).* The model is based on the observations of generations of pyroclastic flows at Unzen volcano, Japan. Fig. 4.19 demonstrates three stages of the collapse of an unstable tip of the dome at Unzen volcano, Japan with the formation of a pyroclastic flow with a volume equal to $5.7 \times 10^4$ $m^3$, recorded by video and seismic instruments: the lava dome collapse (stage A), the downfall of lava blocks onto the slope (stage B), and the generation of pyroclastic flow (stage C). This process may be described by the function representing a single-force system that is shown in Fig. 4.20 (Uhira et al., 1994). The use of the shape of this function for construction of a seismic source time function allows the calculation of synthetic seismograms similar to the observed seismic signals produced by the pyroclastic flow (Fig. 4.21).

The inversion gives the magnitudes of the vertical component of single force corresponding to the collision of fragments with the slope and the horizontal component representing the deceleration of the flowing fragments along the slope.

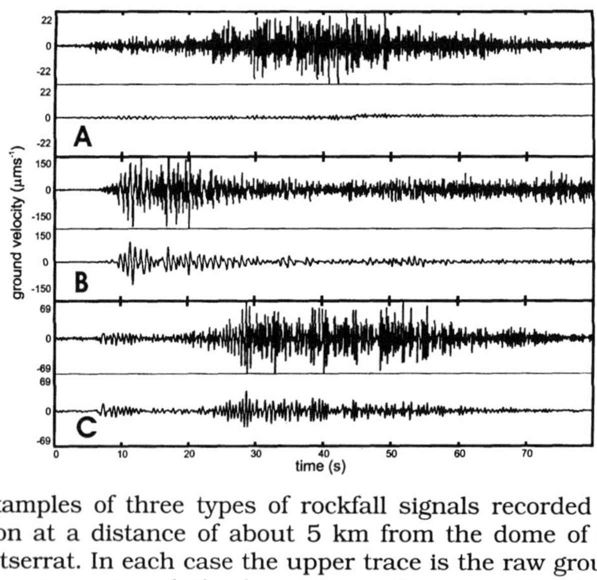


Fig. 4.22. Examples of three types of rockfall signals recorded by broadband seismic station at a distance of about 5 km from the dome of Soufriére Hills volcano, Montserrat. In each case the upper trace is the raw ground velocity on the vertical component and the lower trace the same signal after low-pass filtering at 2 Hz to show the long-period component. A, a rockfall signal with little energy below 2 Hz; B, a rockfall that contains energy below 2 Hz; and C, a long-period rockfall with energy below 2 Hz present as a precursory wavelet, in addition to throughout the rest of the signal. From Luckett et al. (2002).

*Gas-venting + rock-moving model (Luckett et al., 2002).* This model is based on the broadband seismic observations at Soufriére Hills volcano, Montserrat. It consists of two subsources that produce the seismic signal. The first is the resonance of a fluid-filled conduit, which contributes rather long periods (frequencies between 1 and 2 Hz). The long-period resonance is associated with gas escaping into the atmosphere. The second is a combination of forces resulting from rocks moving on the surface of the dome, which contributes the higher-frequency signal. Fig. 4.22 illustrates the presence of long-period (1-2 Hz) component in seismic signals.

# Chapter 5
# Volcano-tectonic Earthquakes at Basaltic Volcanoes

*Basaltic volcanoes* have a characteristic lava of low viscosity ($10-10^2$ poises), high temperature (1100-1600°C), with a small amount of volatiles (0.2-2 wt%), and a high velocity of movement. They are characterized by the effusive type of eruption and produce extensive lava flows.

The eruptions of basaltic volcanoes may occur from the central crater of the volcano, but flank or fissure eruptions are very popular too. The low viscosity and high temperature magmas are very mobile and able to form the passageway to the surface along or out of the central conduit of a volcano. These properties of magma constrain the possible seismic reaction to the magma movement. The fracturing of rock during the forming of new volcanic vents on the flanks of a volcano or along a new fissure produce volcano-tectonic earthquakes. These earthquakes can be observed also during the movement of magma along the volcanic conduit before the central eruptions. The lava effusion from central and flank (fissure) vents is accompanied by volcanic tremor and explosions.

## *5.1. Volcano-tectonic earthquakes associated with shield volcanoes*

Shield volcanoes are broad and have gentle slopes. A few shield volcanoes have persistent lava lakes in their summit calderas. Among them are Mount Erebus (Antarctica) and Nyiragongo (Congo). The most active shield volcanoes are Kilauea in Hawaii and Mt. Etna in Sicily.

### 5.1.1. Kilauea volcano, Hawaii

Kilauea is a tholeitic shield volcano, and the eruptions of a very liquid magma occur mainly in the summit caldera or from the Southwest and East Rift (ERZ) zones (See Fig. 3.6). Formation of vertical dikes appears to be the dominant process in rift-zone intrusion. Kilauea´s volcanic earthquakes occur beneath the rift zones and summit caldera, at a depth of less than 5 km (Klein and Koyanagi, 1989).

Volcanic earthquakes at Kilauea typically occur in swarms. The history of Kilauea´s earthquake swarms that occurred in 1963-1983 (Klein et al., 1987) shows that the swarms can be divided into fast, intense swarms lasting for a few hours to a couple of days (they precede nearly all eruptions of summit caldera and the rift zone intrusions) and slow swarms lasting several days to a few weeks between the eruptive events.

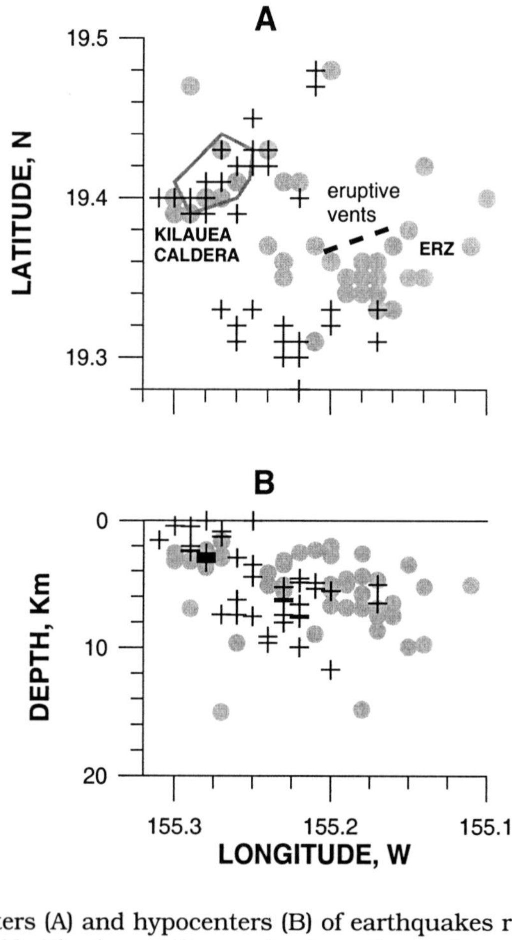


Fig. 5.1. Epicenters (A) and hypocenters (B) of earthquakes recorded from 15 to 22 February 1969. The heavy diamonds show the early events from the East Rift Zone (ERZ) recorded from 15 February to 09:16 h of 16 February; the crosses show the events recorded between 00:42 h of 17 February and 00:49 h of February 22 (inflation swarm); the diamonds show the events recorded from 05:23 to 16:33 of 22 February (eruptive swarm). The outline of caldera is shown, as well as the new ERZ eruptive fissures according to Klein et al. (1987). Data from the catalog of Hawaiian Volcano Observatory.

I selected three episodes of volcanic activity: the 1969 rift-zone intrusion; the 1974 summit caldera eruption with a rift-zone intrusion; and the 1982 summit caldera eruption. These three episodes were preceded by seismic activity. The detailed descriptions of the events are presented in Klein et al. (1987). I use the catalog of Hawaiian Volcano Observatory for an analysis of the seismic process.

*The 1969 rift-zone intrusion* occurred between 22 and 28 of February within the ERZ. The eruptive vents were formed at a distance of about 8 km from the summit caldera (Fig. 5.1). Seismic activity was represented by two seismic sequences: an inflation swarm and an intrusive swarm (Fig. 5.2). The first swarm began on February 17 and was associated with a summit inflation recorded by a deformation network from February 17 to 21, but was not followed by a summit eruption.

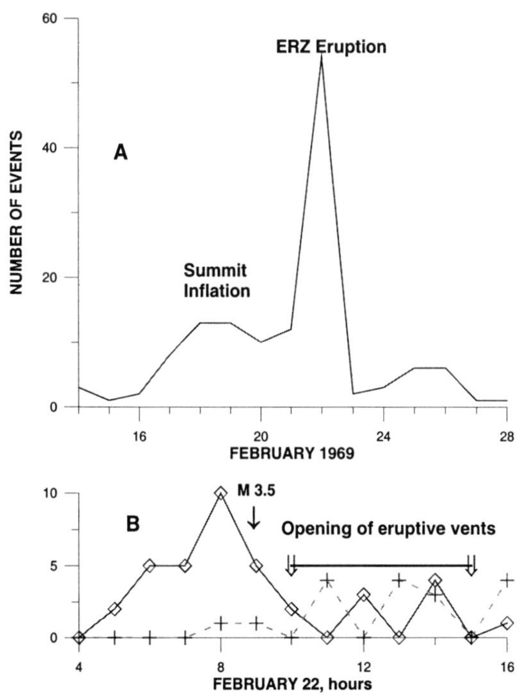


Fig. 5.2. Temporal distribution of total earthquakes recorded before and during the February 22, 1969 ERZ eruption (A) and a structure of the eruptive swarm of February 22 (B). A single arrow indicates the maximum earthquake; the double arrows indicate the period of the opening of the eruptive vents. The dash ed line with crosses shows the summit earthquake variations; a line with diamonds shows the ERZ earthquake variations. Data from the catalog of Hawaiian Volcano Observatory.

The next, more intensive swarm began on February 22 at 5:23 h and lasted for 10 hours. The maximum number of earthquakes per hour and the maximum magnitude event (M 3.5) of this swarm were recorded in three hours after the onset of the swarm. The number of events then gradually decreased. The intrusion began 5 hours after the swarm beginning. Three systems of eruptive vents were opened along the new fissures during about 5 hours (Klein et al., 1987).

Seismic events of the both swarms occurred simultaneously in two zones: beneath the summit caldera and near the new eruptive vents in the ERZ (Fig. 5.1). The events that occurred near the new eruptive vents during the first swarm were more distant to the south than the events of the second swarm that clustered very close by. Analysis of the temporal variations of both groups of events during the second swarm (Fig. 5.2B), shows that the ERZ earthquakes dominated during the first 5 hours prior to the opening of the eruptive vents. With the opening of eruptive vents, the summit and ERZ earthquakes were comparable in number and accompanied the 5-hour opening of fissures.

*The 1974 summit caldera eruption with a rift-zone intrusion* occurred on July 19 and lasted for 4 days. The 9-hour swarm of about 200 located earthquakes (Fig. 5.3) was recorded at depths from 1 km to 4 km between the eruptive vents that were formed within the central caldera and the near-caldera craters of the ERZ (Fig. 5.4). Magma reached the surface about two hours after the swarm had ceased.

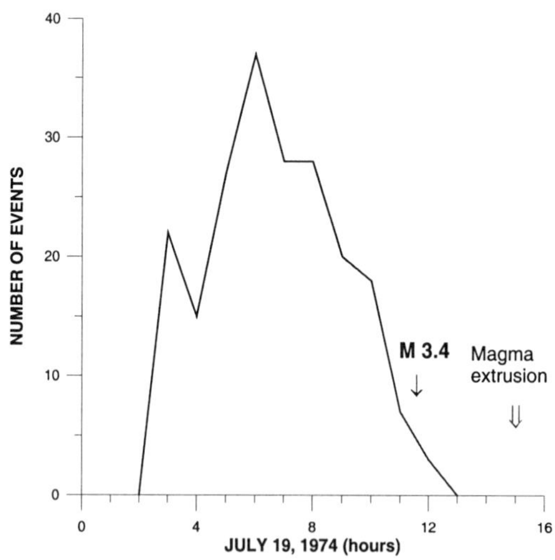


Fig. 5.3. Temporal distribution of earthquakes before the July 19, 1974 Kilauea-ERZ eruption. A single arrow indicates the moment of maximum earthquake; the double arrow indicates the beginning of magma intrusion. Data from the catalog of Hawaiian Volcano Observatory.

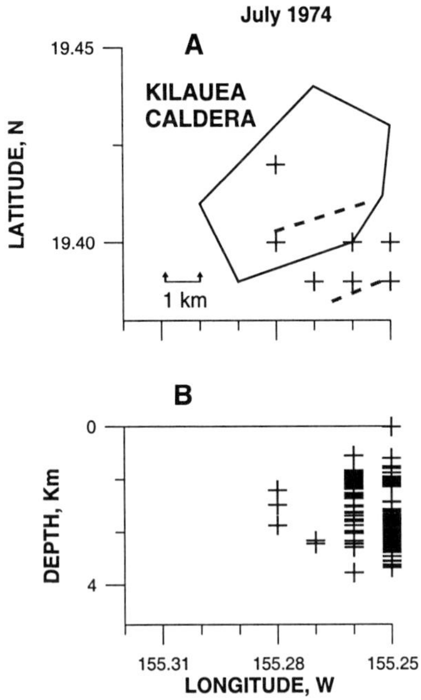


Fig. 5.4. Epicenters (A) and hypocenters (B) of earthquakes recorded on 19 July 1974. The crosses show the events recorded between 03:02 h and 12:28 h of July 19 prior to the magma intrusion. The outline of the caldera (solid lines) and the new eruptive fissures (dashed lines) are shown according to Klein et al. (1987). Data from the catalog of Hawaiian Volcano Observatory.

The swarm developed in two stages. The first peak was recorded during the first hour of activity; the next peak, more intensive, was observed 5 hours after the beginning of the swarm. The maximum magnitude event (M 3.4) was recorded during the final part of the activity, when the number of the events sharply decreased.

*The 1982 summit caldera eruption* occurred on April 30 and continued for one day. The short swarm of earthquakes began about 2.5 hours before the eruption and continued up to the beginning of the volcanic manifestations (Fig. 5.5).

The majority of earthquakes were recorded during the first hour of activity; then the number of events gradually decreased. The maximum magnitude event (M 3.2) was recorded just before the beginning of the eruption.

The epicenters of all 47 events recording between 08:54 h and 11:41 h were located within the central caldera, where the eruptive fissures

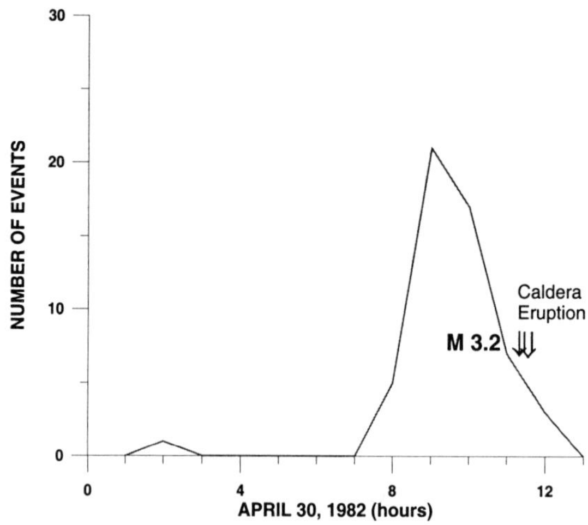


Fig. 5.5. Temporal distribution of earthquakes before the April 30, 1982 Kilauea caldera eruption. A single arrow indicates the moment of maximum earthquake; the double arrow indicates the beginning of eruption. Data from the catalog of Hawaiian Volcano Observatory.

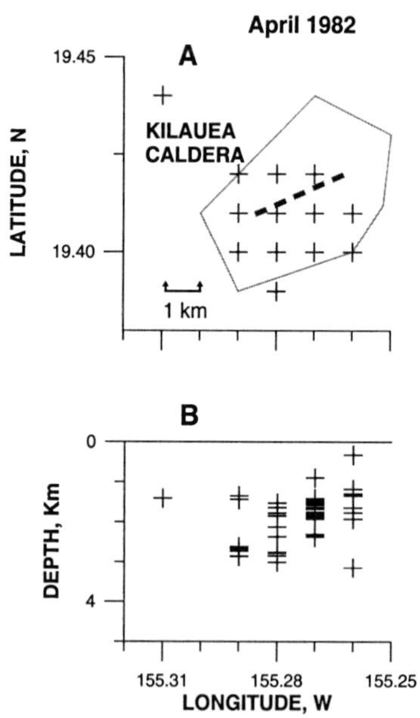


Fig. 5.6. Epicenters (A) and hypocenters (B) of earthquakes recorded on 30 April 1982. The crosses show the events recorded between 08:54 h and 11:41 h of April 30 prior to the caldera eruption. The outline of the caldera (solid lines) and the new eruptive fissures (dashed lines) are shown according to Klein et al. (1987). Data from the catalog of Hawaiian Volcano Observatory.

opened (Fig. 5.6A). Earthquake depths were estimated to be in the range of 1-2 km beneath the caldera (Fig. 5.6B).

5.1.2. Mt. Etna, Sicily

Mt. Etna is a basaltic shield volcano with an elevation of 3350 m. It is the most active volcano in Europe. Etna activity is characterized by persistent activity at the summit craters (central, NE and SE craters) as well as episodic flank eruptions (Gasperini et al., 1990). Geodynamic behavior of the eastern and western sides of the volcano is different (Gresta et al., 1990). Detailed seismic studies began here in 1974, and during 1974-1999 thirteen flank eruptions and eight summit eruptions from the central, NE and SE craters were recorded (Gresta and Patané, 1987; Cristofolini et al., 1988; Simkin and Siebert, 1994; La Delfa et al., 2001). Earthquake swarms preceded and accompanied the majority of flank eruptions. The summit eruptions as usual were not preceded by any seismic crisis (Gresta and Patané, 1987). Below we discuss the relationships between volcano-tectonic seismicity and eruptive process for two largest episodes of 1974-1999.

*The lateral eruption of March 28, 1983* occurred on the southern flank of

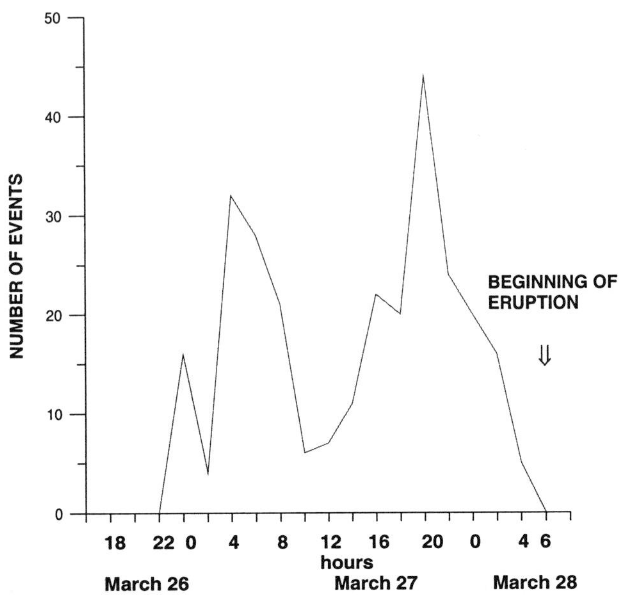


Fig. 5.7. Temporal variations of number of earthquakes before the March 28, 1983 Mt. Etna eruption. The number of earthquakes recorded at least by 3 stations during 2-hour-interval is shown. The double arrow shows the beginning of eruption. From Patané et al. (1984).

the volcano from a short single fracture at the altitude from 2450 to 2250 m (Gresta and Patané, 1987). It continued for 132 days. About $75x10^6$ $m^3$ of volcanic products (mainly lava) were erupted (Gasperini et al., 1990).

A swarm of volcano-tectonic earthquakes began about 31 hours before the eruption and continued for 28 hours (Patané et al., 1984; Fig. 5.7). It consisted of three impulses. The number of events gradually increased from impulse to impulse during first 18 hours and reached its maximum about 10 hours before the eruption. Then the number of events sharply decreased for 6 hours; the eruption began when seismic activity ceased.

Seismic foci were distributed below the SE slope of the volcano at a depth range from 3 km ASL to 15 km BSL but mainly within the volcanic edifice (Fig. 5.8). The majority of events were concentrated in an area along and on the east of the eruptive fissure (Patané et al., 1984). Patané et al. (1984) consider that only the shallowest events

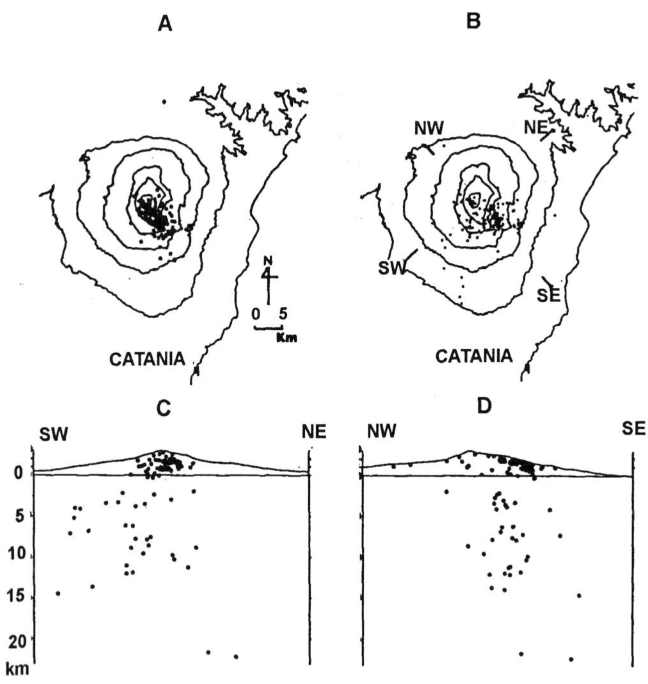


Fig. 5.8. Epicenters (A, B) and hypocenters (C, D) of earthquakes recorded before the March 28, 1983 Mt. Etna eruption. A, Epicentral distribution of shallowest and smallest earthquakes; B, Epicentral distribution of largest earthquakes; C and D, foci distribution along the SW-NE and NW-SE cross-sections, respectively. From Gresta and Patané (1987).

within the volcanic edifice can be related to fracturing processes linked to the opening of eruptive fissures.

During the March-August eruption about 200 volcano-tectonic earthquakes were recorded, only about 74 of them were located (Glot et al., 1984). A significant part of these earthquakes were recorded during the June 3-4 swarm that occurred at a depth from 7 km to 36 km on the upper southern and northwestern flanks of the volcano.

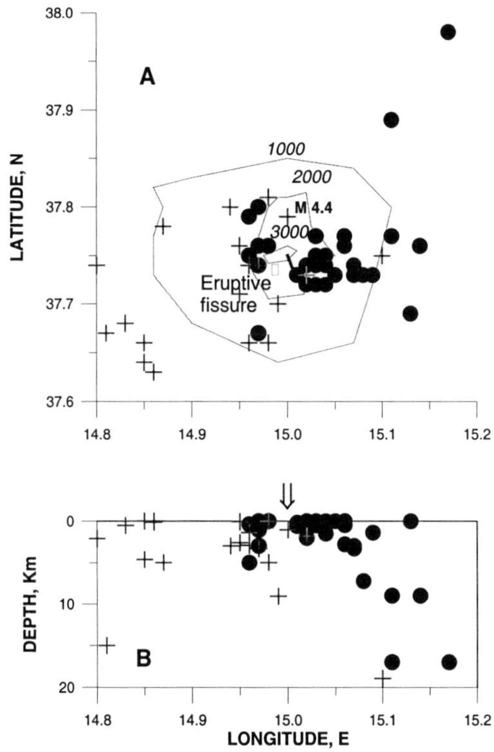


Fig. 5.9. Epicenters (A) and hypocenters (B) of the earthquakes associated with the 1991 flank eruption of Mt. Etna. Filled circles show the events recorded during the re-opening of the 1989 fissures (02:01 h to 18:41 h, December 14); crosses show the events recorded during the lava flow from the vents (from 05:52 h, December 15 to the end of December). A fissure system on the western slope of the volcano is shown by heavy line; its position on Fig. B is shown by the double arrow. The moment of the largest earthquake of the swarm (M 4.4; catalog of S. Gresta) is shown, as are the contours of the volcanic relief (1000, 2000 and 3000 m ). It is seen that the seismic events associated with the opening of the fissures were recorded mainly on the western slope of Mt. Etna while the events recorded during the lava flow from the vents clustered on the eastern slope of the volcano. Data from the catalog prepared by S. Gresta.

*The eruption of April 27 – October 16, 1984.* "As usual for subterminal eruptions, no seismic crisis preceded the long 1984 SE crater eruption" (Gresta and Patané, 1987). It produced the total lava volume of $15 \times 10^6$ $m^3$.

*The eruption of December 1991 – March 1993* was a remarkable event in the recent eruptive history of Mt. Etna, considering both its duration of 473 days and the great volume of emitted products (about $250 \times 10^6$ $m^3$). This eruption took place in the southern flank of the volcano (Lombardo et al., 1996; Fig. 5.9). On December 14, 1991 at about 2:00 h, the uppermost segment of 7-km-long radial fracture that was formed in 1989 (Ferrucci and Patané, 1993) began to re-open on the flanks of the SE crater, accompanied by a swarm of earthquakes (Figs. 5.9 and 5.10). The majority of earthquakes (about 200 events of magnitude from 1 to 3.3; 17 of them were located by S. Gresta, Fig. 5.9) were recorded during the first two hours of the propagation of the fissure below the crater and the western slope of the volcano at the depth from 0 to 10

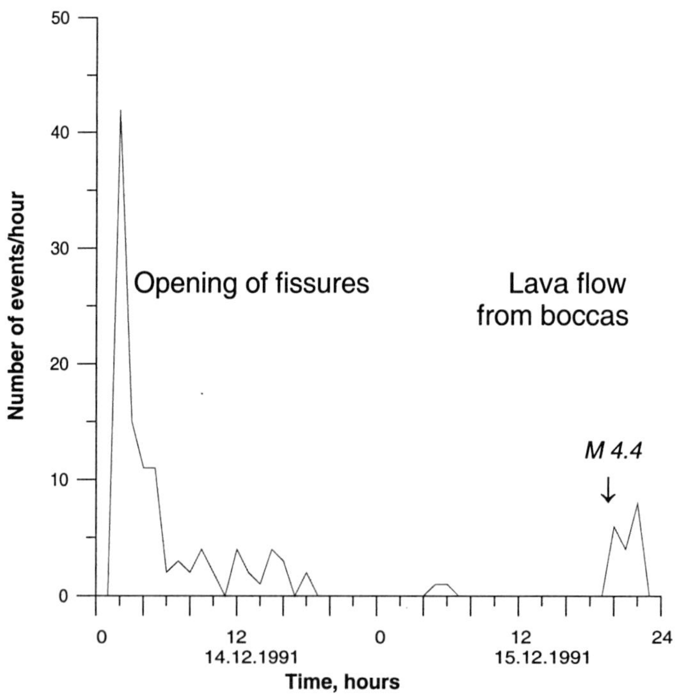


Fig. 5.10. Temporal variations of earthquakes associated with the 1991 flank eruption of Mt. Etna. The arrow shows the onset of the largest earthquake (M 4.4; catalog of S. Gresta) of the swarm. From Ferrucci and Patané (1993).

km. Then seismic activity remained at a low level of 0-3 events per hour.

Early in the morning on December 15, eruptive vents opened at the lower termination of the fracture at a height of 2400 m. A slight increase in seismic activity was observed during the night of December 15 when a sequence of earthquakes occurred below the crater and the eastern slope of the volcano at a depth of 0 to 15 km. This hypocentral zone continued its activity up to the end of December. The largest earthquake of the 14-15 December swarm (M 4.4) was recorded during the lava flow from the new vents, 42 hrs after the beginning of the swarm.

## *5.2. Volcano-tectonic earthquakes associated with stratovolcanoes*

Stratovolcano is a volcano constructed of alternating layers of lava flows and pyroclastic rocks. Stratovolcanoes are steeper than shield volcanoes and tend to a conical form. Among the most active basaltic stratovolcanoes are Oshima in Japan, Hekla in Iceland and Klyuchevskoy in Kamchatka. Many stratovolcanoes have cinder cones and craters of parasitic vents on their flanks (Walker, 2000).

### 5.2.1. Oshima volcano, Izu islands, Japan

Oshima volcano is a basaltic stratovolcano that forms an island in Sagami Bay, about 100 km SSW from Tokyo. The recent eruption of Oshima volcano started on November 15, 1986 within the central cone of Oshima caldera (Mihara-yama). A new cone, A, was formed on the wall of the caldera producing a lava flow. On November 21, a fissure opened on the caldera floor at the NE foot of Mihara-yama (vent B); one hour and half later, a new series of vents (vents C) were opened outside the caldera rim, on the northern slope of the volcano. The activity of new craters continued with some interruptions up to November 18, 1987 (Aramaki, 1988).

Seismic activity of volcano-tectonic earthquakes preceded and accompanied all stages of this one-year sequence of central-flank eruptions (Watanabe, 1988; Yamaoka et al., 1988). Fig. 5.11 shows the epicenters of earthquakes located before the summit eruption of November 15 (A), during this eruption (B) and just before the lateral fissure eruption of November 21 (C).

Before the beginning of summit eruption, the seismic activity of the caldera region was very low and stayed at the same level during the eruption. No preceding swarm was recorded. In contrast, two hours before the fissure eruption a swarm of very shallow (depth < 5 km) earthquakes began at the northern part of caldera, and the seismic foci

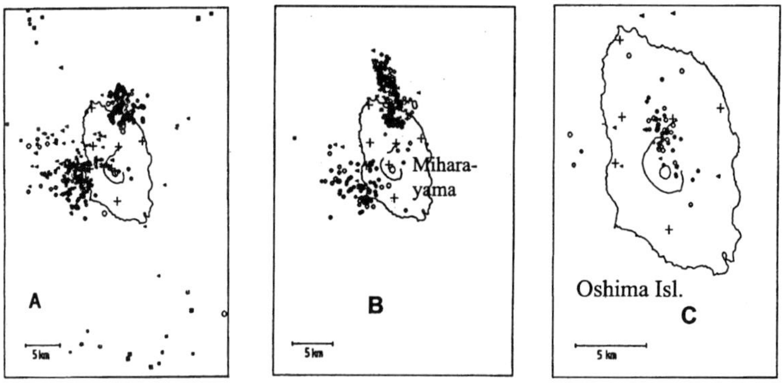


Fig. 5.11. Epicenters of earthquakes located before the November 15, 1986 summit eruption (A, from April 20 to November 15, 1986), during the eruption (B, from November 15 to 20) and just before the November 21, 1986 fissure eruption (C) of Oshima volcano. From Yamaoka et al., (1988).

migrated toward the north (Fig. 5.11 C). The chain of craters B and C opened along the western part of the swarm area.

A very intensive swarm accompanied the fissure eruptions (Figs. 5.12 and 5.13). It occupied a large area of about 40 km². The earthquake foci were distributed at shallow depths (0 to 5 km) below the island and formed the clusters at the depths from 5 km to 10 km to the SE (about 10-15 km from the volcano) and to the NW (about 5-10 km from the volcano). Yamaoka et al. (1988) noted that the foci of events located in these two clusters propagated during the first day of fissure eruption simultaneously to the SE and NW directions from the eruptive site with the velocity of 11-16 m/min. These two hypocentral

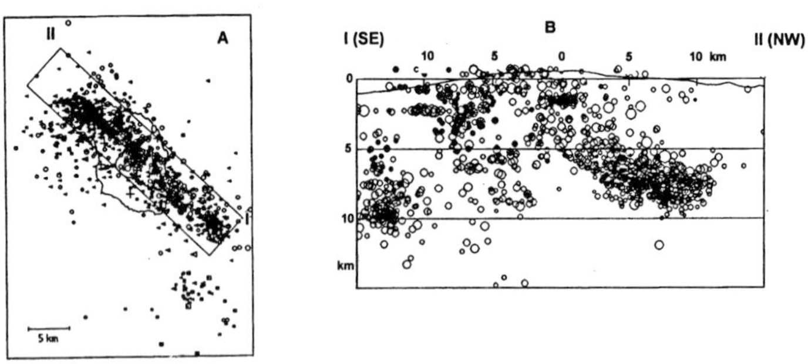


Fig. 5.12. Seismic activity accompanied the fissure eruption of Oshima volcano from 16:15 on November 21 to November 30, 1986. A, epicenters of earthquakes; B, hypocenters of earthquakes. I-II is a zone of cross section of B. From Yamaoka et al. (1988).

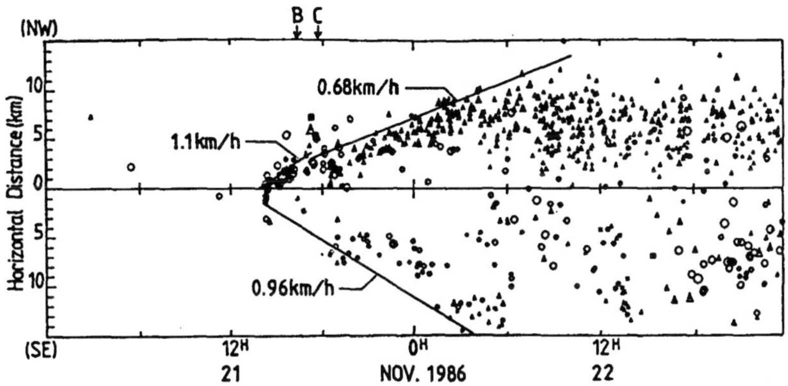


Fig. 5.13. Temporal variations of epicenters from November 21 to 22 in the region shown in Fig. 5.12 A. The focal depths are shown by circles (0-5 km), triangles (5-10 km) and squares (>10 km). The difference in magnitude is shown by the difference in the size of the symbol. Arrows B and C denote the time of B and C craters eruptions. From Yamaoka et al. (1988).

zones were active up to the end of March 1987. The largest earthquake of the swarm (Mw 5.9) occurred on November 22, about 18 hours after the beginning of fissure eruptions, off the SE coast of the island.

### 5.2.2. Klyuchevskoy volcano, Kamchatka, Russia

Klyuchevskoy stratovolcano is the most active basaltic volcano of the Kurile-Kamchatka volcanic zone. It is the highest active volcano in Eurasia; its altitude is 4750 m. Its activity is represented by paroxysmal and small summit eruptions and frequent flank eruptions. The summit eruptions of Klyuchevskoy occur with the formation of cinder cones within the summit crater following by explosions and lava flows.

The paroxysmal eruption of 1945 was preceded by significant seismic activity; the shocks were felt at a distance of 32 km from the volcano a few hours before the eruption of January 1 (Piip, 1956). The small summit eruptions that occurred during 1949-1986 were not preceded by seismic activity that could be recorded at a distance of 16 km where the nearest seismic station was installed; only volcanic tremor was observed before and during the summit eruptions (Tokarev, 1966; Khrenov et al., 1991).

The twelve lateral (flank) eruptions that occurred on the flanks of Klyuchevskoy from 1932 to 1987, at altitudes from 450 to 3900 m (Khrenov et al., 1991) were preceded by seismic activity. We discuss here the seismic activity associated with two flank eruptions, of 1951 and 1966.

*The November 19, 1951 Bylinkina flank cone eruption* occurred at the height of 950 m with forming of two new cones followed by lava flow (Tokarev, 1981). Seismic activity began 6 days before the beginning of eruption (Fig. 5.14). Number of earthquakes increased rapidly during first 30 hours, then gradually decreased for three and half days. The strongest earthquake of magnitude 3.1 occurred in two days after the beginning of seismic activity. The eruption began in one day after the cease of seismic activity. Only one station KLY was in operation, so no locations were made.

*The October 6, 1966 Piip flank cone eruption* occurred at an altitude of 1900-2100 m with the formation of a sequence of explosion craters followed by a lava flow (Tokarev et al., 1968). Seismic activity began 5 days before the eruption with a sharp increase in number of earthquakes; the largest earthquake (M 3.7) occurred during the first hour of activity (Fig. 5.15).

Two peaks in activity were recorded during first 3 days of the swarm. Then the number of events decreased rapidly and the eruption occurred above the background of a termination of seismic activity. Fig. 5.16 shows the distribution of epicenters of 22 earthquakes of the swarm (M > 2). They formed an elongated area of about 20 km of length with the site of eruption in the center of the area. The largest

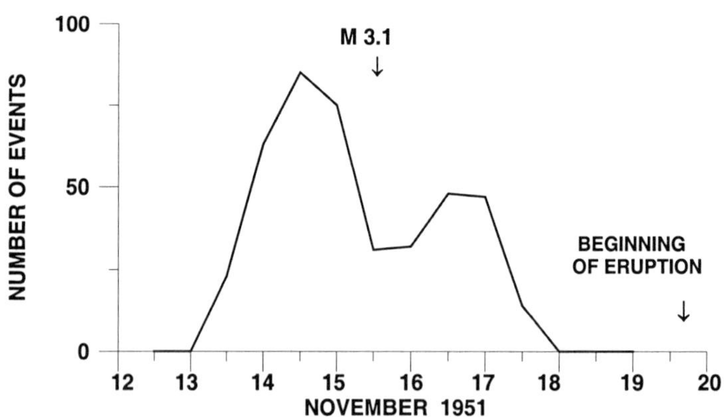


Fig. 5.14. Temporal distribution of 12-hrs number of earthquakes (M > 1) prior to the November 19, 1951 eruption of the Bylinkina flank cone of Klyuchevskoy volcano, recorded at seismic station KLY (see Fig. 5.16), at a distance of 22 km. The moment of the largest event of magnitude M 3.1 is shown. Data from Tokarev (1981).

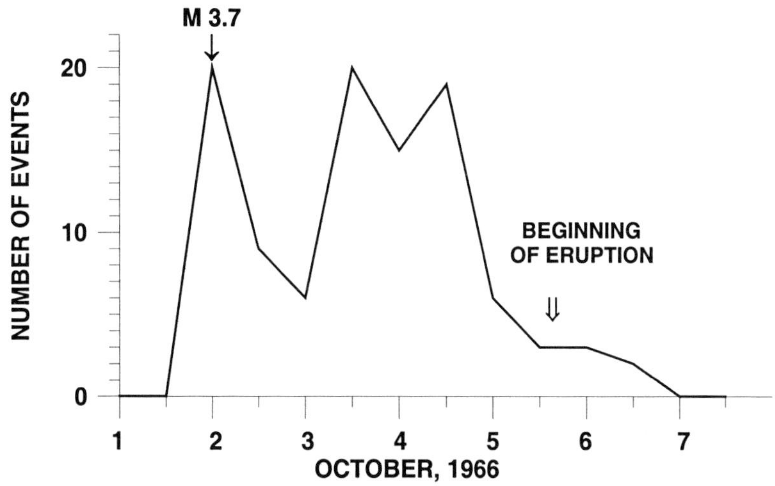


Fig.5.15. Temporal distribution of 12-hrs number of earthquakes (M > 1) prior to the October 6, 1966 eruption of the Piip flank cone of Klyuchevskoy volcano recorded at a seismic station KLY (Fig. 5.16) at a distance of 25 km. The moment of the largest event of magnitude M 3.7 is shown. Data from Tokarev (1981).

earthquakes of magnitude 2.5-3.7 form a lineal group, including the eruptive fissure within. The first three largest earthquakes that were recorded during the first hour of activity occurred at depths of 10-20 km. All remaining events were very shallow.

## 5.3. *Volcano-tectonic earthquakes associated with fissure eruptions*

A fissure eruption is an eruption that takes place from an elongate fissure, rather than from a central vent (Schmidt and Schmincke, 2000). The most famous recent fissure eruptions were of Parícutin in Mexico (1943-1952), Krafla in Iceland (1975-1983) and New Tolbachik volcanoes in Kamchatka (1975-1976).

### 5.3.1. New Tolbachik Volcanoes, Kamchatka, Russia

A chain of new volcanoes was formed from 6 July 1975 to 10 December 1976 along the new 12-km-length fissure that was opened at the distance of about 18 km SW from Plosky Tolbachik volcano (Fig. 5.17). The new eight volcanoes were basaltic cinder cones with a height up to 300 m; 1.2 $km^3$ of lava was produced by the cones, which covered an area of about 45 $km^2$. The total volume of pyroclastic material was about 1 $km^3$.

The preceding swarm of earthquakes (73 events of magnitude M > 2) began on June 27. It is possible to discriminate two stages in seismic activity development (Fig. 5.18): from 27 June to 30 June and from 1 July to 5 July. During the first stage, a sharp increase in the number of events was observed on June 27, and then the number gradually decreased for three days. The epicenters were distributed within an ellipse-shaped area just to the north from the site of forthcoming eruption of the cone I (Fig. 5.17A) and covered the space where the following 6 cones appeared. The depths of earthquakes varied from 15 km to the surface (Fig. 5.17B).

The second impulse in seismic activity was characterized by lower number of events and ceased one day before the July 6 eruption. The epicenters occupied more compact area than during the first stage, nearer to the forthcoming eruption site. They were shallower and were distributed from 8 km of depth to the surface. The largest earthquake of swarm (magnitude Mw 4.9) was observed during the second stage, on July 2 at a depth of 11 km. Its epicenter was located below the cones II

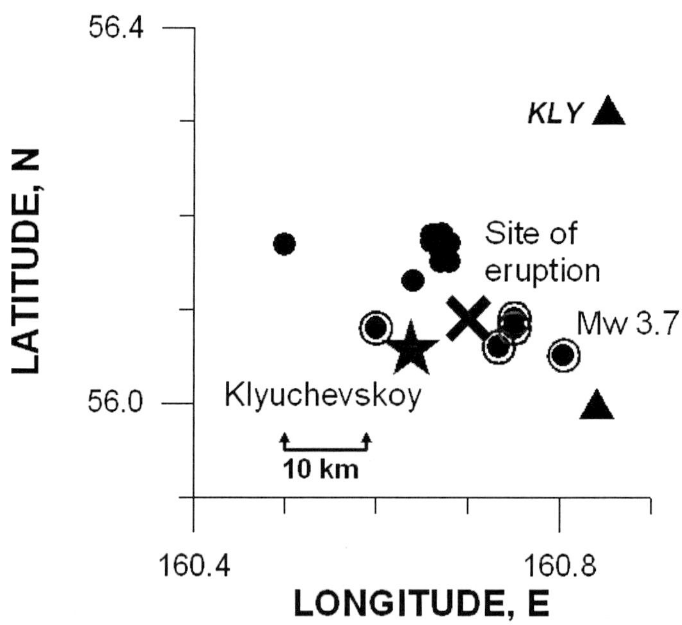


Fig. 5.16. Epicenters of earthquakes (M > 2) recorded before the October 5, 1966 eruption of the Piip flank cone of Klyuchevskoy volcano. Star shows Klyuchevskoy volcano; cross shows a site of new eruptive center. The two nearest seismic stations are shown by triangles. The large (magnitude M > 2.5) earthquakes are shown by additional circle. Taken from the Kamchatka Regional Earthquake Catalog.

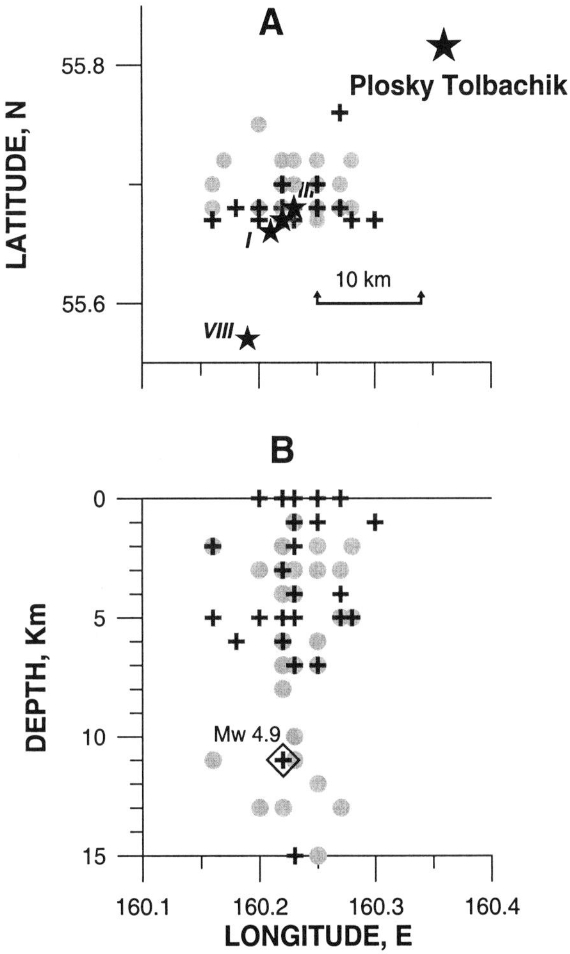


Fig. 5.17. Epicenters (A) and hypocenters (B) of the earthquakes (M > 2) preceding the July 6, 1975 eruption of the 1st cone of New Tolbachik volcanoes. The epicenters of events recorded from 26 to 30 June are shown by filled circles, the events recorded from 1 to 6 July are shown by crosses. The largest earthquake is shown by diamante. Active volcano Plosky Tolbachik (PT in Fig. B) is shown by large star; new cones I, II, II and VIII are shown by stars. Data from the Kamchatka Regional Earthquake Catalog.

and III. During the following eruptive activity with the formation of the new cones in August-September, 1975, a few seismic sequences were observed (Fig. 5.19). One sequence (11 events of M > 2) occurred from 6 to 8 of August and preceded the birth of cone II on 9 August. The epicenters were distributed within the southern part of epicentral zone of the first swarm of June 26-July 6, on the level of the forthcoming site of cone II. The next sequence of small earthquakes (35 events of M > 2) occurred from 10 to 14 August, before the forming of the third and fourth cones on August 17. These seismic events connected the new eruptive zone with the old basaltic Plosky Tolbachik volcano (Fig. 5.19) where reda caldera collapse occur in the middle of August (Dvigalo et al., 1984) marking a clear relation between the old volcano and the new active cones.

The last seismic swarm (12 events of magnitude M > 2) was occurred from 16 to 18 September 1975, preceding the birth of the 8$^{th}$ cone VIII (Fig. 5.19). The location of these events continued the active line from Plosky Tolbachik volcano through the North Zone of the cones I-VII (they finished their activity on September 15) to the south up to the site of cone VIII that was born on September 18.

The total process of migration of epicenters along the 30-km line Plosky Tolbachik – Cone VIII from August 6 to September 18 is seen in Fig. 5.20. During the initial part of the post-I-cone seismicity (Aug. 6-8), the epicenters were clustered on the position of the forthcoming cone II. Then, after forming of cone II on August 9, a NE migration of foci was observed to Plosky Tolbachik volcano, that continued from 10 to 14 August and coincided with the caldera collapse of Plosky Tolbachik. The last SW focal movement was from the site of cones I-VII to the site of the forthcoming cone VIII.

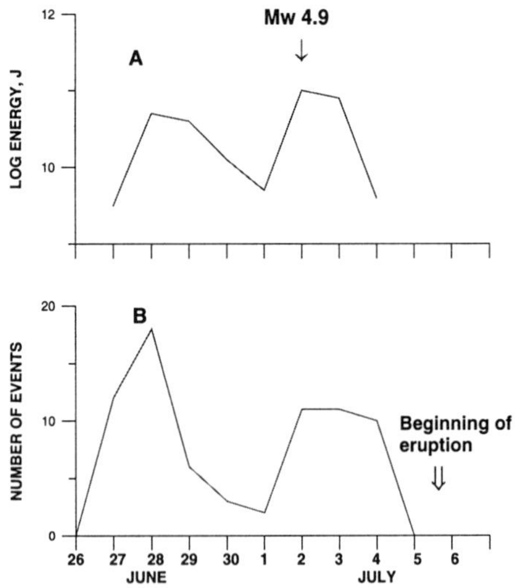


Fig. 5.18. Temporal variations of daily number of earthquakes (M > 2) preceding the July 6, 1975 eruption of the 1st cone of New Tolbachik volcanoes (B) and their maximum daily energy (A). The moment of largest event is shown. Data from the Kamchatka Regional Earthquake Catalog.

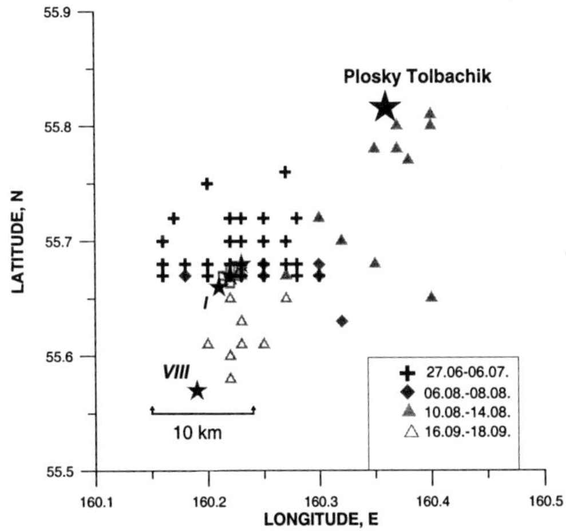


Fig. 5.19. Epicenters of earthquakes (M > 2) that were observed during June-September, 1975 forming of the eight new Tolbachik cones. Four groups of events are shown: 26 June - 6 July (crosses); 6-8 August (diamonds); 10-14 August (black triangles) and 16-18 September (open triangles). Cones I to VIII are shown by gray stars; active volcano Plosky Tolbachik is shown by black star. Data from the Kamchatka Regional Earthquake Catalog.

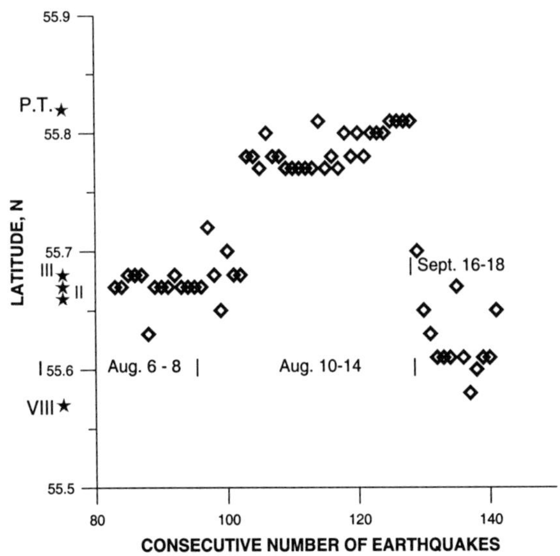


Fig. 5.20. Migration of earthquake foci (M > 2) along the NE line Plosky Tolbachik volcano – Cone VIII from August 6 to September 18. Stars along the latitude axis show the positions of new cones (II, III, VIII) and Plosky Tolbachik volcano. Data from the Kamchatka Regional Earthquake Catalog.

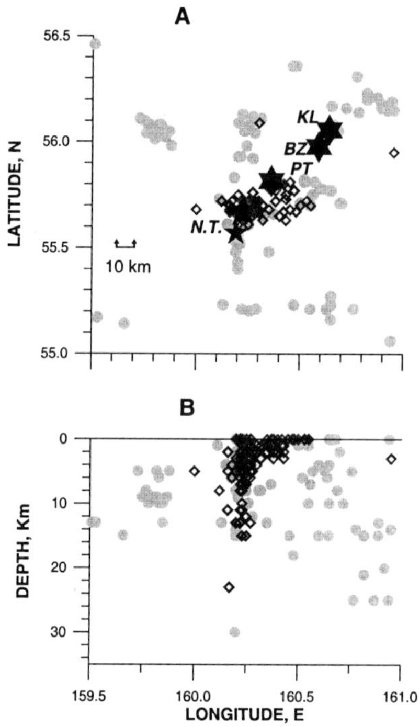


Fig. 5.21. Post-eruption (16.12.1976-1978) seismic activity in the region of New Tolbachik volcanoes. A, epicentral distribution of earthquakes; B, hypocentral distribution of earthquakes. Open diamonds are the earthquakes recorded during the active period (27.07.1975-15.12.1976); filled circles are the earthquakes recorded from December 16, 1976 to the end of 1978. The 5-ray stars show the new Tolbachik volcanoes; the active volcanoes of Klyuchevskaya group (KL, Klyuchevskoy; BZ, Bezymianny; and PT, Plosky Tolbachik) are shown by 6-ray stars. N.T. is a site of New Tolbachik volcanoes. Data from the Kamchatka Regional Earthquake Catalog.

After the end of activity of the eighth cone on December 15, 1976, seismic events began to spread along the line of the new volcanoes, to the north and south from the active centers (Fig. 5.21). During two weeks, a line of 90-km length was outlined by the epicenters. In 1977 and 1978, seismic activity was recorded at the periphery of the active volcanic zone at the distance of 40-60 km from the new volcanoes.

### *5.4. Volcano-tectonic earthquakes associated with caldera collapse*

A caldera is a gigantic basin with steep walls at the summit of a volcano; larger than a crater. It is usually formed by collapse (Decker and Decker, 1999). The multiple stages of collapse following evacuation of an underlying magma chamber are characteristic for Kilauea (Hawaii) and Fernandina (Galapagos Islands). The great 1883 eruption of Krakatau (Indonesia) and the 1912 eruption of Katmai (Alaska) are the best geologically documented examples of a caldera-forming events.

### 5.4.1. Fernandina volcano, Galápagos Isl.

Fernandina is a large basaltic shield volcano on the western edge of Galápagos platform (Fig. 5.22). At its summit, some 1500 m above sea level, an elliptical caldera of 4 x 6.5 km$^2$ of area is situated. On 21 May 1968, the eruption started, and a lava flow was observed on the east flank of Fernandina. A sequence of strong explosions began on June 8 and then declined on June 12. The aerial reconnaissance of June 21 revealed that the 8-km$^2$ caldera floor subsided in a series of short drops along an elliptical boundary fault. The maximum subsidence was approximately 350 m with a resulting collapse volume of 2 km$^3$ (Filson et al., 1973; Francis, 1974).

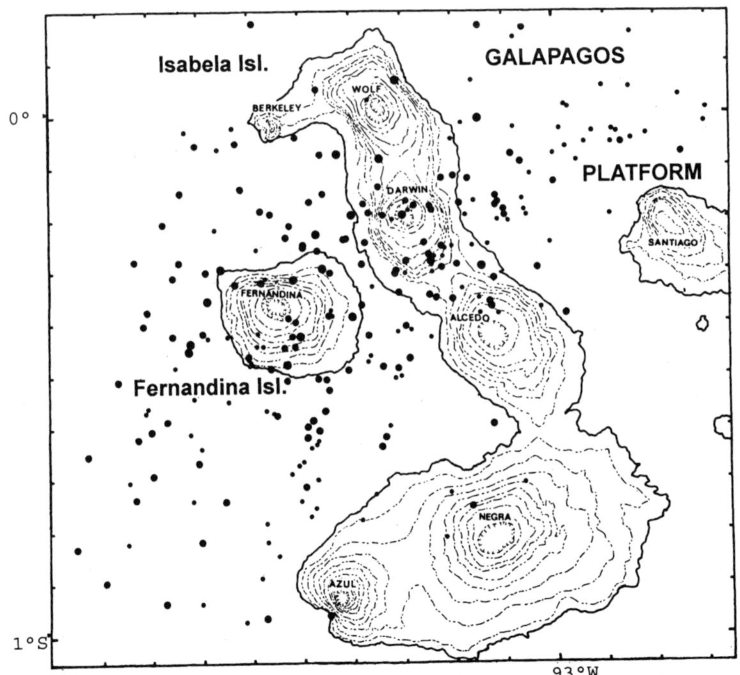


Fig. 5.22. Epicenters of Galapagos swarm earthquakes located by NOS network. The large distance of the nearest stations from the epicenters (about 400 km) produced a large error in locations; therefore, the great area of distribution of earthquakes may not reflect the real situation. From Filson et al. (1973).

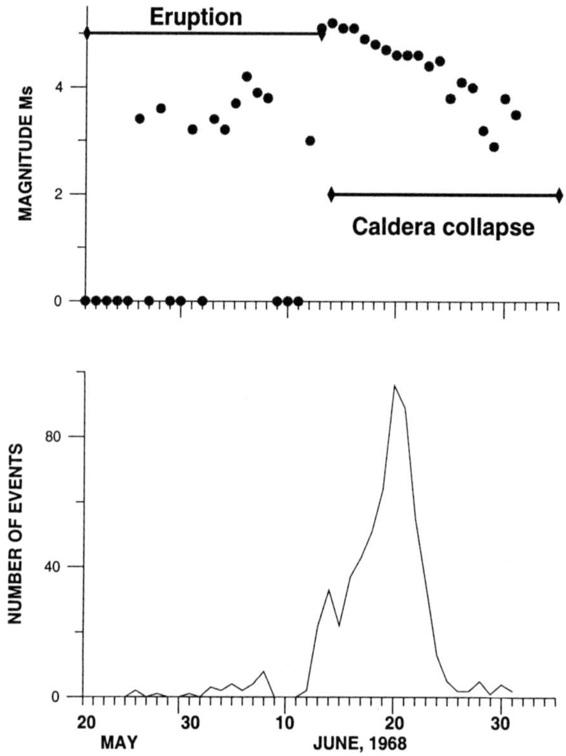


Fig. 5.23. Distribution of maximum magnitudes Ms (A) and daily number of earthquakes (B) from May 20 to June 30, 1968. Data from Filson et al. (1973).

Minor seismic activity (Ms < 4.2) was recorded during the first week of June (Fig. 5. 23B). On June 12, a sharp increase in seismic activity began and reached its maximum on 19 June. Then the number of earthquakes declined rapidly, and the swarm practically stopped after June 23. This seismic swarm was characterized by a great number of high-magnitude events that had practically the same form (Filson et al., 1973; Kaufman and Burdick, 1980). Seismic activity developed in two stages. The first stage was associated with the eruptive activity of the volcano (May 20 – June 11). The maximum magnitude of events recorded during this stage was 4.2. The second stage began on June 12, when the magnitude and number of events began to sharply increase. The maximum daily magnitudes 5.1-5.2 were recorded during four days, 12 to 15 of June (Fig. 5.23 A). The total number of events of magnitude Ms ≥ 5.0 was 5 (Filson et al., 1973), which is very large for any volcanic earthquake swarm. Filson et al. (1973) suggested that the caldera collapse occurred during the second stage of seismic activity, and the earthquakes were resulted from the dropping of the caldera block. All epicenter locations were made by the global seismic network (Fig. 5.22) that produced large errors in location. It made it difficult to do any tracing of epicenters and comparison with the volcanic structures.

*5.5. Volcano-tectonic earthquakes associated with submarine eruptions*

Submarine eruptions vary between its birth on the deep sea-floor and its emergence above sea level. Evidence of deep submarine eruptions is rarely visible at the surface, because the steam they produce is quickly condensed as it rises through cold seawater. At the same time, if the eruption is vigorous and the vent is within 500 m or so of the surface, the water may become turbulent and cloudy (Bardintzeff and McBirney, 1999). The best-documented seismological example is the 1989 Teishi Knoll eruption near Japan.

5.5.1. Teishi Knoll volcano, 1989. Izu Islands, Japan

On June 30, 1989, an earthquake swarm started off eastern Izu Peninsula in Japan (Figs. 5.24 and 5.25). This was then culminated in the submarine eruption on July 13. The hourly numbers of earthquakes that were observed at station KMT, situated at a distance of 6.8 km from the eruption site, show (Fig. 5.26; Yamasato et al., 1991) that three main stages in development of the swarm were observed: (1) from 30 June to 4 July; (2) from 4 July to 8 July and (3) from 8 July to 13 July.

During the stage 1 seismic activity was very low; the great increase in the number of events began from 4 July, then gradually decreased during the next three days (stage 2). The next large impulse in seismic activity (stage 3) began on 8 July and reached its maximum on July 9, when the M 5.5 earthquake, largest in this sequence, occurred. Then the seismic activity gradually declined and the eruption began on 13 July with a low background level.

Fig. 5.24 demonstrates that a regularization of the epicentral zone of events during the swarm was observed. At the 1st and 2nd stages, the epicentral zone had an elliptical form (Fig. 5.24b). At the 3rd stage, the epicenters clustered in a lineal submeridional zone including the site of forthcoming eruption (Fig. 5.24c and d). After the beginning of the eruption, the seismic events continued to cluster within this lineal zone. The majority of seismic events of the swarm were located within the depth range from 5 to 10 km. Two hypocentral subzones were formed: the initial large subzone just below the eruption site that was active during stage I and II, and a narrow smaller size subzone that was

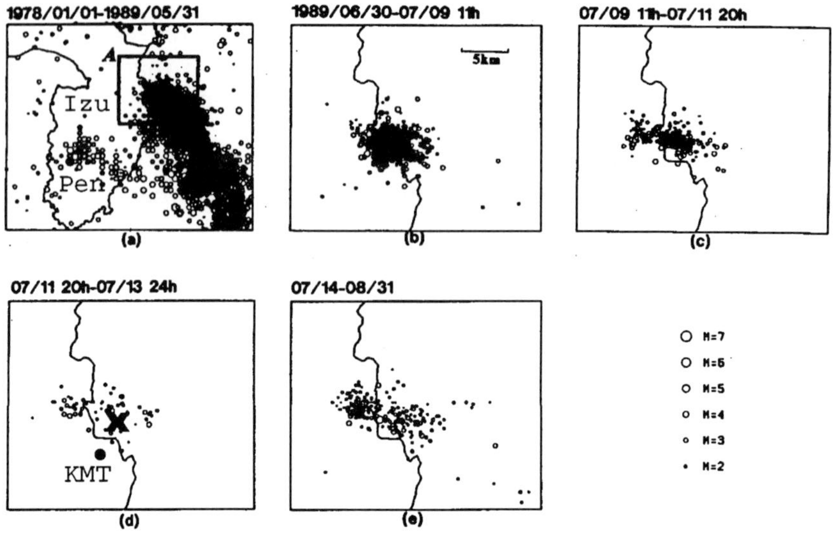


Fig. 5.24. Spatial-temporal development of the 1989 Teishi-knoll earthquake swarm. a, Seismic activity of the zone of eruption before the swarm (1978-31/05/1989); b, Seismicity during the first and the second stages of the swarm (June 30 – July 9, 11 h); c, Seismicity during the third stage of the swarm (July 9, 11 h – July 11, 20 h); d, Seismicity just before the eruption (July 11, 20 h – July 13, 24 h); e, Seismicity during the eruption (July 14 – August 31). Insert A in Fig. 5.24 a shows the area of study in Figs. 5.24 b-e. Cross in Fig. 5.24 d shows the site of eruption. KMT is a seismic station. From Yamasato et al. (1991).

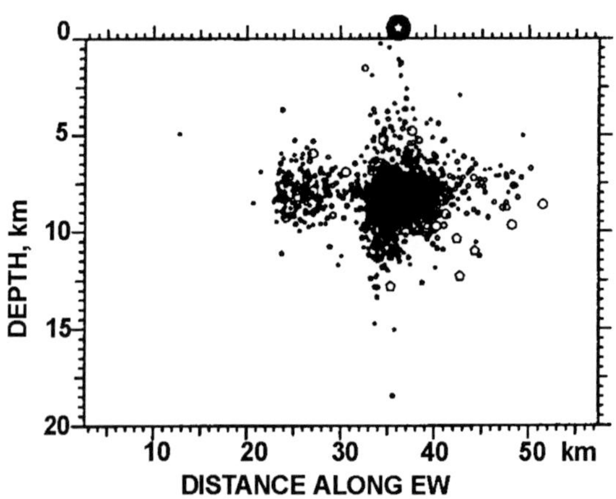


Fig. 5.25. Depth distribution of the 1989 Teishi Knoll earthquake swarm events along the EW direction. A black circle shows the site of eruption. From Matsumura et al. (1991).

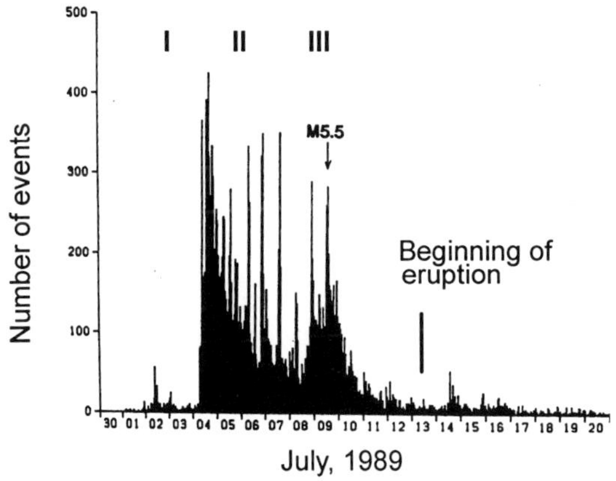


Fig. 5.26. Temporal variations of the number of earthquakes during the 1989 Teishi Knoll swarm according to data of station KMT (See Fig. 5.24d). From Yamasato et al. (1991).

formed during stage III of the swarm activity. A thin lineal zone of a few events connects the large hypocentral subzone with the site of eruption (Fig. 5.25).

# Chapter 6
# Volcano-tectonic Earthquakes at Andesitic Volcanoes

*Andesitic volcanoes* produce lava with a viscosity in the range $10^3$-$10^5$ poises at temperatures of about 1000-1300°C; the volcanic activity is of explosive and effusive types; they are characterized by higher level of volatiles. Andesitic volcanoes usually have the form of stratovolcanoes. The most famous among them are Bezymianny in Kamchatka, Merapi in Indonesia and Colima in Mexico.

The eruptions of andesitic volcanoes occur mainly from the central crater of the volcano and rather seldom on its flanks. The magmas are rich in volatiles, mobile, able to produce phreato-magmatic or phreatic explosions and can form lava domes in the crater of the volcano. These eruptions are accompanied by lava and pyroclastic flows and numerous rockfalls. The normal cycle for andesitic volcanoes consists of the formation of a lava dome and then its destruction with its collapse or an explosion. The rock fracturing during the forming of the passageway for the magma to the surface produces volcano-tectonic earthquakes. The lava effusion from central and flank (fissure) vents may be accompanied by volcanic tremor.

I discuss here the volcano-tectonic seismicity associated with three types of activity of andesitic volcanoes: seismic activity related (1) to the great "directed blasts" at volcanoes, (2) to the phreatic or phreatio-magmatic explosions and (3) associated with block-lava extrusions. These three types of eruptions may form also subsequent stages of the eruptive activity of an andesitic volcano.

## *6.1. Volcano-tectonic earthquakes associated with volcanic "directed blasts"*

The "directed blasts" of andesitic and dacitic volcanoes are within the most "large-scale" eruptive events. During an eruption of this type, a great part of volcanic cone can be exploded horizontally and end up distributed over a great area (Gorshkov and Bogoyavlenskaya, 1965). Typical eruptions: Bezymianny, Kamchatka, 1956; Sheveluch, Kamchatka, 1964; Mt. St. Helens, Cascades, USA, 1980.

### 6.1.1. Bezymianny volcano, Kamchatka, Russia

The March 30, 1956 explosion of Bezymianny volcano became a classic example of a "directed blast" eruption (Gorshkov and Bogoyavlenskaya, 1965). This stratovolcano with a height of 3085 m was considered as dormant, but on 29 September 1955 the first earthquake in the region of the volcano was recorded. From October 6, a swarm of volcano-tectonic earthquakes (Fig. 6.1) began, and the first explosion was recorded on October 22, 1955. A strong ash eruption continued for one month, then it changed to moderate explosive activity that culminated in the paroxysmal explosion of March 30, 1956. This explosion destroyed the SE part of volcanic cone and pyroclastic material was erupted to the SE at the angle of 30-40° to the horizon. The ash cloud rose up to 40 km. An area of 500 $km^2$ at a distance of 30-32 km to the SE of the volcano was covered by about 3 $km^3$ of erupted material. The height of the volcano was diminished by about 180 m. Later, in April 1956, a new dome began to form in the crater. This process of the formation and destruction of the dome continued for many years, after 1965 accompanied by the extrusion of block lava flows (Bogoyavlenskaya and Kirsanov, 1981; Bogoyavlenskaya et al., 1991).

Fig. 6.1 shows the development of seismic activity in 1955-1956. The only seismic station KLY was situated at a distance of 42 km from the volcano and was equipped by short and mediate period three-component seismometers with magnification of 5000 and 300, respectively. It failed to localize the events but it was used to count the number of events and to estimate their energy. One month before the eruption the number and energy release of seismic events gradually increased up to the first explosion. During next month, when numerous strong ash explosions were observed, the number of events recorded was up to 300-400 quakes per day. They were mainly of small magnitude. With the beginning of moderate explosion stage, the number of events sharply decreased (about 100-200 events per day). In March 1956, only 40-100 events per day were observed; the maximum magnitude of events decreased too. The paroxysmal explosion of March 30 occurred on the background of the low level in seismic activity and was associated with an earthquake of magnitude Mw 5.3.

Gorshkov and Bogoyavlenskaya (1965) noted that the variations in the number of seismic events correlated well with the manifestations of eruptive activity while the variation in the maximum magnitude of earthquakes had no such dependence. Larger earthquakes formed a permanent background from October 18, 1955 (before the eruption) to the end of February 1956; their occurrence did not depend on the eruptive manifestations.

After the paroxysmal explosion, a dome growth in the crater was observed. The partial destruction of the dome that was accompanied by explosions and lava flows was preceded by an increase in seismic activity 2-5 days before the eruption (Tokarev, 1966; 1981; Bogoyavlenskaya et al., 1991).

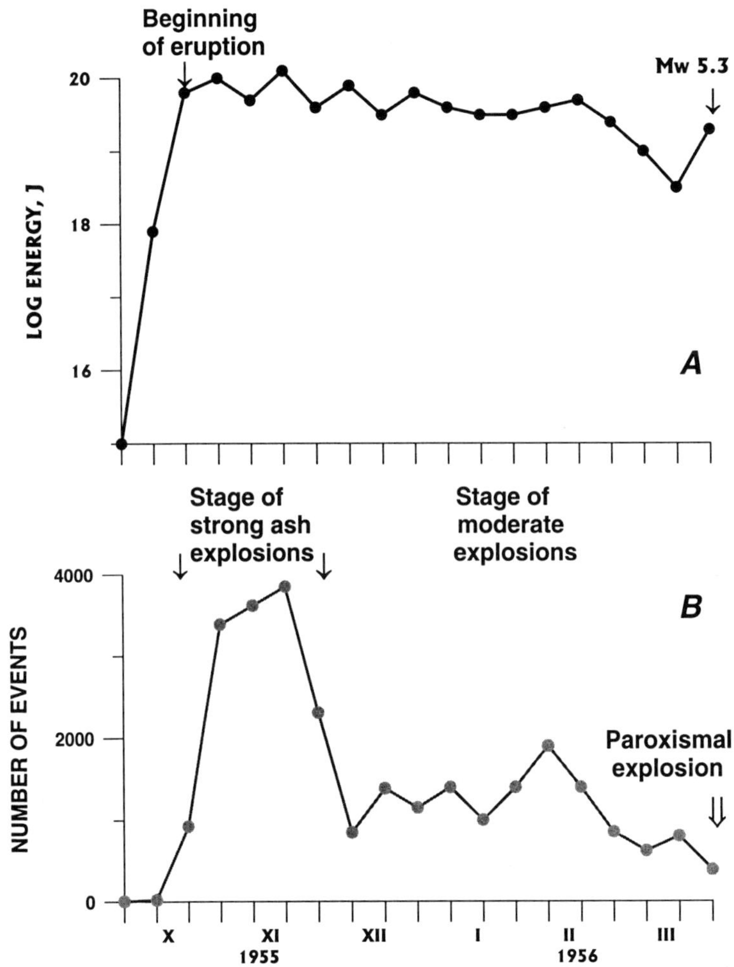


Fig. 6.1. Variations in seismic energy release (A) and in the number of earthquakes of magnitude M > 0 (B) that were observed at KLY station at a distance of 42 km from the Bezymianny volcano in October 1955 – March 1956. The main stages in eruption development up to the paroxysmal explosion of March 30, 1956 are shown. From Gorshkov and Bogoyavlenskaya (1965).

Two typical variations of seismic activity before the explosions of the dome in 1959 and 1960 are shown in Fig. 6.2. The number of events was counted at seismic station KLY (Tokarev, 1981). In 1959, an increase in seismic activity began on March 26 (Fig. 6.2A). The number of events sharply increased up to the moment of the March 28 explosion. After the explosion, the number of events gradually decreased. The majority of seismic events had magnitudes from 0 to 1. The maximum event of magnitude 1.3 was recorded on March 28.

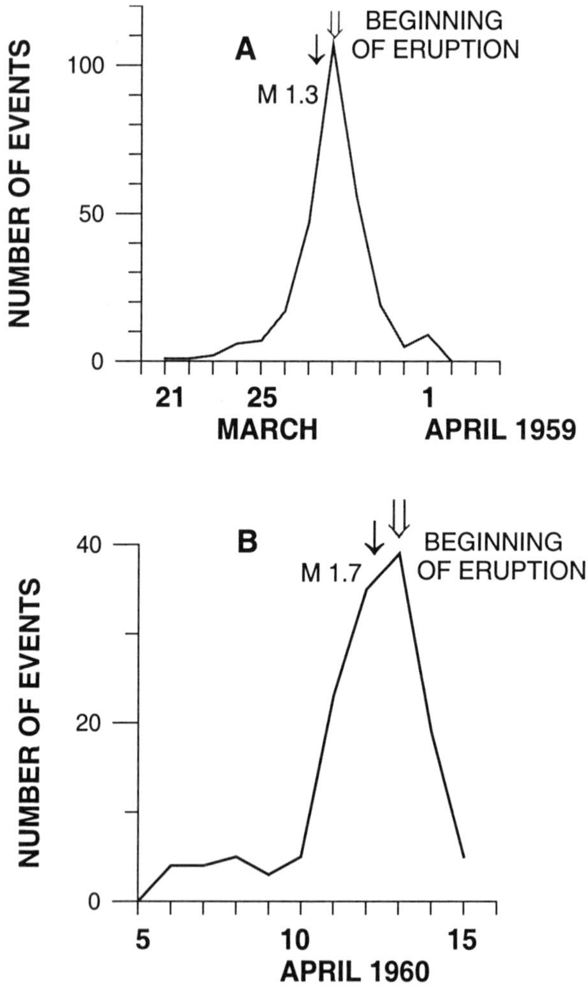


Fig. 6.2. Variations in the number of earthquakes before two explosions of Bezymianny volcano in 1959 (A) and 1960 (B) that were observed at KLY station at a distance of 42 km from the Bezymianny volcano. Data from Tokarev (1981).

A similar situation was observed before the April 13-14, 1960 eruption (Fig. 6.2B). An increase in seismic activity began 3 days before the explosion; the explosion occurred at the maximum level of seismic activity. After the explosion, the number of events gradually decreased. The majority of seismic events had magnitudes from 0 to 1. The maximum event of magnitude 1.7 was recorded on April 12.

6.1.2. Sheveluch volcano, Kamchatka, Russia

Sheveluch volcano is the northernmost active andesitic volcano of Kamchatka. It is located at the junction of the Kurile-Kamchatka and Aleutian subduction zones. It consists of two parts: the old edifice with an altitude of 3283 m and the young (active) edifice with an altitude of 2800 m. During the last 10 000 years, the volcano has been characterized by two types of eruptions: the great eruptions of the "directed blast" type (they have continued from hours to a few days) and the slight to moderate eruptions associated with the growth and destruction of the crater dome (they have continued for several years) (Bogoyavlenskaya et al., 1991).

*1964 eruption.* The "directed blast" eruption took place on November 12, 1964 after 15 years of calm. It continued for only 70 minutes. As a result, about 1.5 $km^3$ of resurgent material and 0.8 $km^3$ of juvenile material were ejected; the summit crater and five periphery domes were destroyed by the explosion. A new crater was formed. Fig. 6.3 shows the seismic history of the eruptive process (Tokarev, 1967; Gorshkov and Dubik, 1970).

The first alert pronouncing the volcanic unrest was initiated by the April-May, 1964 swarm of 37 small shallow earthquakes located near the volcano. This swarm was not followed by any volcanic manifestations. The next swarm of volcano-tectonic earthquakes began on 17 October (Fig. 6.4). About 500 earthquakes of magnitude M 2 to 5 were recorded. Epicenters of all the earthquakes were located by three seismic stations of Klyuchevskaya Volcano Station and occurred within the volcanic edifice at depths of less than 5 km. The number and energy of the earthquakes (according to the data of the nearest seismic station KLY situated at a distance of 45 km to the south of the volcano) gradually increased up to the moment of the paroxysmal explosion of November 12. The maximum magnitude earthquakes were observed on November 12 just, and associated with, the paroxysmal eruption, their magnitudes Mw were 5-5.2 (Tokarev, 1967; Zobin, 2001). The airwaves of the November 12 explosions were recorded by barographs and microbarographs at a distance up to a few thousand kilometers. After

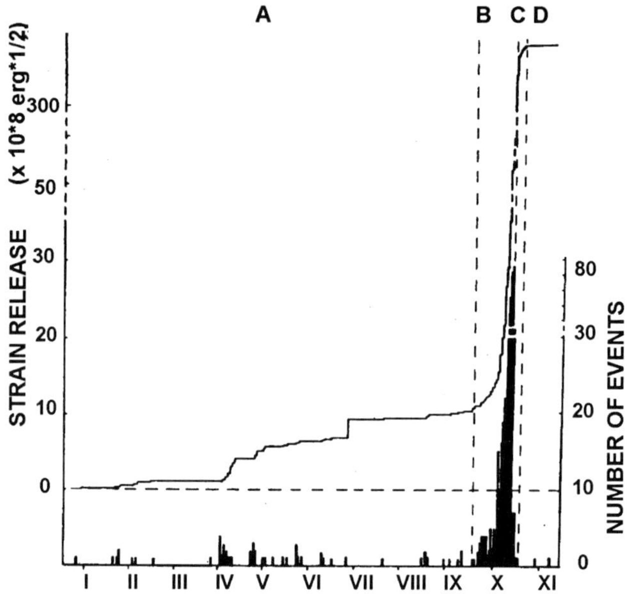


Fig. 6.3. Characteristics of seismic activity at Sheveluch volcano in 1964. The strain release during the seismic activity (as a cumulative value of quadratic root of seismic energy in Erg) and variations in the number of earthquakes according to the records at seismic station KLY situated at a distance of 45 km to the south of the volcano are shown. A-C, the stages in seismo-volcanic activity: A, the preliminary stage (January to the middle of October 1964); B, the preliminary seismic swarm (October 17 to November 12); C, the paroxysmal eruption on November 12; and D, the post-eruptive stage. From Tokarev (1967).

the 12 November explosion, seismic activity sharply decreased and ceased in four days.

*1993 eruption.* A new dome began to grow in 1980 within the crater that was formed by the 1964 explosion, without any preceding seismic events. Its 1980-1993 growth was accompanied by infrequent Vulcanian-type and phreatic explosions and culminated in the April 21, 1993 paroxysmal explosion of the dome that produced about 0.5 $km^3$ of pyroclastic material (Gorelchik et al., 1995; 1997).

Fig. 6.5 shows the temporal variations in the number of earthquakes preceding the eruption. A seismic swarm began three weeks before the eruption. The majority of earthquakes occurred within a distance of less than 5 km from the crater dome; their depths did not exceed 3 km. The number of earthquakes gradually increased up to the moment of the eruption. After the eruption, the number of earthquakes sharply decreased.

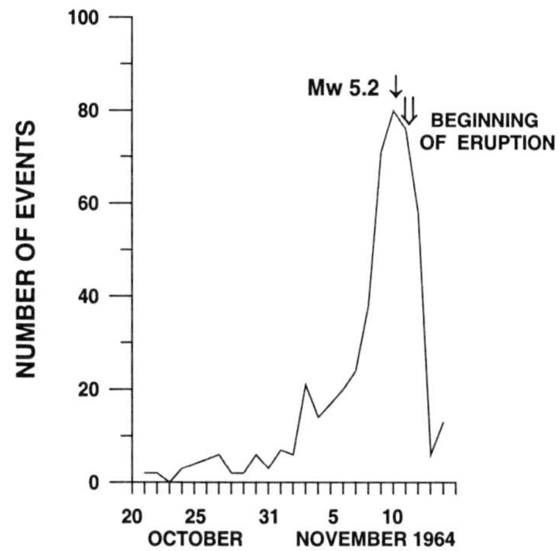


Fig. 6.4. Variations in the number of earthquakes before the November 11, 1964 eruption of Sheveluch volcano. Data from the Kamchatka Regional Earthquake Catalog.

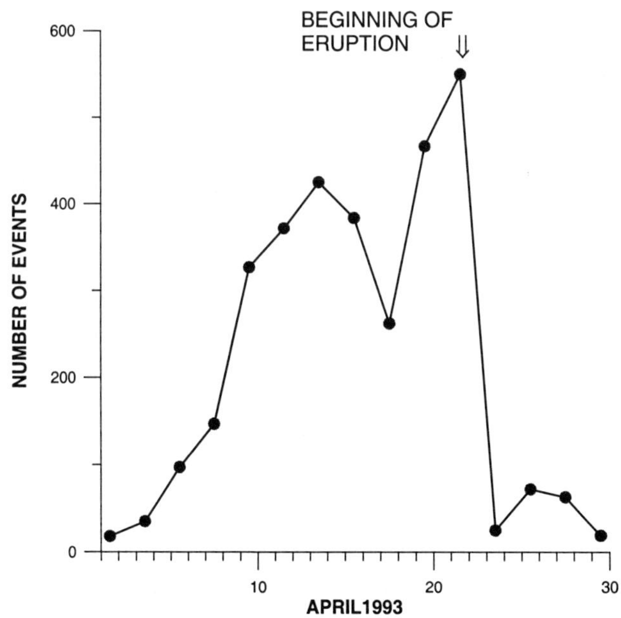


Fig. 6.5. Variations in the two-days number of earthquakes before the 1993 eruption of Sheveluch volcano according to the records of seismic station SVL situated at a distance of 8 km from the crater. A double arrow shows the beginning of eruption. Data from Gorelchik et al. (1997)

### *6.2. Volcano-tectonic earthquakes associated with phreatic and phreatomagmatic explosions*

#### 6.2.1. El Chichón volcano, Mexico

El Chichón volcano is an andesitic stratovolcano in southern México. It erupted in March 1982, after about 600 years of quiescence (Simkin and Siebert, 1994), and produced a sequence of 6 large phreatomagmatic explosions from March 29 to April 4 that destroyed the summit dome and formed a 1-km-wide crater (SEAN, 1982).

The eruption was preceded by a 28-day seismic swarm, which began on March 1, 1982 (Figs. 6.6-6.8). Seismic activity was recorded before the beginning of eruption by a 14-station seismic network of UNAM that was installed in this region in 1979-1980 (Havskov et al., 1983; Yokoyama et al., 1992). The swarm began with a sharp increase in the number of earthquakes during the first five days of activity. The maximum magnitude event (M 4.0) was recorded on the fifth day of the swarm on the background of a decreasing number of events. The activity increased again about 2 days before the eruption. The first explosion was recorded with the cease in seismic activity (Fig. 6.6).

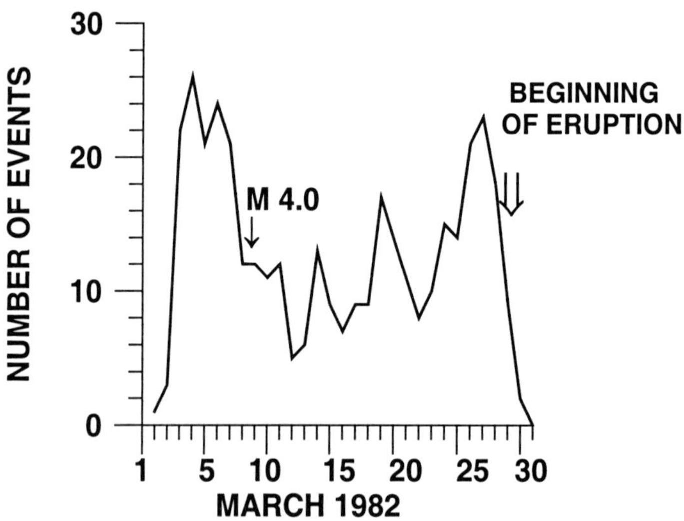


Fig. 6.6. Temporal variations in the number of earthquakes before the 1982 El Chichon eruption. The maximum magnitude earthquake is shown by a single arrow; the beginning of the first explosion is shown by a double arrow. From Havskov et al. (1983).

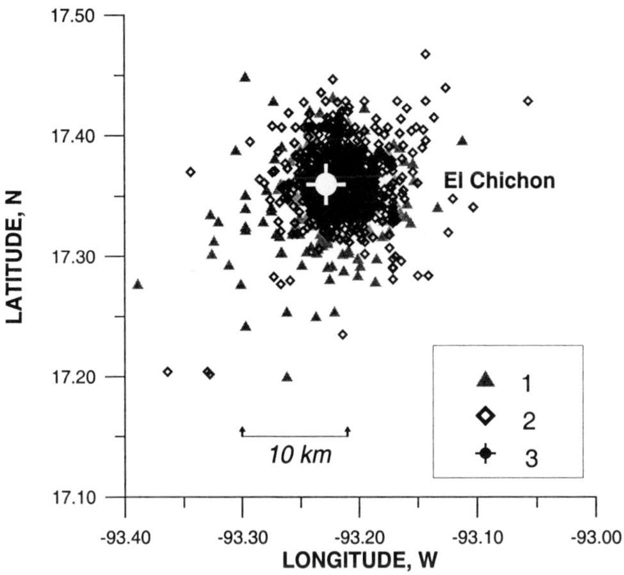


Fig. 6.7. The epicenters (A) and hypocenters (B) of earthquakes that occurred before the March 29, 1982 El Chichon eruption (diamonds) and after the eruption (circles). El Chichon volcano is shown by a 6-ray star. The catalog of earthquakes was prepared by J.M. Espíndola (personal communication, 2002).

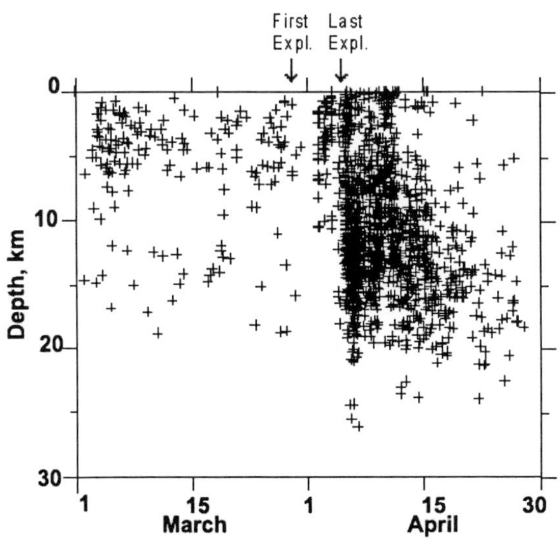


Fig. 6.8. Temporal variations of earthquake focal depths in March and April 1982. The interval between the first and last explosions is shown by arrows. From Jiménez et al. (1999).

The majority of earthquakes were located within the 5-km radius zone at depths from 2 to 8 km; a small group of foci was located also at the depths from 10 to 18 km (Fig. 6.7). During the time interval between the first and the last explosions, the seismic events of the second, deep group disappeared.

After the last explosion, seismic activity continued one month more. The depth of events went down up to 20-25 km (Fig. 6.8). The earthquake foci filled the depth interval from 0 to 25 km (Fig. 6.7) and stayed at this depth in April (Jimenez et al., 1999).

6.2.2. Volcán de Colima, México

The andesitic stratovolcano Volcán de Colima (height of 3,860 m) is one of the most active volcanoes in Mexico. The eruptions of Volcán de Colima during last 480 years involved a wide spectrum of eruption styles, including small phreatic explosions, major block-lava emissions, and large explosive events (Breton González et al., 2002).

*The phreatic explosion of 1994* occurred at 20:13 on 21 July. It destroyed the 1991 crater dome and produced avalanches and ash fall. The explosion formed a shallow crater approximately 140 m in diameter and 50 m in depth. The explosion was heard at a distance of 20-35 km to the south of the volcano (Jiménez et al., 1995). The eruption was preceded by a seismic sequence from 13 to 21 of July that was located within the volcanic edifice.

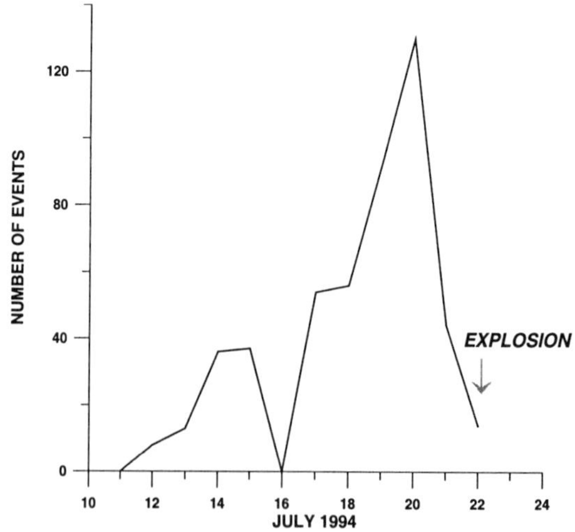


Fig. 6.9. Variations in the 12-hour number of events recorded before the 1994 Volcán de Colima explosion.

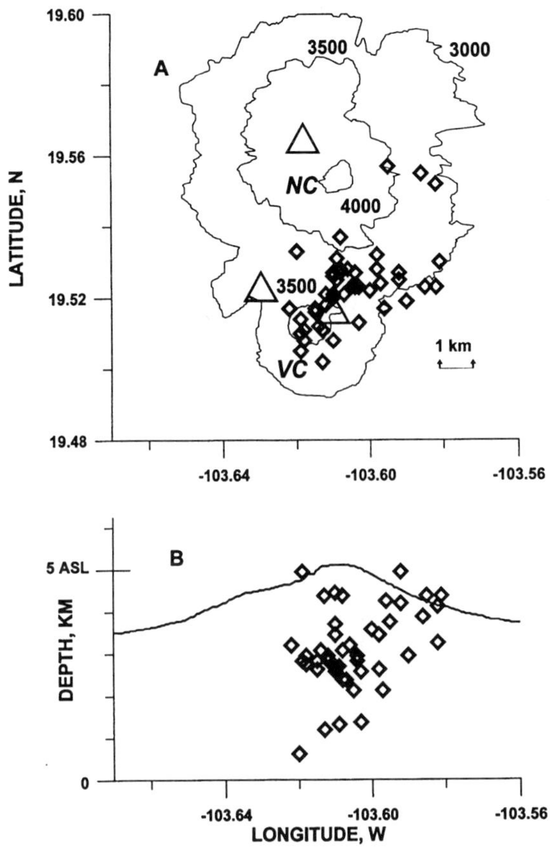


Fig. 6.10. The distribution of epicenters (A) and hypocenters of earthquakes (B) located during the 1994 seismic crisis at Volcán de Colima. The nearest seismic stations are shown as triangles. Shown are contours of the height above sea level for 2500 m, 3000 m, and 3500 m. VC is Volcán de Colima; NC is Nevado de Colima.

About 450 seismic events were recorded by the nearest seismic station (a distance from the crater of 1.7 km) during the 1994 swarm, 45 of them were located. Fig. 6.9 shows the temporal variations in the number of events. The seismic activity developed in two stages, from July 12 to July 15, and from July 17 to July 22. The maximum magnitude Mw 1.2 event was recorded during the peak activity of the second stage.

The events formed a compact group within an area of about 6 $km^2$ beneath the active crater. They clustered at shallow depths in the range of 0 km to 4 km above sea level (Fig. 6.10). The detailed analysis of depth variations within the 1994-swarm shows that this swarm began with the relatively deep events (July 13-15) but then, from July 17, the swallowing of events was observed. The very shallow events were observed up to the moment of the phreatic explosion.

### 6.2.3. Soufriére Hills volcano, Montserrat

Andesitic stratovolcano Soufriére Hills (SHV) with a height of 915 m is situated in the southern part of Montserrat Island, West Indies. Its recent eruption, after about 370 years of repose (Simkin and Siebert, 1994), began on 18 July 1995 with the steam and ash venting from the northwest crater. The eruption activity passed through the phreatic and magmatic explosions, the growth and collapse of lava domes accompanied by pyroclastic flows, and a great directed explosion. It continued for about 4 years.

About 10 000 volcano-tectonic and hybrid earthquakes of magnitudes from –0.5 to 2 were located from 28 July 1995 to the end of

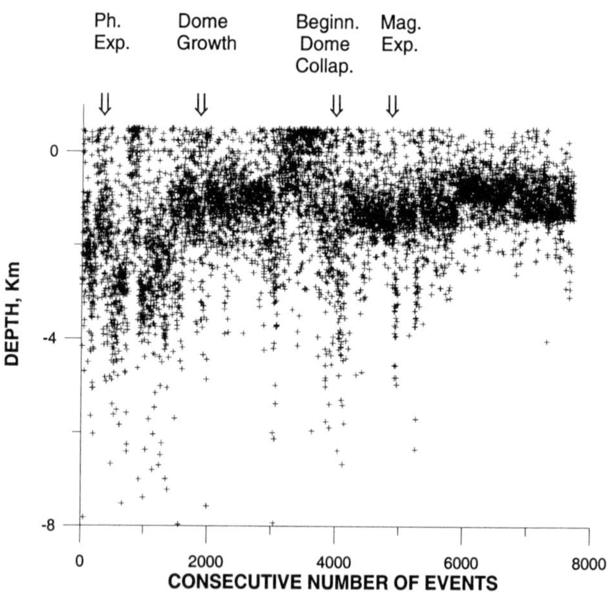


Fig. 6.11. Depth variations of earthquakes that were located beneath Soufriére Hills volcano from July 28, 1995 to February 28, 1997. Depth 0 km marks the sea level. Arrows show the main episodes in volcanic activity (Ph. Exp., phreatic explosion of August 21, 1995; Dome Growth and Beginn. Dome Collap., the initial growth and the beginning of dome collapses; Mag. Exp., magmatic explosion of September 17, 1996). Data from the catalog of Montserrat Volcano Observatory (MVO).

February 1997 with a high precision (Aspinall et al., 1998; Miller et al., 1998). They followed all main events in volcanic activity. Earthquakes occurred in three epicentral zones, beneath Soufriére Hills volcano, along the structure extending to the NE of the volcano, and beneath St. George Hill (SGH) that is situated about 5 km to the NW from the volcano. All events were recorded at the shallow depths, mainly from 5 km BSL to the surface. Fig. 6.11 shows the temporal variations in the depths of seismic events during 1995-1997 for the epicentral zone of the volcano together with the main events in volcanic activity. The tendency in shallowing of earthquake foci during the first year of activity from the phreatic explosion up to the beginning of dome collapses is clear seen.

We discuss here the variations in seismic activity preceding two important moments in volcanic activity of SHV: the phreatic explosion of 21 August 1995 and the magmatic explosion of 17 September 1996.

*Phreatic explosion of 21 August 1995.* It was the first large phreatic explosion in the 1995-1997 eruptive sequence. It blanketed the nearest city Plymouth, situated at a distance of about 6 km from the volcano, in a thick ash cloud and caused darkness for about 15 minutes (Chronology..., 2002).

Fig. 6.12 and 6.13 show the temporal variations in earthquake number and the distribution of epicenters and hypocenters of earthquakes from 1 to 25 of August. An increase in seismic activity began with a two-day swarm of earthquakes along the NE extend of the SHV from 5 to 6 of August, then a new two-day swarm was recorded beneath SGH that continued from 12 to 13 August. Seismicity beneath the volcano was low up to 20 August. During two days, 20 and 21 of August about 140 earthquakes were recorded. The explosion occurred at 08:03 a.m. (L.T.) of 21 August (B. Voight, personal communication, 2002) during the maximum level of seismic activity below the volcano. Just after the explosion, the number of seismic events sharply decreased.

The seismic events recorded beneath the volcano were clustered within a vertical column shape volume extending from 3 km BSL to the summit of the volcano. The foci tended to be shallower before the phreatic explosion. Fig. 6.14 shows that the focal depths of earthquakes changed from 1-3 km BSL on 1-19 August to the depth range from 2 km BSL to the summit of the volcano on August 20-21. The events associated with the extension of the NE structure of the volcano, as well as the SGH earthquakes, were more deep and located at depths from 5 km to 3 km BSL.

*Magmatic explosion of September 17, 1996.* A series of dome collapses that began in July 1996 (See Fig. 6.11) lead to the September

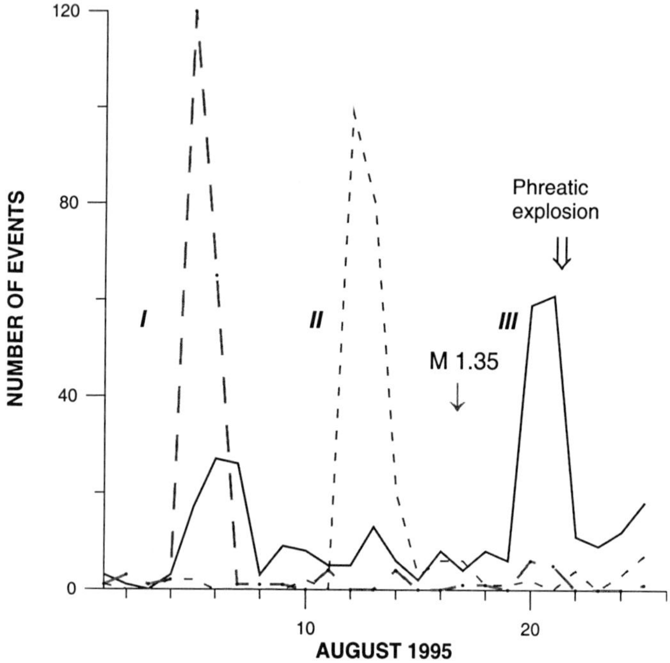


Fig. 6.12. Variations in daily earthquake number before the August 21, 1995 phreatic explosion. I, II, and III are the earthquake sequences that occurred within the NE extend from the volcano (I), beneath SGH (II), and beneath the volcano (III). The largest earthquake is shown by a single arrow; the phreatic explosion is shown by double arrow. Data from the catalog of MVO.

17 (23:46 LT) magmatic explosion. Rocks and pumice fell in the south of the island, and houses were destroyed the nearest village. A major ash plume rose to about 12 km, and about 600 000 tons of ash were deposited in southern Montserrat (Chronology..., 2002).

The seismic activity that preceded the September 17 explosion is shown in Fig. 6.15-6.17. The earthquake foci located during about 40 days of activity are clustered mainly beneath the SHV forming the vertical column from the depth of 3 km bsl to the summit of the volcano. The zones that were active before the phreatic explosion of August 21, 1995, now were calm.

The swarm of earthquakes had no clear beginning; it began on the background of continuous seismic activity due to the process of partial collapsing of the dome that began in July 1996. Nevertheless, we can mark the gradual increase in earthquake numbers beginning from

August 29; therefore, the swarm continued for about 23-24 days. All earthquakes were of low magnitude; the largest event of magnitude 1.9

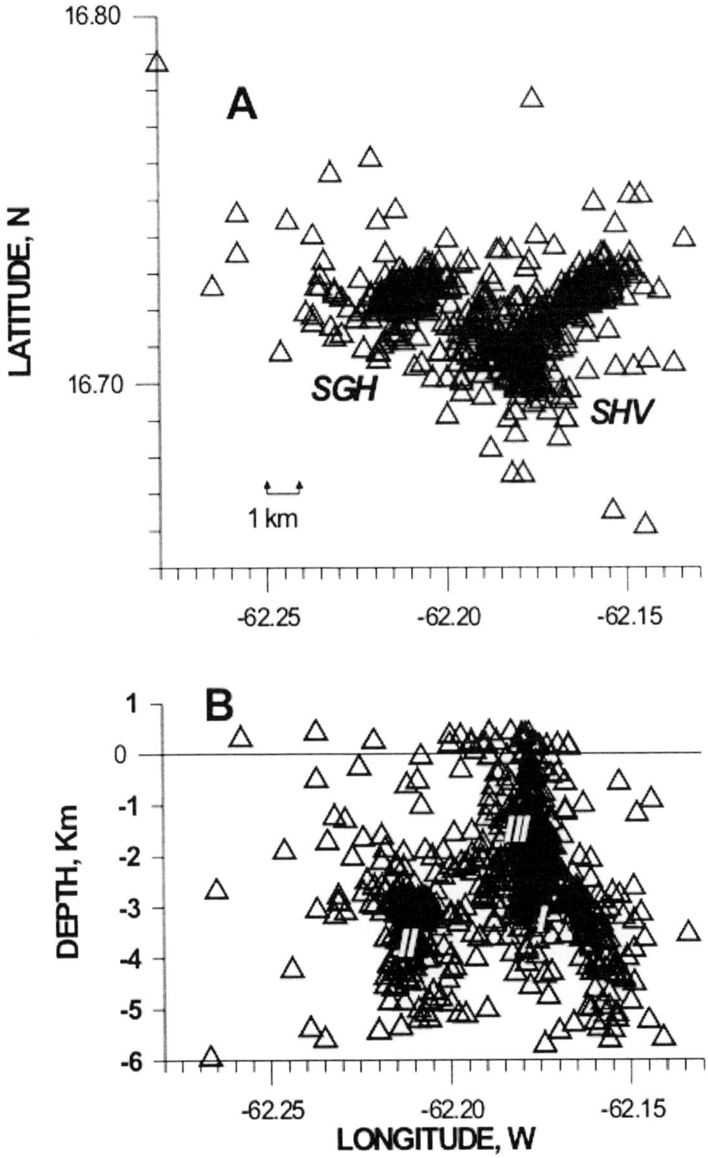


Fig. 6.13. Distribution of epicenters (A) and hypocenters (B) of earthquakes located before the August 21, 1995 phreatic explosion. Square is SGH, star is SHV. The zero-line of sea-level depth is shown. Data from the catalog of MVO.

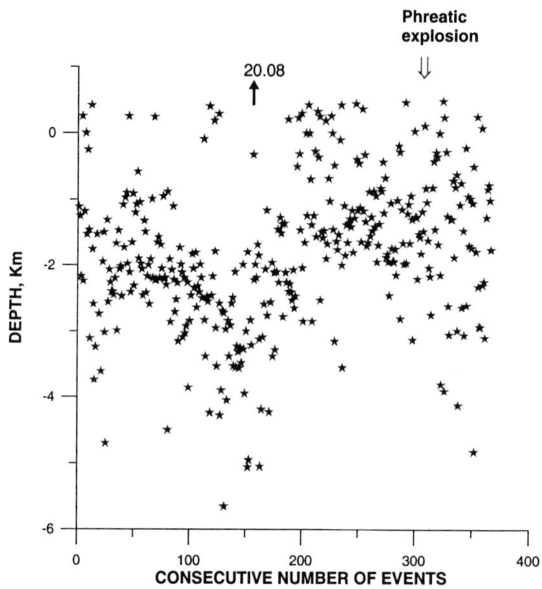


Fig. 6.14. Temporal variations in earthquake depths beneath the volcano before the August 21, 1995 phreatic explosion. The phreatic explosion is shown by double arrow. The beginning of the 20-21 August swarm is shown by a single arrow. Data from the catalog of MVO.

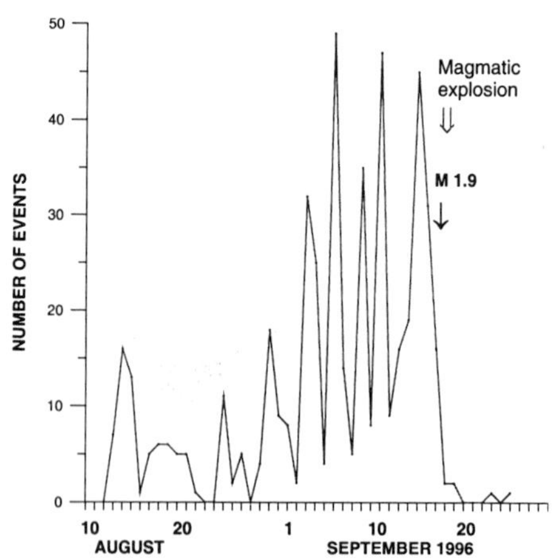


Fig. 6.15. Variations in daily earthquake numbers before the September 17, 1996 magmatic explosion. The largest earthquake is shown by a single arrow; the magmatic explosion is shown by double arrow. Data from the catalog of MVO.

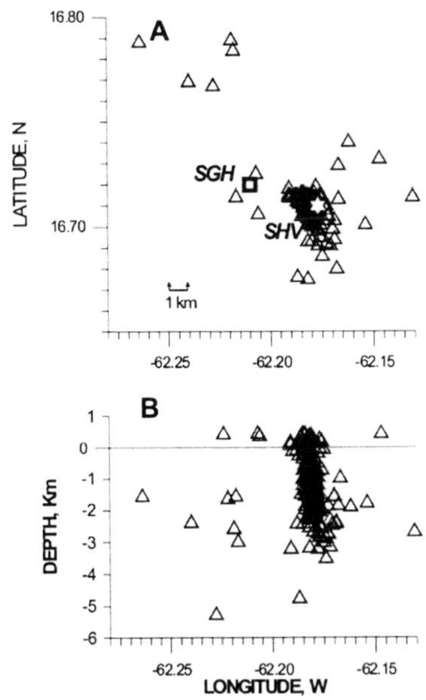


Fig. 6.16. Distribution of epicenters (A) and hypocenters (B) of earthquakes located before the September 17, 1996 magmatic explosion. A square marks St. George´s Hill (SGH), a star marks Soufriére Hills volcano (SHV). Data from the catalog of MVO.

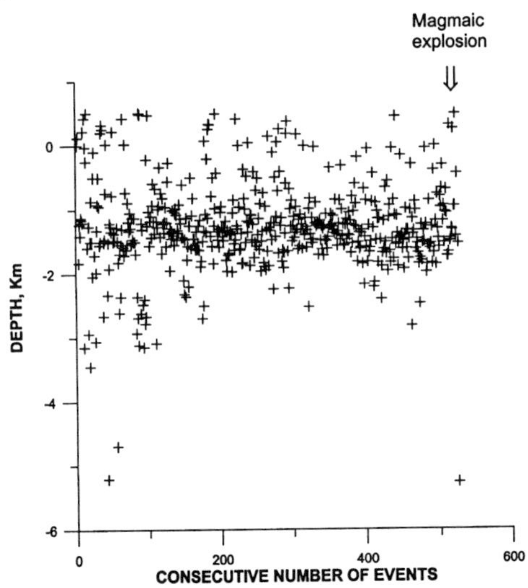


Fig. 6.17. Temporal variations in earthquake depths beneath the volcano before the September 17, 1996 magmatic explosion. The magmatic explosion is shown by double arrow. Data from the catalog of MVO.

occurred just before the explosion of September 17. The explosion occurred during the apparent decrease in number of seismic events; after the explosion, the number of earthquakes sharply decreased.

Fig. 6.17 shows that the seismic process developed permanently at the depths between 1 and 2 km. No focal depth migration to the surface was noted before the explosion.

### *6.3. Volcano-tectonic earthquakes associated with lava extrusions*

#### 6.3.1. Volcán de Colima, México

An eruption occurred on 20 November 1998, when andesitic block lava began to extrude in the summit crater. The dome quickly spilled over the SW crater rim and fed block-lava flows that descended in three branches on the southern slope of the volcano, to a maximum distance from the crater of 3.8 km; the advance of the block lava flows was accompanied by pyroclastic flows. (Zobin et al., 2002a).

Fig. 6.18 shows the temporal development of seismic activity before the 1998 eruption. About 15,000 small earthquakes were recorded. The eruption was preceded by five earthquake swarms from November 28,

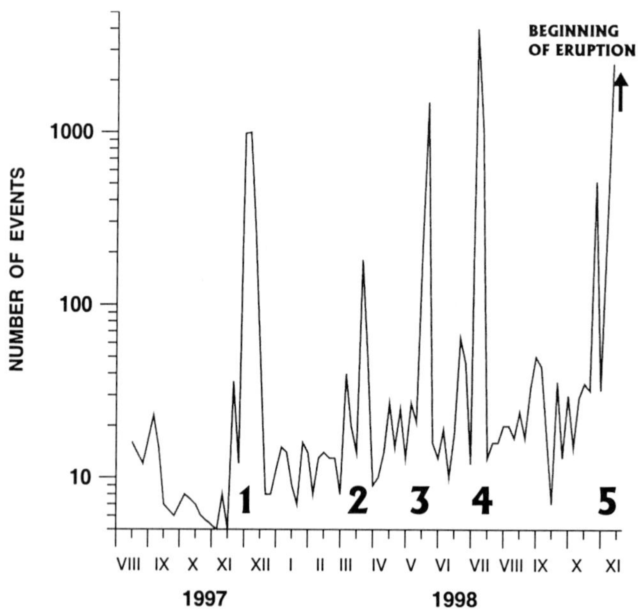


Fig. 6.18. Variations in number of events recorded in 1997-1998. Counted events are those recorded by a minimum four stations over 5-day intervals. Numbers 1 to 5 identify individual swarms. From Zobin et al. (2002a).

1997 to November 20, 1998. Each swarm consisted of a sequence of volcano-tectonic microearthquakes.

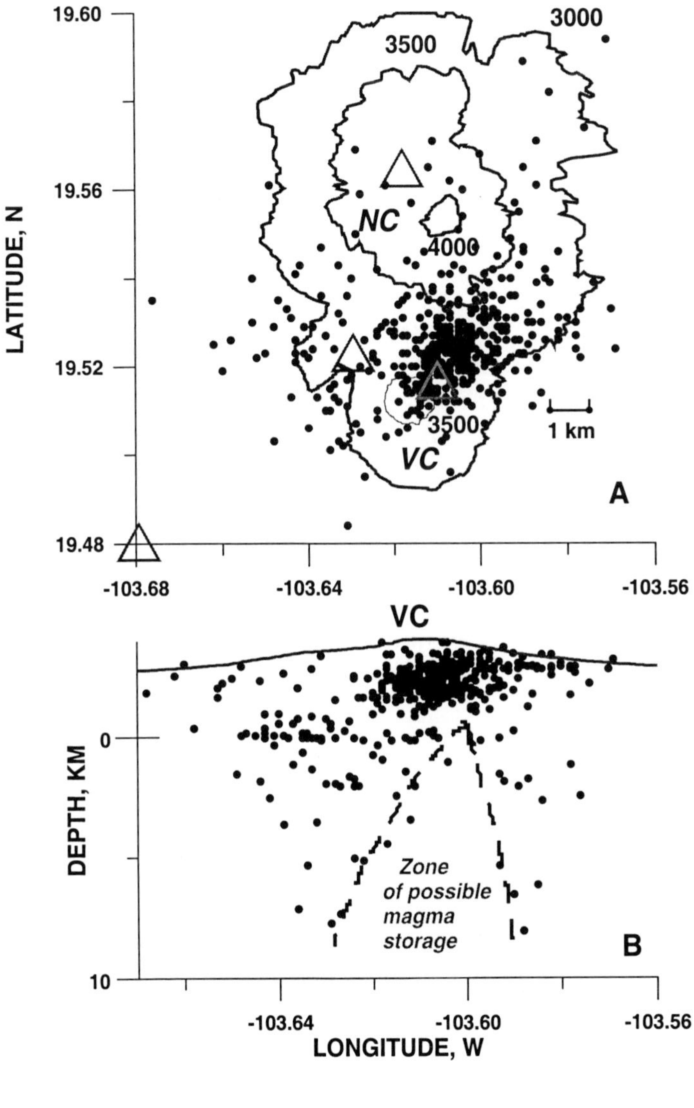


Fig. 6.19. The distribution of epicenters (above) and hypocenters of earthquakes (below) located during the seismic crisis of 1997-1998. The contours of 3000, 3500 and 4000 m show the relief of Colima Volcanic Complex. VC is Volcán de Colima; NC is Nevado de Colima. The seismic stations are shown as triangles. An earthquake-free zone is outlined in the cross-section. From Zobin et al. (2002a).

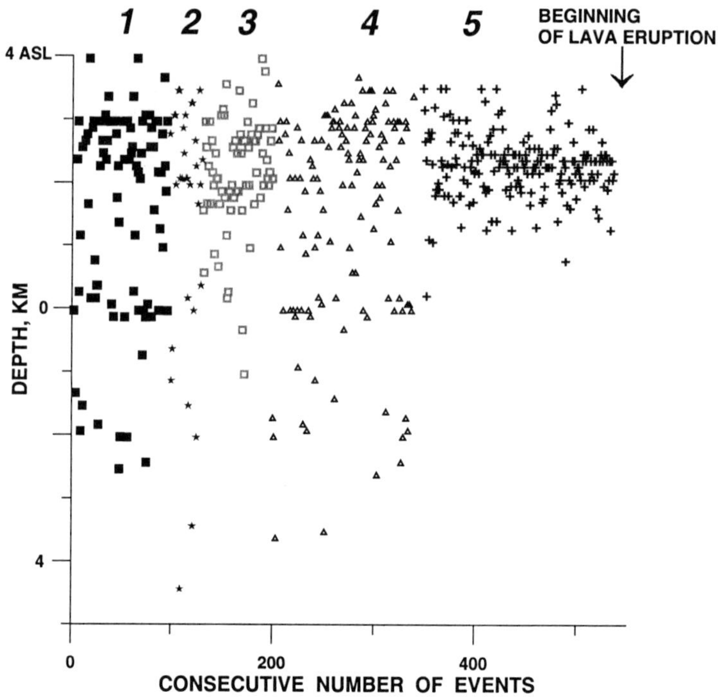


Fig. 6.20. Temporal variations of event depths during the five swarms (numbers 1 to 5). Events in each swarm are marked by particular symbols. From Zobin et al. (2002a).

The distribution of epicenters and hypocenters of earthquakes of magnitude Mw from –0.5 to 2.7 recorded from November 1997 to November 1998 is shown in Fig. 6.19. We plotted here only the earthquakes located within the area of the seismic network, as those outside the network had larger location uncertainties. The epicenters were located within the Colima volcanic complex (CVC) and covered an area of about 50 $km^2$. The majority of events clustered in the area of Colima volcano´s active crater and between the volcanic edifices of Colima and Nevado volcanoes.

The majority of hypocenters were located at depths ranging from 10 km (mainly from 5 km) below sea level (BSL) to the crater zone of the volcanoes, or up to heights of about 4000 m. A dense clustering of hypocenters was observed within a narrow depth interval of 1 to 4 km above sea level (ASL) within the volcanic constructions. Fig. 6.19B shows a space free of earthquake foci just below the volcano at depths of 0 km to 10 km BSL. This free space could correspond to a zone of possible magma storage.

Fig. 6.20 shows the depth variations of earthquakes in time. It is difficult to note any variations in depth for the events within the individual swarms. However, there were substantial variations in foci depth from swarm to swarm. It is possible to distinguish two stages in depth variations. The first stage includes the first three swarms from November 1997 to May 1998; the depths gradually decreased from swarm to swarm. In June-July (swarm 4), the depths again increased, with their distribution similar to the depth distribution of the first swarm in November-December 1997. The last swarm (swarm 5) that occurred just before the eruption was shallower, with events occurring only from 1 to 4 km ASL.

### *6.4. Volcano-tectonic earthquakes associated with flank eruption*

#### 6.4.1. Sakurajima volcano, Japan

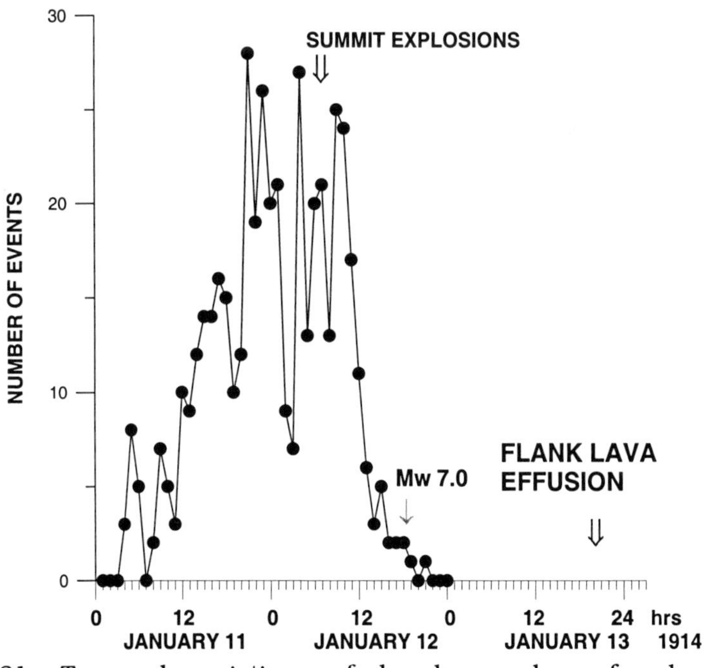


Fig. 6.21. Temporal variations of hourly number of volcano-tectonic earthquakes of Sakurajima volcano recorded by the Gray-Milne-Ewing seismic station in Kagoshima at the distance of 10 km from eruptive centers. A single arrow shows the largest earthquake; double arrows show the eruptive events. From Omori (1914).

Sakurajima volcano is an andesitic stratovolcano (altitude of 1110 m) that is situated in the southern part of Aira caldera, Southern Kyushu. The volcano had repeated flank and summit eruptions during historic times (Ishihara, 1990). The 1914 eruption of the volcano was one of the largest volcanic events in the twentieth century in Japan and it was one of the first volcanic eruptions that were studied using instrumental seismic observations (Omori, 1914; Yokoyama, 1997).

The eruption began on January 12, 1914 at about 08 hrs (local time) with the emissions of white smoke from the summit crater, and 2 hrs later explosions occurred at the western and eastern slopes, where the vents were opened. Just after midnight of January 13, the strongest explosions were observed and at about 20 hrs lava began to effuse from the craterlets on both the eastern and western slopes. The eruption continued until May 1915 (Yokoyama, 1997).

Seismic activity near the volcano was recorded in Kagoshima by local seismic station beginning January 11 (Omori, 1914). The recording drum of the low-sensitivity seismographs (magnification 5 to 10) was triggered into motion by intense shocks over a certain threshold (Fig. 6.21). Their magnitudes were 4.8 to 5.2, according to Yokoyama (1997). The higher sensitivity seismic station in Nagasaki (epicentral distance was about 148 km, but the magnifications were of 20 and 120) began to record the swarm at the lower threshold on January 9. Earthquakes increased in number and magnitude, with the maximum activity at midnight on January 12. After the opening of flank vents in the morning of January 12, earthquake activity rapidly decreased. Then, at 18:30 of January 12, with a background of seismic calmness, the Mw = 7.0 (Zobin, 2001) shock occurred. About 26 hrs later the great lava flank extrusion began. About 1.6 $km^3$ of lavas overflowed from the volcano, which connected the volcanic island with Oosumi Peninsula (Yokoyama, 1997).

# Chapter 7
# Volcano-tectonic Earthquakes at Dacitic Volcanoes

The eruptions of *dacitic volcanoes* may occur from the central crater and from the new vents on the flank of volcano. The high viscosity and relatively low temperature magmas with a great amount of volatiles are characterized by low mobility and high explosivity. The magmas may construct a lava plug dome, cryptodome or lava spine in the crater of the volcano. The large volume of volatiles can produce numerous explosions of different sizes. Volcano-tectonic earthquakes can be observed during the movement of the magma to the surface before the lava extrusion. The lava effusion from central and flank vents may be accompanied by volcanic tremor. Dacitic volcanoes have a low altitude and a stratovolcano shape.

## *7.1. Volcano-tectonic earthquakes associated with central eruptions*

The central eruption of dacitic volcanoes may consist of the growth of a dome in the crater with possibly the formation of lava flows and (or) a great explosion or a sequence of numerous small explosions. The most famous dacitic volcanoes with central eruptions are Mt. St. Helens in the U.S.A, Usu in Japan, Pinatubo in the Philippines, and Karymsky in Kamchatka. Volcano-tectonic earthquakes are related to lava dome eruptions and magmatic explosions.

### 7.1.1. Mt. St. Helens, Cascades, U.S.A

Mount St. Helens (MSH) is a stratovolcano of 2,950 m of height. It is situated within the Cascade Range in the western state of Washington, U.S.A. The eruptive history of MSH began about 40,000 years ago with dacitic volcanism. The products of eruptive activity during the past 2,500 years were represented mainly by dacite-andesite-dacites. The volcano had been apparently dormant since 1857 (Mullineaux and Crandell, 1981).

The first 1980 steam-blast eruption from its summit crater was observed on March 27, and was followed by a sequence of small explosive eruptions up to the climactic eruption on the morning of May

18. This explosive eruption of dacitic magma formed a vertical Plinian column more than 25 km high. The northward-directed lateral blast of the bulging north flank of the volcano erupted about 2 $km^3$ of cold and juvenile material and devastated an area of nearly 600 $km^2$. A new crater, with an area of 1.5x3.0 $km^2$ and of depth of about 500 m, was formed in the northern flank of the volcano. The height of the volcano diminished by about 400 m. Several smaller magmatic eruptions occurred then during 1980 producing pyroclastic flows, ash falls and several lava domes (Christiansen and Peterson, 1981). Lava domes began to grow in the crater beginning from June 1980 (Moore et al., 1981). The eruption terminated in 1986.

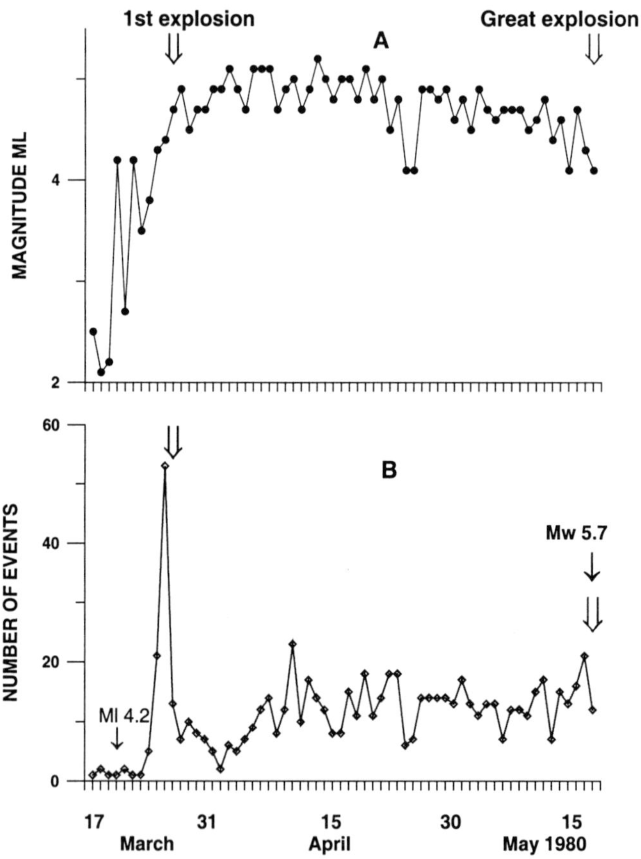


Fig. 7.1. Variations in daily maximum magnitude (A) and the number of earthquakes (B) located during the seismo-volcanic activity of MSH in March to May 1980. The double arrows show eruptive events; the single arrow shows the maximum magnitude earthquake. Data from the Pacific Northwest Seismic Network (PNSN) catalog.

The seismic activity of MSH was monitored by a regional network (Endo et al., 1982). Fig. 7.1 shows the temporal variations in daily number of events of magnitude $M_L$ greater than 3.2. The 1980 swarm of volcano-tectonic shallow earthquakes began on March 16 with small events; the first significant ML 4.2 event was recorded beneath Mt. St. Helens on March 20. The number and magnitude of earthquakes sharply increased starting March 25 with peak number on March 27, when the first explosion occurred. After the beginning of the eruption, the number of earthquakes gradually decreased during one month; then their number was more or less stable up to May 18, when the cataclysmic explosion occurred. The largest earthquake of Mw 5.7 (Zobin, 2001) occurred on May 18 on the background of low-level seismic activity, practically simultaneously with the great directed blast of the volcano.

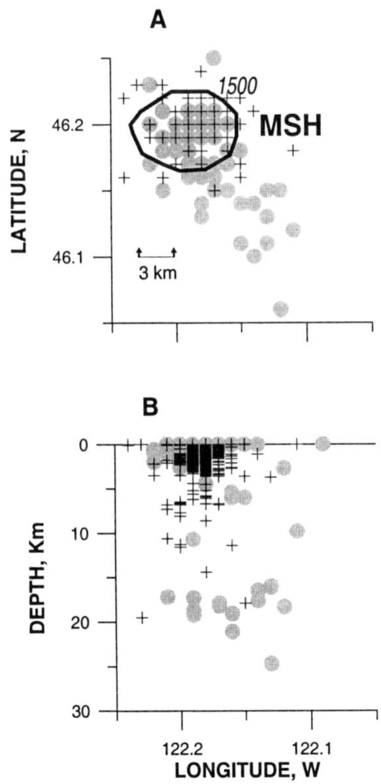


Fig. 7.2. Epicenters (A) and hypocenters (B) of earthquakes located during the seismo-volcanic activity of MSH in March to May 1980. The earthquakes located before March 27 (first explosion) are shown by circles; the earthquakes located from March 27 to May 18 are shown by crosses. The position of volcano MSH is shown by the 1500 m contour. Data from the PNSN catalog.

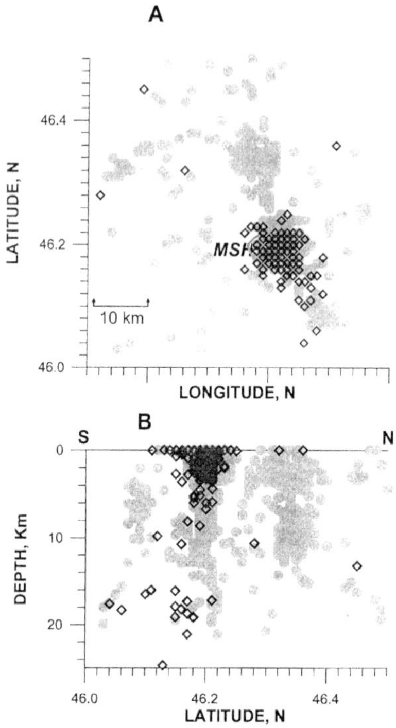


Fig. 7.3. Post-May 18, 1980 seismic activity in the region of Mt. St. Helens observed from May 19, 1980 to August 31, 1981. A. Epicenters of earthquakes; B, hypocenters of earthquakes (N-S cross section). The earthquakes recorded before the May 18 explosion are shown by diamonds; the earthquakes recorded after the May 18 explosion are shown by circles. The 1500-m contour of Mt. St. Helens (MSH) is shown. Data from the PNSN catalog.

Fig. 7.2 shows the location of earthquakes. Before the March 27 explosions, 93 earthquakes were located beneath MSH and to its SE. The seismic events formed two clusters: the shallow, at the depth range of 0 to 5 km, and the deep cluster, at the depth range from 18 to 25 km. During the active period from March 27 to May 18, 627 located earthquakes occurred beneath the volcanic edifice, mainly at the depth range from 0 to 10 km; the deep cluster disappeared. The largest earthquake of May 18 was located at the depth of 2.8 km beneath the NW flank of MSH.

After the great May 18 explosion, seismic activity beneath the crater continued, and was associated with a few magmatic eruptions of the volcano on May-August, 1980 (Weaver et al., 1981; Malone et al., 1981). At the same time, the regions to the north and to the south of the volcano became seismically active (Fig. 7.3). The 50-km-length submeridional linear structure with the volcano in its middle section was activated. Seismic foci to the south of the volcano clustered at the depth range from 20 to 1 km; the northern foci formed a cluster with the length of about 15 km and within the same depth range from the surface to 20 km.

The seismic foci just beneath the volcano occupied the space between the deep (16-22 km) and shallow (from depths of 10 km to the surface) groups of earthquakes recorded before the May 18 explosion.

### 7.1.2. Usu volcano, Hokkaido, Japan

*Eruption of 1977-1978.* The eruption occurred in the southeastern part of the summit crater (Fig. 7.4). Four new craters were formed during a sequence of magmatic explosions on August 7-14. The volume of ejected pumice material was about 0.01 km$^3$ (Yokoyama et al., 1981).

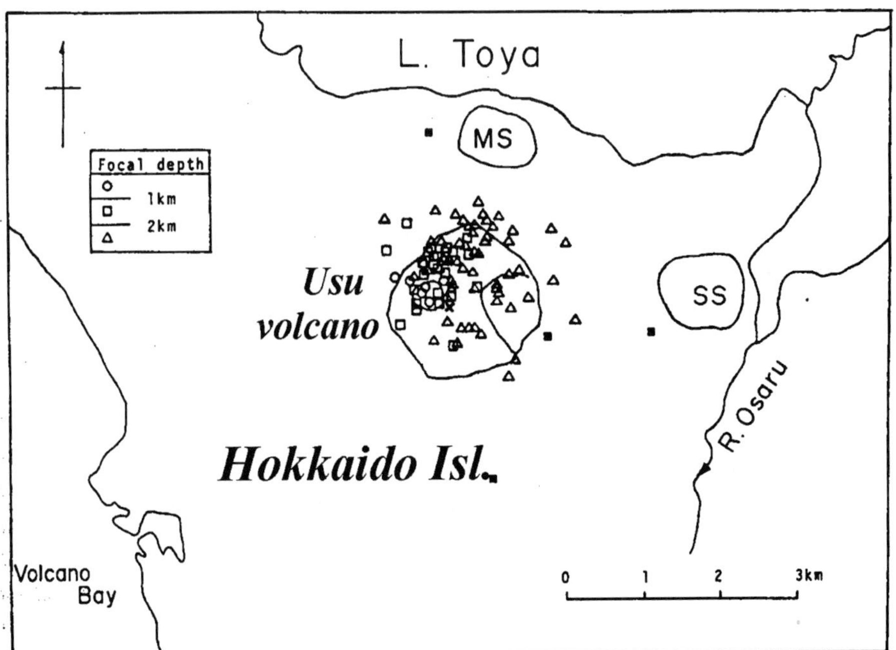


Fig. 7.4. Epicenters of earthquakes located just before the August 7, 1977 eruption of Usu volcano. MS is the 1910 cryptodome Meiji-shinzan; SS is the 1943-1945 lava dome Showa-shinzan. Solid squares show the position of seismometers; the cross indicates the position of crater 1. From Yokoyama et al. (1981).

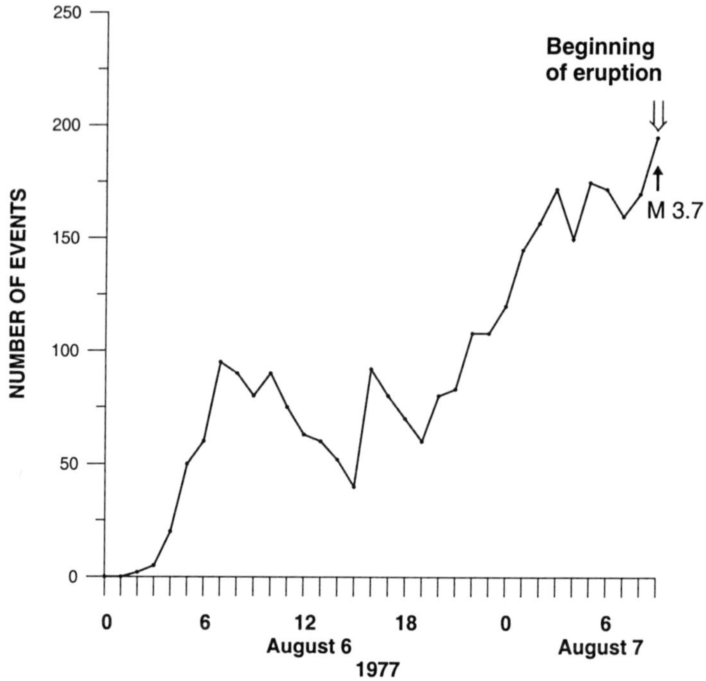


Fig. 7.5. Temporal variations in hourly number of earthquakes recorded at the station JMA-A located on the somma of Usu 32 hours before the eruption. The double arrow shows the beginning of the eruption; the single arrow shows the maximum magnitude earthquake. From Yokoyama et al. (1981).

The eruption was preceded by 32 hours of seismic activity (Fig. 7.5). The earthquake swarm consisted of about 3000 small (magnitude 1-2) volcano-tectonic earthquakes that were located beneath the summit crater at a depth from 0 to 2 km (Fig. 7.4). The number of earthquakes gradually increased with time, and the eruption began when the frequency of events reached its maximum. The maximum magnitude M 3.7 earthquake was observed half an hour before the beginning of eruption (Yokoyama and Seino, 2000; Yokoyama, personal communication, 2001). The new eruptive craters were situated within the epicentral zone of earthquakes. The alarm was issued by JMA, and there were no victims (Yokoyama et al., 1981).

During the stage of pumice explosions in August 8-14, the strong explosions were preceded and followed by a sharp decrease in seismic activity (Fig. 7.6). The stage of calm after the explosions continued for 3 to 10 hours, then seismic activity gradually increased again (Suzuki and Kasahara, 1979).

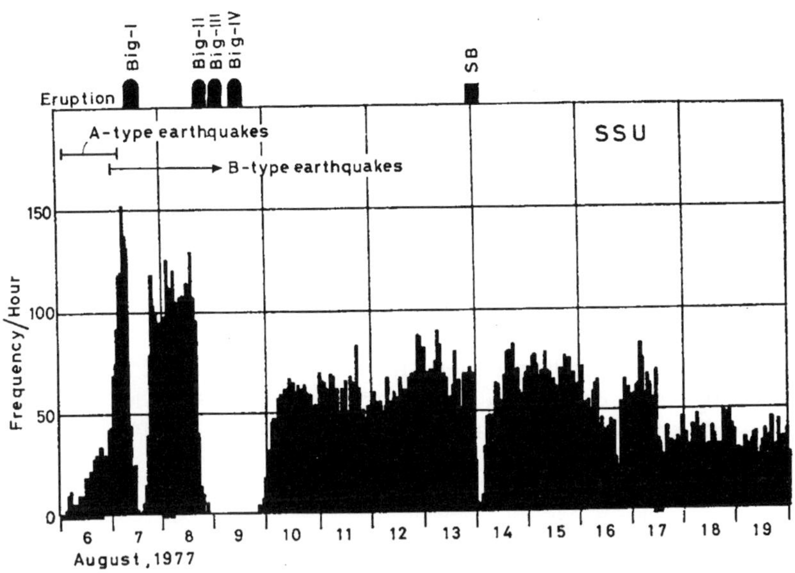


Fig. 7.6. Variations of hourly earthquake frequency at the station SSU situated at a distance of 2.5 km east from the crater (threshold M = 0.8) from August 6 to 19. It is clearly seen that there is an absence of seismic events just after the large explosions Big-I, II, III, IV and SB. From Suzuki and Kasahara (1979).

### 7.1.3. Unzen volcano, Kyushu, Japan

Unzen volcano is a composite volcano of 1359 m of height and consists of the multiple volcanic cones of dacite to andesite. Following 198 years of dormancy, a small phreatic eruption started at its summit on November 17, 1990. During February-April 1991, a sequence of phreatomagmatic eruptions followed a resumed phreatic eruption on February 12; they developed into a dacitic dome eruption on May 20, 1991. The growth of the dome continued for about four years accompanied by repeating partial collapses and pyroclastic flows. The total volume of erupted materials was $2.1 \times 10^8$ $m^3$ (Nakada et al., 1999).

Seismic activity began one year before the phreatic eruption (Nakada et al., 1999; Umakoshi et al., 2001). Fig. 7.7 shows the temporal variations in the number of earthquakes. Five swarms of volcano-tectonic earthquakes were recorded up to the moment of lava extrusion. The first two swarms were not followed by volcanic activity. Swarm 1 occurred at a distance of about 20 km to the west of the volcano at depths from 10 to 20 km below the ancient Chijiwa caldera (Fig. 7.8A). It began on November 21, 1989 and continued for four days. About 600 earthquakes were located with a maximum event of M 3.3.

Swarm 2 that occurred during July-August of 1990 was shallower, it occurred within the depth range from 15 to 5 km. The largest earthquake of magnitude 3.9 was observed on July 7. This swarm occupied the territory between the first swarm epicenters and the volcano (Fig. 7.7A).

The next three impulses of seismic activity occurred in November 1990 and during February-March and May 1991. They were associated with eruptive events. Their hypocenters were clustered, practically repeating the locations from swarm to swarm, within the long inclined zone that began at the depth of 15 km at the distance of about 20 km from the volcano and terminated at the surface about 2 km from the summit of the volcano (Fig. 7.8C and D).

Swarm 3 (Fig. 7.8B) preceded the phreatic eruption of November 17, 1990. It began on November 13 with a sharp increase in the number of events and continued for about 25 hours (Fig. 7.9A). The phreatic eruption occurred after the cease of seismic activity.

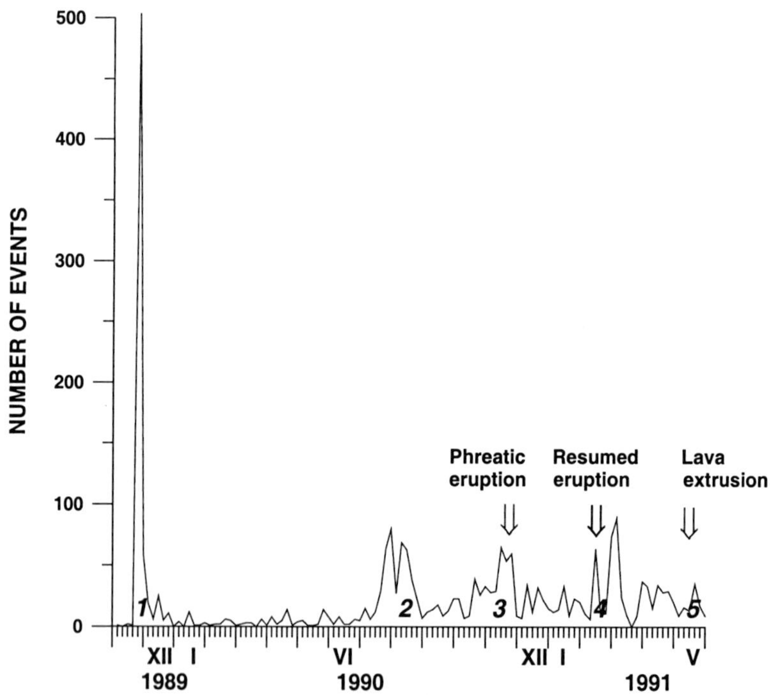


Fig. 7.7. Temporal variations of located volcano-tectonic earthquakes (number of events per 5 days) before and during the 1990-1991 eruption sequence at Unzen volcano, Japan. Arrows show the main eruptive events. Numbers correspond to the swarm of earthquakes. Data from the catalog of Shimabara Earthquake and Volcano Observatory.

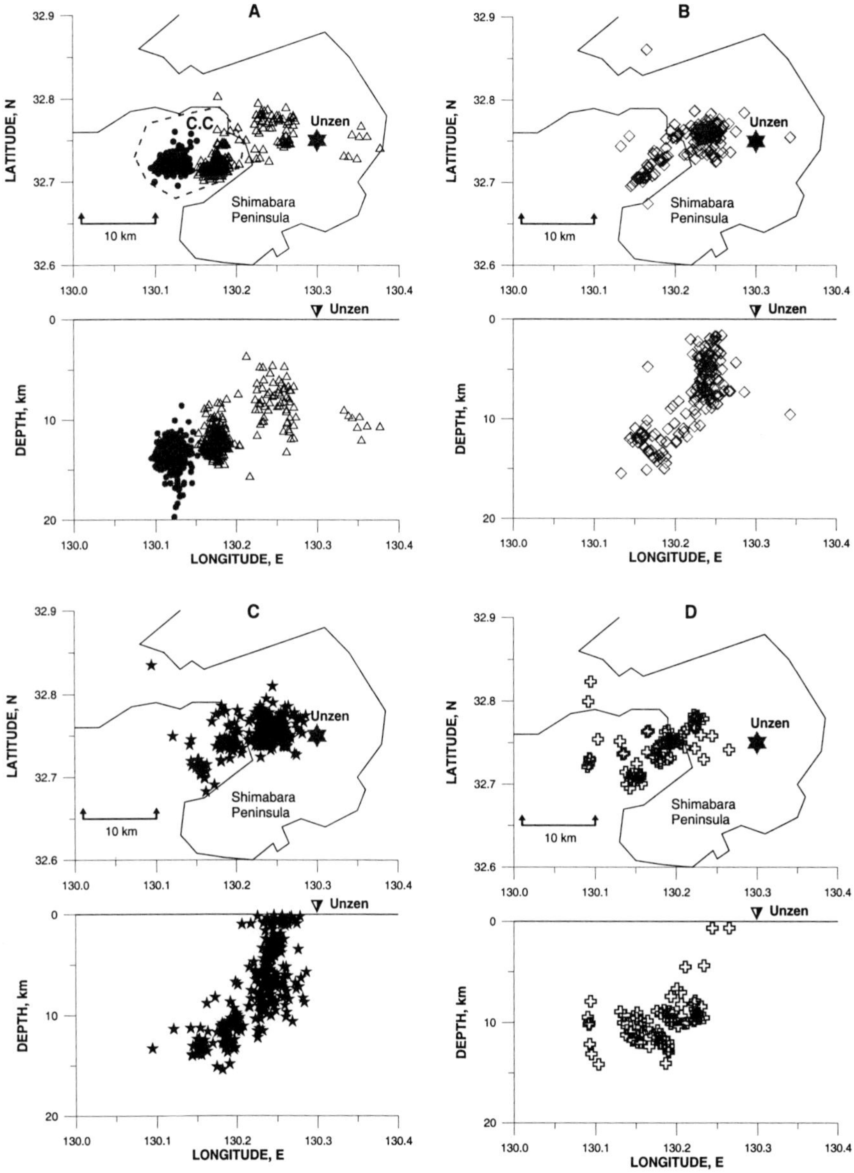


Fig. 7.8. Epicenters and cross-sections for the swarms of volcano-tectonic earthquakes that occurred during 1989-1991 at Unzen volcano, Japan. The swarms that were not followed by volcanic activity are shown in Fig. A (circles, swarm 1; triangles, swarm 2). The swarms associated with volcanic eruptions are shown in Fig. B (swarm 3), C (swarm 4) and D (swarm 5). Chijima caldera (C.C.) is shown by dashed line. Data from the catalog of Shimabara Earthquake and Volcano Observatory.

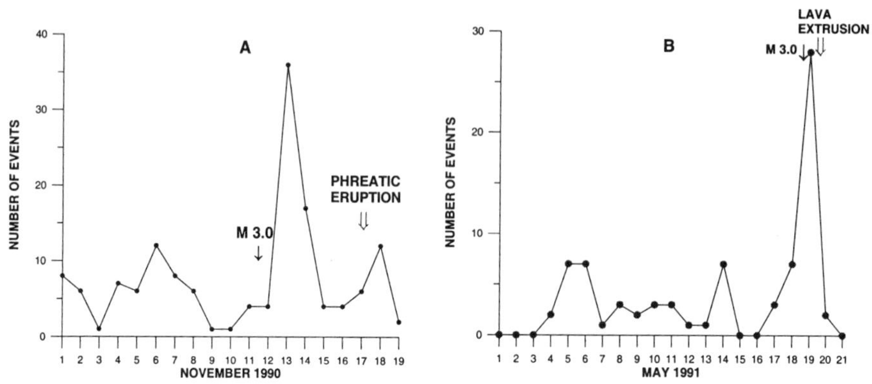


Fig. 7.9. Variations of the daily number of earthquakes before the November 17, 1990 phreatic explosion (A) and the May 20, 1991 lava extrusion (B) at Unzen volcano, Japan. Data from the catalog of Shimabara Earthquake and Volcano Observatory.

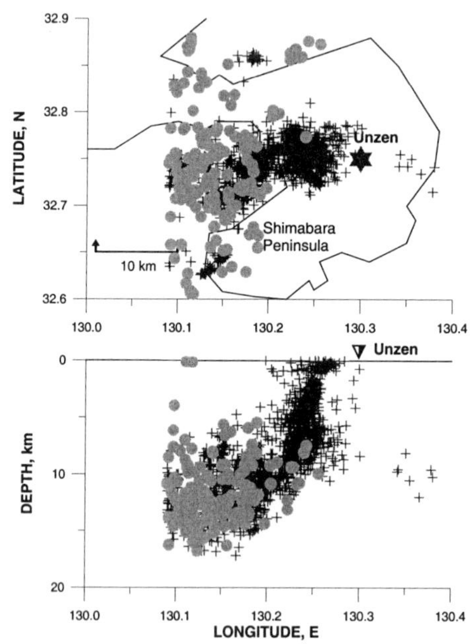


Fig. 7.10. The epicenters (above) and hypocenters (below) of the post-eruption volcano-tectonic earthquakes (circles) at Unzen volcano, recorded from February 1995 to the end of 1999. Crosses show the earthquakes recorded during the eruptive activity. Data from the catalog of Shimabara Earthquake and Volcano Observatory.

The resumed eruption, that began on February 12, 1991 and was represented by a sequence of phreatic and phreatomagmatic explosions, was not preceded by volcano-tectonic seismicity. Swarm 4 began after the phreatic explosion and accompanied a sequence of phreatomagmatic explosions.

Swarm 5 occurred just before the lava extrusion of May 20, 1991. It began on May 18 and reached its peak in the number of events on May 19, 1991. The lava extrusion was observed just after the ceasing of the swarm (Fig. 7.9B). This swarm was the smallest of five seismic sequences; it was practically at the background level of seismicity (Fig. 7.7). It was characterized by a small number of shallow seismic events; the majority of earthquake foci were located at depths from 6 to 15 km (Fig. 7.8D).

The characteristic feature of the Unzen seismicity was the absence of volcano-tectonic events just beneath the volcano crater. Umakosi et al. (2001) suggested that it was a zone of high-temperature, ductile bodies, or shallow magma chamber.

After the termination of lava activity in February 1995, volcano-tectonic earthquakes were observed along the submeridional structure crossing the focal zone of swarm 1 at a distance of about 20 km from the volcano (Fig. 7.10). They were located at the depth range from 5 to 17 km.

### 7.1.4. Pinatubo volcano, Philippines

Mount Pinatubo is a composite volcano of 1745 m elevation and is situated in the Luzon volcanic arc. Its recent unrest began after about a 500-year calm on April 2, 1991 after about 20 days of locally felt earthquakes related to steam explosions. The unrest culminated 10 weeks later in the great explosion of June 15 that produced about 10 $km^3$ of porous, pumiceous deposits and formed a caldera of 2.5 km in diameter. Within the twentieth century, it was the second largest eruption after the 1912 Novarupta-Katmai (Alaska) eruption (Wolfe and Hoblitt, 1996).

Seismic observations began only in May; therefore, the preceding earthquake swarm was not recorded. Nevertheless, the seismic activity, that preceded the June 7 shallow lava extrusion at the summit and the following June 15 great explosion, was located by a seismic network. It allowed forecasting the June 15 climactic eruption (Harlow et al., 1996).

Figures 7.11 to 7.13 show the development of seismic activity from May 7 to June 12, 1991. Seismic events were recorded within two clusters: just beneath the summit (I) and 5 km NW of the summit (II) (Fig. 7.11). The second group of seismic foci occurred mainly within the depth range from 7 to 3 km while the second group of events was more shallow, within the range 4 km bsl to 1 km asl. The events from the distant zone were dominant from 21 May to 5 June. Seismicity of zone I beneath the volcano began to increase from May 30, and an intensive swarm of low magnitude (from 0 to 1) was recorded there from June 6 to 7 (Fig. 7.12). This swarm culminated in the June 7 lava dome extrusion. Just after the beginning of dome growth, the number of earthquakes sharply decreased, and the June 12 explosion occurred on the background of low-level seismicity for both seismic zones.

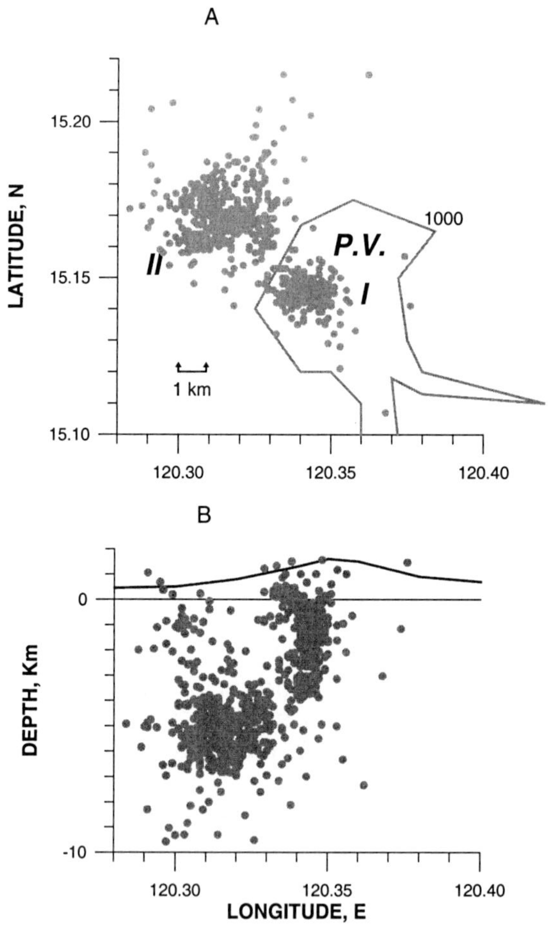


Fig. 7.11. Epicenters (A) and hypocenters (B) of earthquakes located from May 7 to June 12, 1991 near Pinatubo volcano. The 3-D velocity model was used for location (Harlow et al., 1996). P.V. is Pinatubo volcano. I and II mark two clusters of earthquakes. The 1000-m contour of the volcano's relief and the zero-line of sea-level depth are shown. Data from Mori et al. (1996b).

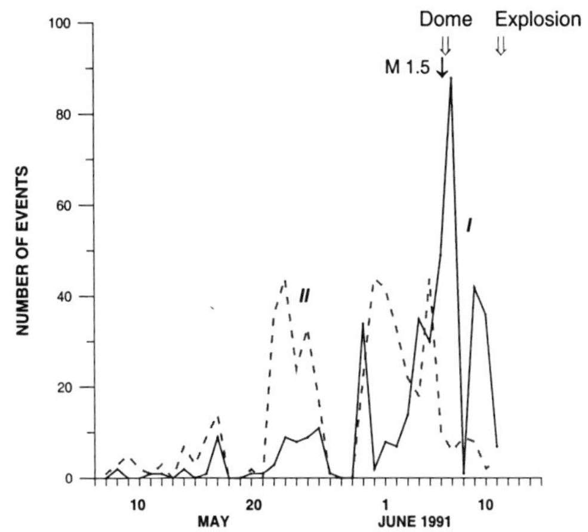


Fig. 7.12. Temporal variations in the daily number of earthquakes located from May 7 to June 12, 1991 beneath the Pinatubo volcano (I) and 5 km to NW from the volcano (II). The double arrows show the eruptive events; the single arrow shows the maximum magnitude earthquake. I, II are two clusters of earthquakes. Data from Mori et al. (1996b).

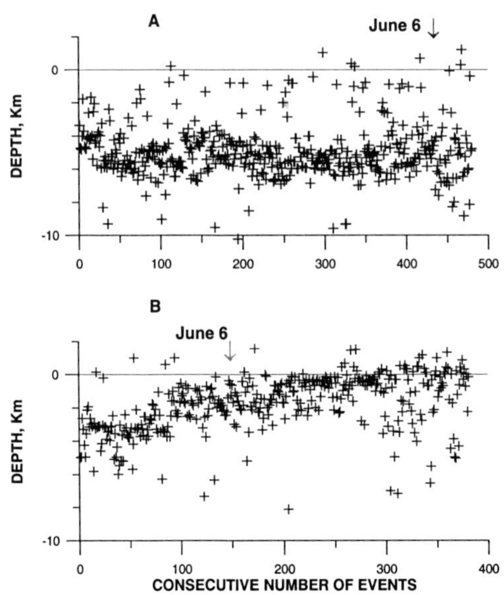


Fig. 7.13. Variations in the focal depths of the events from cluster II situated 5 km to NW from the volcano (A) and from cluster I beneath the volcano Pinatubo (B) from May 7 to June 12, 1991. Zero-line of sea-level depth is shown. The beginning of the June 6-7 swarm is shown by the arrow. Data from Mori et al. (1996b).

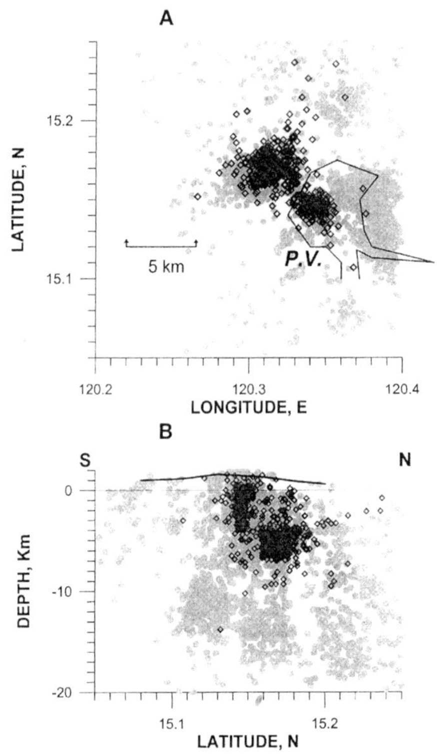


Fig. 7.14. Post-June 15, 1991 seismic activity in the region of Pinatubo observed from June 26 to August 18, 1991. A. Epicenters of earthquakes; B, hypocenters of earthquakes (N-S cross section). The earthquakes recorded before the June 15 explosions are shown by diamonds; the earthquakes recorded after the June 15 explosions are shown by circles. The 1000-m contour of Pinatubo (P.V.) is shown. Data from Mori et al. (1996b).

Fig. 7.13 shows that the seismic foci, that occurred beneath the volcano from May 8 to June 5, became shallower before the June 6-7 swarm (and the lava dome extrusion), changing the depths from 3-5 km to 0-2 km.

The intense seismic activity, which occurred from June 12 to 15, was practically off the scale. Several of the seismic stations were destroyed, the remaining were very often saturated (Harlow et al., 1996). The 9-hour climactic sequence of explosions of June 15 was accompanied by large volcano-tectonic earthquakes. Two of them, which were recorded at 18:41h and 19:15 h, reached magnitude Mw 5.7 and 5.6 respectively (Harvard, 1996).

The location of seismic events was again possible only from June 29. The post-explosion distribution of seismic foci sharply changed from the previous one that was shown in Fig. 7.11. The seismically active zones extended in space and depth (Fig. 7.14). A zone beneath the volcano spread to the west within volcanic edifice and occupied all beneath-volcano space filling it up to depths of 25 km, about 20 km deeper than before the June 15 explosion. The NW seismic zone became also deeper and larger in area.

### *7.2. Volcano-tectonic earthquakes associated with flank eruptions*

We have only two cases; both are related to the activity of Usu volcano, Hokkaido.

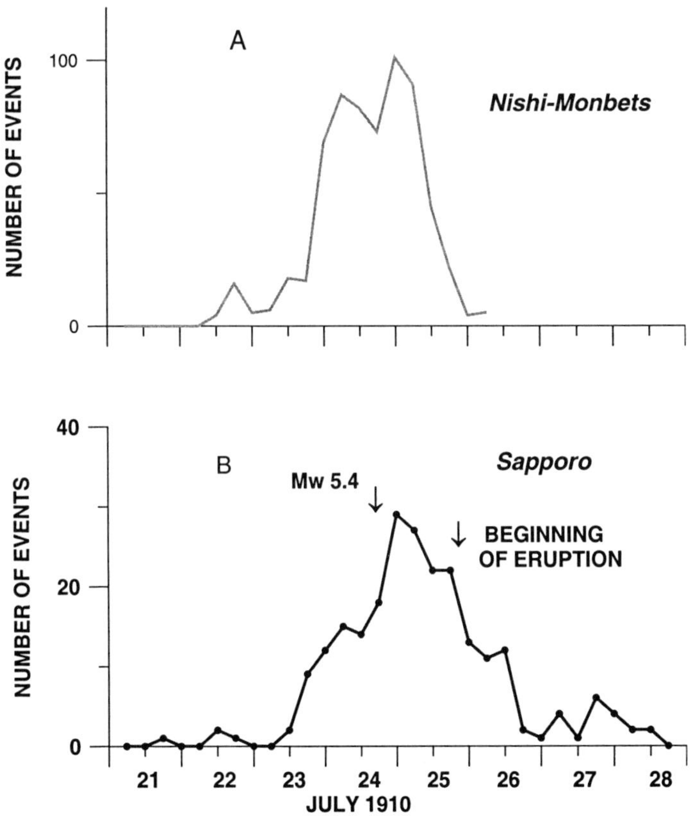


Fig. 7.15. Variations in the number of earthquakes preceding the 1910 flank eruption of Usu volcano from macroseismic data recorded at a distance of 8.5 km from the new volcanic site (A) and recorded by a seismic station at a distance of 70 km from the new volcanic site. Data from Omori (1911-1913).

*Eruption of 1910.* The eruption of the cryptodome Meiji-shinzan began at the northern foot of the volcano on July 25 (see Fig. 7.4), after four days of precursory earthquakes (Fig. 7.15). Nearly 40 craterlets were formed along the arcade line approximately 3 km long during ten days of phreatic explosive activity. The eruption caused an upheaval of approximately 90 m above the original terrace (Yokoyama and Seino, 2000).

The felt shakes began on July 21. On July 22, 25 shocks were felt at a distance of 8 km from the volcano. Fig. 7.15B shows the 6-hour frequency of earthquakes recorded by horizontal pendulum seismograph (magnification 30) that was installed at the meteorological observatory of Sapporo, at a distance of 70 km from the volcano. The number of earthquakes increased rapidly and reached its maximum 13 hours before the beginning of eruption. The strongest earthquake (Mw 5.4; Zobin, 2001) was recorded 30 hours before the beginning of eruption. The eruption began on the background of a gradual decrease in the number of earthquakes (Omori, 1911-1913). Fig. 7.15 shows also the number of earthquakes that were felt at Nishi-Monbets village, at a distance of 8.4 km from the volcano (Fig. 7.15A). The remarkable similarity between the curves is seen.

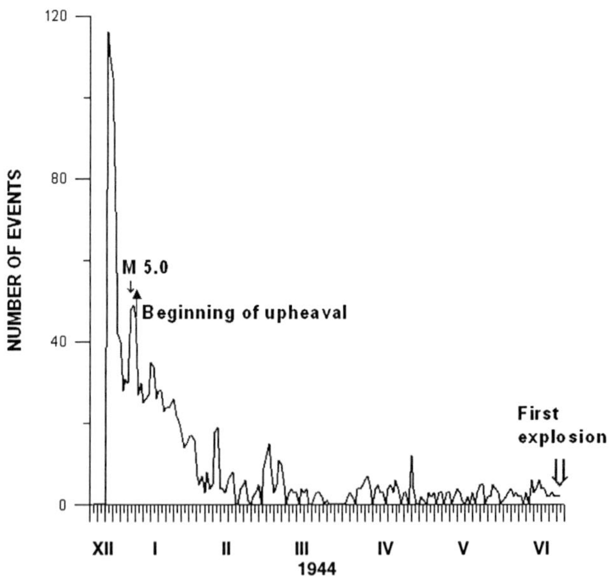


Fig. 7.16. Variations in daily number of earthquakes that were felt at a distance of 3 km from the volcanic site. The double arrow shows the beginning of the eruption; the single arrow shows the maximum magnitude earthquake. Data from Mimatsu (1995).

*Eruption of 1944.* It was a birth of a new volcano Showa-Shinzan on the flank of Usu (See Fig. 7.4). The seismic activity and deformations began about 6 months before the first explosion of June 23, 1944. Fig. 7.16 shows the variations in number of felt earthquakes recorded at a distance of about 3 km from the new volcano site by M. Mimatsu (Mimatsu, 1995). I consider the list of Mimatsu-san is complete enough for analysis of temporal variations in seismicity.

Seismic activity began with numerous earthquakes; the maximum number of events was recorded during the first three days of activity on December 29-31, 1943. Then the number of earthquakes began to gradually decrease during January-February, 1944 and stayed at a background level up to the first explosion of June 23. The maximum magnitude event (M 5; Yokoyama and Seino, 2000) occurred in 12 days after the beginning of activity on January 9 and was located by a three-station network at a depth of 11 km (Kizawa, 1958).

# Chapter 8
# General Properties of Volcano-tectonic Earthquake Swarms

This chapter discusses the general properties of volcano-tectonic earthquake swarms. At first, we generalize the earthquake swarm properties inferred from the data presented in Chapters 5 to 7. Then these data would be supplemented and compared with the published data in this field.

## *8.1. Properties of volcano-tectonic earthquake swarms inferred from the data of Chapters 5 to 7*

We passed through a sequence of seismo-volcanic episodes in Chapters 5-7. Table 8.1 collects the characteristics of seismic sequences of volcano-tectonic earthquakes that were observed in the twentieth century at 17 volcanoes of the world during their 33 eruptions. These data are not representative for the Earth´s volcanic activity for that century; it is only a small sample showing the most studied or (and) most remarkable examples of seismic activity at volcanoes. Nevertheless, we can discuss the general properties (or tendencies) of seismic swarms, associated with volcanic activity, using these observations. We select for discussion the following characteristics: temporal variations in seismic activity, spatial distribution of seismic foci, post-eruption seismicity; duration of seismic activity before volcanic event, position of volcanic event in seismic sequence, and dependence of these characteristics on the type of magma feeding the eruption.

### 8.1.1. Temporal variations

The temporal variations in seismic activity were of three types: one-peak (I), multi-peak sequences with the maximum peak at the beginning of the sequence (II) or with maximum peak at the end of sequence (III) (Fig. 8.1). The first type of sequences is rare (Fig. 8.1B), it was observed only for six from 30 cases. The most popular is the third type of variations when the maximum peak is observed at the end of seismic sequence; it was observed for 14 cases. The second type of sequences

**Table 8.1**
**Data on volcano-tectonic earthquake sequences**

| Volcano, Region | Type of magma | Year | Type of erupt. | VEI | Duration of earthq. seq., hrs | Dep., km | Range of magnitudes | Type of eart. seq. | Posision of erup. event |
|---|---|---|---|---|---|---|---|---|---|
| Kilauea Hawaii | Basaltic | 1969 | 4 | 0 | 5 | 2-10 | 0-3.5 | 1 | 4 |
| | | 1974 | 1,4 | 0 | 9 | 0-4 | 0-3.4 | 3 | 4 |
| | | 1982 | 1 | 0 | 2.5 | 1-3 | 0-3.2 | 1 | 3 |
| Etna, Sicily | Basaltic | 1983 | 2 | 1 | 28 | (-)3-15 | | 3 | 4 |
| | | 1984 | 1 | 2 | 0 | | | - | - |
| | | 1991 | 2 | 2 | 14 | 0-10 | 0-4.4 | 2 | 1 |
| Oshima, Izu isl. | Basaltic | 1986 | 1 | 3 | 0 | | | | |
| | | 1986 | 2 | 3 | 2 | 0-5 | 0-5.9 | - | 3 |
| Klyuchevskoy, Kamchatka | Basaltic | 1951 | 2 | 2 | 96 | | 0-3.1 | 2 | 4 |
| | | 1966 | 2 | 3 | 84 | | 0-3.7 | 2 | 4 |
| New Tolbachik, Kamchatka | Basaltic | 1975 | 4 | 4 | 216 | 0-16 | 0-4.9 | 2 | 4 |
| Fernandina, Galapagos Isl. | Basaltic | 1968 | 1 | | | | 0-5.2 | - | - |
| Teishi Knoll, Izu Isl. | Basaltic | 1989 | 5 | | 216 | 0-12 | 0-5.5 | 2 | 4 |
| Bezymianny, Kamchatka | Andesitic | 1955 | 1 | 5 | 408 | | | 3 | 2 |
| | | 1956 | 1 | 5 | 4320 | | 0-5.2 | 2 | 3 |
| | | 1959 | 1 | 1 | 48 | | 0-1.3 | 1 | 2 |
| | | 1960 | 1 | 2 | 70 | | 0-1.7 | 1 | 2 |
| Sheveluch, Kamchatka | Andesitic | 1964 | 1 | 4 | 216 | | 0-5.2 | 3 | 2 |
| | | 1993 | 1 | 3 | 500 | | | 3 | 2 |
| El Chichon, México | Andesitic | 1982 | 1 | 5 | 620 | 0-20 | 0-4.0 | 2 | 3 |
| Volcán de Colima, México | Andesitic | 1994 | 1 | | 220 | (-)4-0 | 0-1.2 | 3 | 4 |
| | | 1998 | 1 | | 600 | (-)4-0 | 0-2.7 | 3 | 3 |
| Soufriére Hills, Montserrat | Andesitic | 1995 | 1 | | 30 | (-)0.5-4 | 0-1.4 | 3 | 3 |
| | | 1996 | 1 | | 600 | (-)0.5-3 | 0-1.9 | 3 | 3 |
| Sakurajima, Kyushu | Andesitic | 1914 | 2 | 4 | 80 | | 0-7.0 | 3 | 4 |
| Mount St. Helens, | Dacitic | 1980 | 1 | 5 | 80 | 0-20 | 0-4.4 | 2 | 2 |

| | | | | | | | | | |
|---|---|---|---|---|---|---|---|---|---|
| Cascades | | | | | | | | | |
| | | 1980 | 1 | 5 | 1300 | 0-10 | 0-5.7 | 2 | 3 |
| Usu, Hokkaido | Dacitic | 1910 | 2 | 2 | 108 | | 0-5.4 | 3 | 3 |
| | | 1944 | 2 | 2 | 700 | | 0-5.0 | 2 | 3 |
| | | 1944 | 2 | 2 | 4300 | | | 2 | 4 |
| | | 1977 | 1 | 3 | 32 | 0-2 | 0-3.7 | 3 | 2 |
| Unzen, Kyushu | Dacitic | 1990 | 1 | 1 | 25 | 1-15 | 0-3.0 | 1 | 4 |
| | | 1991 | 1 | 1 | 48 | 0-15 | 0-3.0 | 1 | 3 |
| Pinatubo, Philippines | Dacitic | 1991 | 1 | 6 | 160 | (-)1-1.6 | 0-1.6 | 3 | 2 |

*Note.* Data about the type of eruptions and their VEI are taken from Simkin and Siebert (1994): 1, central eruption; 1a, caldera collapse; 2, flank eruption; 3, radial fissure; 4, regional fissure; 5, submarine eruption. The codes of earthquake sequence (1 to 3) and of the position of volcanic events (1 to 4) see in text. Symbol (-) before the depth means the depth above sea level.

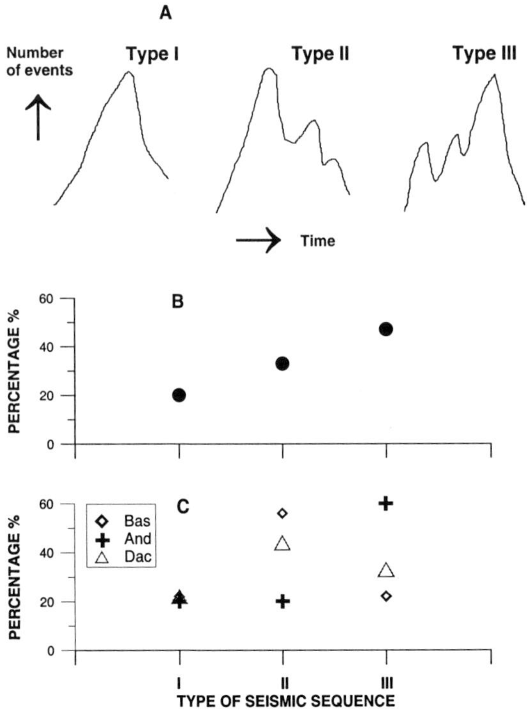


Fig. 8.1. Distribution of three types of volcano-tectonic earthquake swarms for all events of Table 8.1 (B) and for basaltic, andesitic and dacitic volcanoes (C). The three types of volcano-tectonic earthquake swarms are the one-peak (I) and multi-peak sequences with the maximum peak in the beginning of sequence (II) or with maximum peak at the end of sequence (III). The typical curves of swarms are shown in the upper part of Figure (A).

was observed for 10 cases.

For basaltic and dacitic volcanoes, the most popular is a seismic sequence with the maximum peak in the beginning of the sequence, while for andesitic volcanoes, the majority of seismic sequences end with the maximum number of seismic events (Fig. 8.1C).

8.1.2. Spatial distributions

Generally, the seismic events cluster within or below the volcanic edifice within a depth range of 20 km to the summit of the volcano. The seismic events of preliminary swarms occupy a greater area than the site of forthcoming eruption. At the initial stage of seismic activity, the events at the depths 10 to 20 km were observed together with the shallower events. Before an eruption, the earthquakes cluster at shallower depths. It was observed for New Tolbachik volcanoes (Fig. 5.17), Mt. St. Helens (Fig. 7.2), Pinatubo (Fig. 7.11), and Volcán de Colima (Fig. 6.19), among others. Very often the earthquakes originate not only beneath the volcano, but along a system of local faults adjusting the volcano also. The examples of well-located earthquakes show that during the initial stage of seismic activity before an eruption, the seismic events were observed along the local faults, such as St. George´s Hill for Soufriére Hills volcano (Fig. 6.13); the fault 5 km NW of Pinatubo volcano (Fig. 7.11); and the fault SE of Mt. St. Helens (Fig. 7.2). The adjacent fault events had a greater depth than the events originating beneath the volcano. Just prior to an eruption, seismic activity began to concentrate at shallower depths beneath the volcanic edifice.

The important feature of spatial distribution of volcano-tectonic earthquakes is the migration of seismic foci (or seismic clusters) for a distance of 10-20 km. The focal migration was observed before the forming of the eighth cone of New Tolbachik volcanoes (Fig. 5.20). Earthquakes migrated during 6 days along the fault at a distance of about 15 km from the site of the formation of the first 7 cones to the site of the eighth cone.

Multi-peak temporal variations in the number of seismic events may be considered as a result of the migration of seismic clusters. Before the submarine eruption of Teishi Knoll, the two peaks in temporal variations of the number of earthquakes corresponded to the clustering of seismic events within two separated zones (Figs. 5.24 and 5.26). One of these zones was active from 4 to 7 of July 1989, and produced the first peak in frequency of events; the second cluster developed in a longer and narrower zone, and produced the second peak of frequency. The same was observed at New Tolbachik volcanoes (Fig. 5.17 and 5.18). During June 27-30, 1975, the earthquake source zone was larger and deeper, and produced the first peak in earthquake frequency; in July 1-6, the earthquakes occurred within a long narrow zone, and produced the second peak of earthquake frequency.

8.1.3. Post-eruption seismic activity

Very often, we have no volcano-tectonic seismicity corresponding to the beginning of an eruption or following it. Nevertheless, we observed the post-eruption seismicity after the eruptions of New Tolbachik volcanoes, Mt. St. Helens, Oshima, Unzen and Pinatubo. The spatial distribution of the post-eruption volcano-tectonic earthquakes overfills the areas that were active before and during the eruptive events. For New Tolbachik volcanoes (Fig. 5.21), the epicenters of post-eruption earthquakes, that were observed during two years after the eruption, outline the 100-km-length fault that includes the 12-km new-eight-cones fissure. They occurred at the same depths as the events of preliminary earthquake swarm.

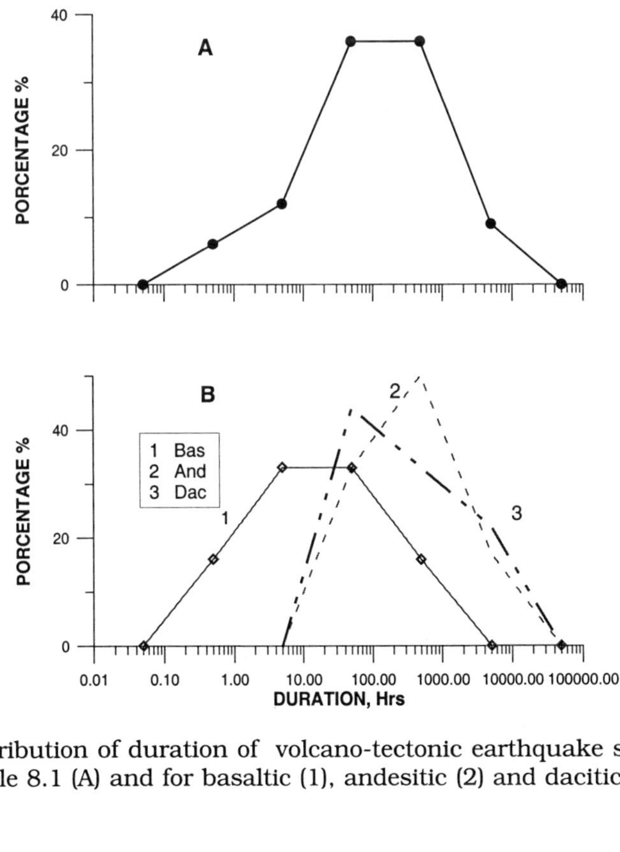


Fig. 8.2. Distribution of duration of volcano-tectonic earthquake swarms for all events of Table 8.1 (A) and for basaltic (1), andesitic (2) and dacitic (3) volcanoes (B).

For Mt. St. Helens, the activated post-eruption fault, whose central part was occupied by the volcano, was outlined to the north and to the south of the volcano. The free space between two seismic clusters that occurred separately at the depths of 0 to 10 km and from 16 to 22 km before and during the eruption was filled by the post-volcanic seismic events during one year of activity (Fig. 7.3).

For Unzen volcano, the post-eruption events traced a 30-km-long structure that intersects at a depth of 20 km the seismic volume where the first preliminary swarm occurred one year before the eruption (Fig. 7.10).

In the case of Pinatubo volcano, the post-eruption seismic events significantly expanded the seismic volume of preliminary events. The epicentral area and the depths of earthquakes were both increased by a factor of two (Fig. 7.14).

The volcano-tectonic earthquakes, that followed the eruption of Oshima volcano, also significantly expanded the seismic volume that was active before the eruption (Figs. 5.11 and 5.12).

These spatial characteristics of the post-eruption seismicity suggest a relationship between the collapses of magma chambers after the magma release during the eruption.

### 8.1.4. Duration of seismic activity before volcanic event

Generally, the duration of a seismic sequence lies between 100 and 1000 hrs before an eruption (Fig. 8.2A). These values mainly reflect the properties of andesitic and dacitic volcanoes (Fig. 8.2B); for basaltic volcanoes, the durations of seismic activity prior to eruption are shorter (from 1 to 100 hrs).

### 8.1.5. Position of a volcanic event according to the stage of volcano-tectonic earthquake swarm

Volcanic events may occur during the different stages of seismic sequence (Fig. 8.3A): at the initial stage of seismic sequence (I), at the maximum frequency of seismic events (II), at the ceasing of seismic events (III), and after ceasing of seismic sequence (IV). The majority of the discussed eruptions occurred while the earthquakes were ceasing (stage III) (Fig. 8.3B), or after the earthquake sequence had ceased (stage IV).

For basaltic volcanoes, the majority of eruptions occur after the end of the seismic sequence, while for andesitic and dacitic volcanoes, the majority of eruptions occur during the maximum number of volcano-tectonic earthquakes (Fig. 8.3C).

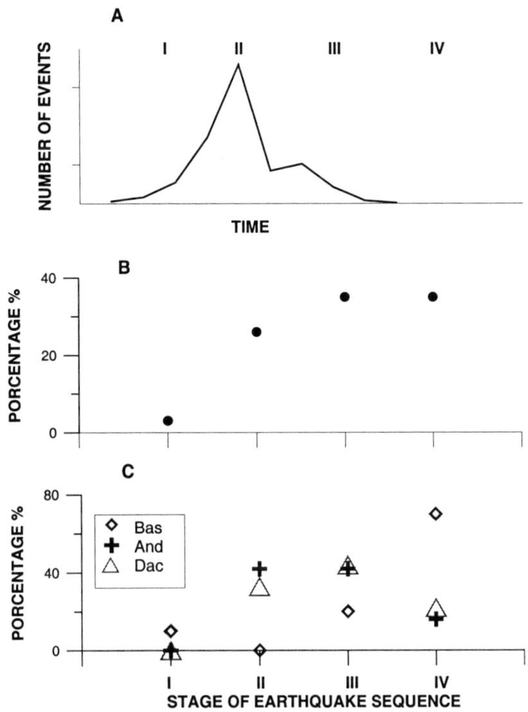


Fig. 8.3. Position of volcanic event according to the different stages in volcano-tectonic earthquake swarms for all events of Table 8.1 (B) and for basaltic, andesitic and dacitic volcanoes (C). The stages are illustrated in Fig. A: the initial stage of seismic sequence (I), stage at the maximum frequency of seismic events (II), at the ceasing of seismic events (III), and after ceasing of seismic sequence (IV).

## *8.2. Additional data about volcano-tectonic earthquake swarm properties*

### 8.2.1. Size of volcano-tectonic earthquake swarm area

The swarm epicentral area is defined as the area of volcano-tectonic earthquake epicenters (magnitude 1-2 and greater) during eruptive activity. Areas of epicentral zones vary for different volcanoes from 25 km for Mount St. Helens to 9450 km for northern Tonga (Zobin, 1996). Zobin (1996) showed that the areas are larger for volcanoes on small islands and smaller for volcanoes on large islands or on continents. Assuming that the crust is thinner for oceanic volcanoes than for continental volcanoes, it was found that a significant negative correlation exists between the epicentral areas of volcanic earthquakes S and the crustal thickness H under the volcano (Fig. 8.4):

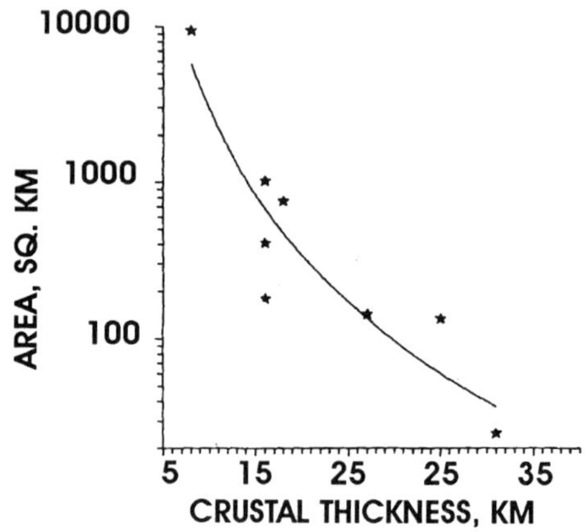


Fig. 8.4. Relationship between the epicentral area of volcano-tectonic earthquake swarms and the crustal thickness for eight selected eruptions. From Zobin (1996).

$$\text{Log S (km}^2) = 3.6 \times 10^5 \text{ H}^{-3.1} \text{ (km)} \quad (8.1)$$

8.2.2. Earthquake swarm duration

Benoit and McNutt (1996) have compiled a Global Volcanic Earthquake Swarm Database, 1979-1989. This database gives the data about swarm durations as reported in the *Bulletin of Volcanic Eruptions* of the Volcanological Society of Japan (Fig. 8.5).

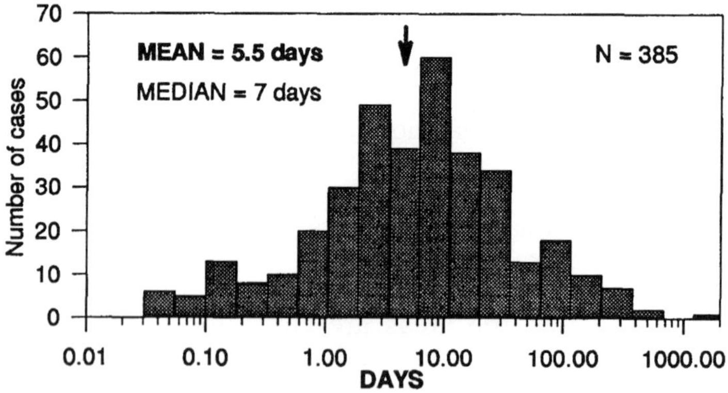


Fig. 8.5. Histogram of the duration of 385 volcanic earthquake swarms. The data form a nearly log-normal distribution with a geometric mean of 5.5 days. From Benoit and McNutt (1996).

This Figure shows that the main part of earthquake swarms had durations from 2 to 20 days (or from 48 to 480 hours). The mean geometric value of swarm duration is 5.5 days (132 hrs). The data of Table 8.1 give for 31 eruptions the mean geometric value of swarm duration of 106 hours (or 4.4 days) that is very close to the Fig. 8.5 estimations.

Benoit and McNutt (1996) made an interesting observation about the relationship between the swarm duration and eruption repose time. Fig. 8.6 shows that the long swarms (longer than 4 months) have a good correlation of their duration with the time of repose; the longer the period between eruptions, the greater duration of seismic activity observed before the eruption. The authors guess that longer lasting swarms may be the result of ascending magma encountering a cooler and stronger crustal environment.

### 8.2.3 Magnitude-frequency relations of events in volcano-tectonic earthquake swarms

The frequency of the occurrence of earthquakes in a given region and for a given period of time can be represented by the equation $\log N = a - b M$, where $N$ is a number of earthquakes with magnitudes of a fixed range around magnitude M, and $a$ and $b$ are constants. The constant $b$ is called the *b*-value and depends on the relation between large and small events in the earthquake sequence. For tectonic earthquakes, *b*-values are between 2/3 and 1 and do not show much regional variability.

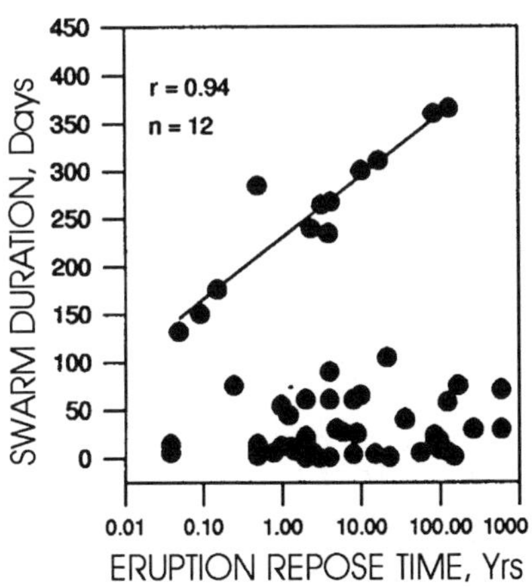


Fig. 8.6. Relationship between the duration of 65 precursory swarms and eruption repose time. The regression line shown is for 12 swarms with a duration of 100 days or greater. The correlation coefficient is 0.94, which is significant at the 99% confidence level. From Benoit and McNutt (1996).

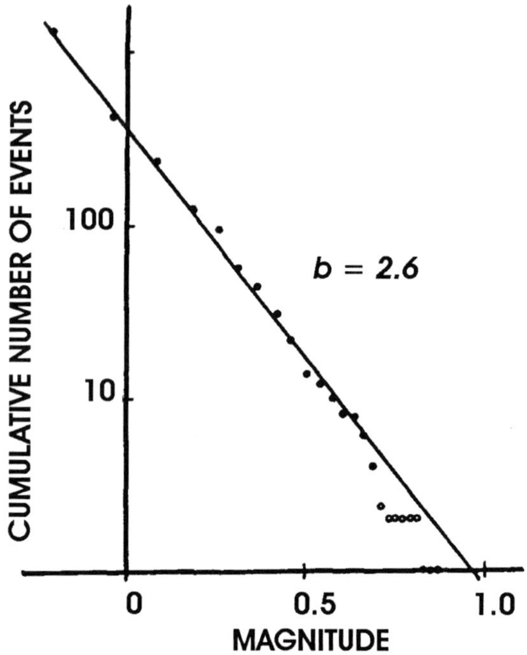


Fig. 8.7. A log-linear magnitude-frequency plot for 1316 B-type volcano-tectonic earthquakes recorded 1 to 8 December 1974 during the warning stages of the period of explosive eruptions at Pavlof volcano, Alaska. It was obtained the relationship Log N = 2.53 – 2.6 M. From McNutt (1986).

For volcanic earthquake swarms, *b*-values were estimated as close to 1 or greater, up to 2.6 (Fig. 8.7), indicating that volcano-tectonic earthquake swarms tend to have an anomalously large number of small events compared to the usual seismic sequence (Gresta and Patané, 1983a, b; McNutt, 1986; Zobin et al., 2002b; etc). Results of McNutt (1986) for Pavlof volcano, Alaska show that *b*-values of volcano-tectonic microearthquake (M = 0-1) swarms do not depend on the stage of volcanic activity.

At the same time, when larger earthquakes participate in the swarm, the magnitude-frequency distribution can vary during the volcanic activity or be non-log-linear (Gresta and Patané, 1983; Okada, 1982). This reflects the unstable relationship between small and large events during the volcano-tectonic earthquake swarm. Gorshkov (1959) has noted that during a course of an earthquake swarm related to the 1956 eruption of Bezymianny volcano, Kamchatka (See Chapter 6), there were observed two independent sequences of large and small events. The large events (magnitude about 4) occurred as a background without any

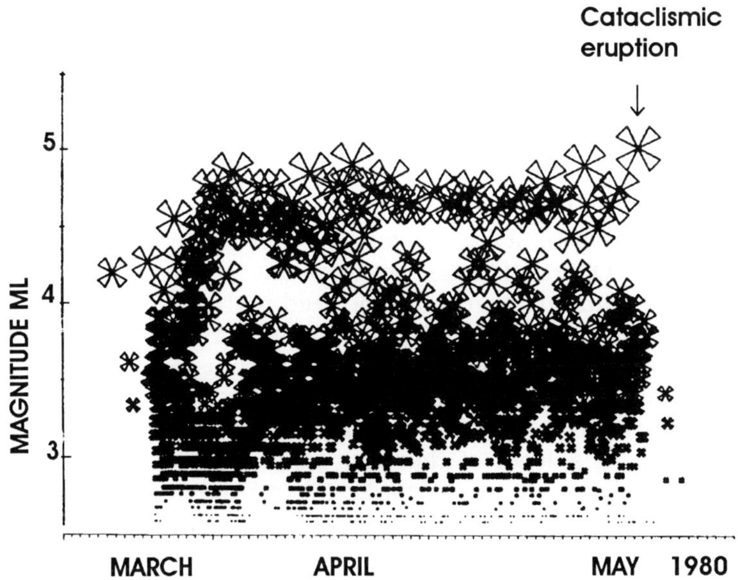


Fig. 8.8. Temporal variations in magnitudes of volcano-tectonic earthquakes preceding the May 18, 1980 directed blast at Mt. St. Helens volcano, Cascades. From Okada (1983).

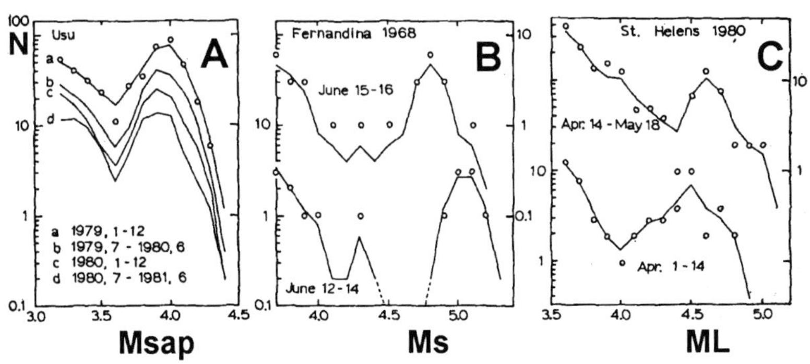


Fig. 8.9. Non-log-linear magnitude-frequency plots for volcano-tectonic earthquakes associated with the 1979-1980 lava dome activity at Usu, Hokkaido (A), the 1968 caldera collapse at Fernandina, Galapagos (B) and the 1980 directed blast at Mt. St. Helens, Cascades (C). N is a number of earthquakes; Msap is a magnitude determined by Sapporo station records. The plots are shown for different stages of volcanic activity indicated by dates. From Okada (1982).

relation to the course of eruption, while the smaller events reflected the variations in volcanic activity. Okada (1982) demonstrated the same characteristics for the earthquake swarm preceded the 1980 cataclysmic eruption of Mt. St. Helens, Cascades volcano (See Chapter 7). Fig. 8.8 shows that the large and small earthquakes develop in time distinctively. There exists an intrinsic lack of earthquakes over a particular magnitude range.

The effect of the lack of earthquakes within the volcano-tectonic earthquake swarms may be a reason for the non-log-linear magnitude-frequency relations. Okada (1982) pointed out this effect for three volcano-tectonic earthquake swarms associated with the great eruptions of Fernandina, Galapagos (1968), Usu, Hokkaido (1977-1978) and Mt. St. Helens, Cascades (1980) (Fig. 8.9). It is seen that for all three cases there is a minimum in magnitude-frequency relations. Okada (1982) guessed that larger earthquakes are related to the large amount of surface deformation in a form of doming, caldera collapse or bulging. Therefore, the volcano-tectonic earthquake swarm may consist of two or more separate swarms reflecting the different-scale processes during great manifestations of volcanic activity.

# Chapter 9
# Source Properties of Volcano-tectonic Earthquakes

In this Chapter, I consider a group of volcano-tectonic earthquakes that includes the earthquakes generated by stresses, which arise during magma migration in the Earth's crust. They correspond to A and B-types events of Minakami's classifications or to high-, mix- and low-frequency earthquakes of McNutt's classification. These earthquakes may be caused by shear failure, tensile failure or fluid pressurization processes and occur during magma movement from depth to the Earth's surface through conduits, dikes, etc, producing the rupture at the tip of magma body and at adjacent fault systems. Here I discuss the focal mechanisms and spectral properties of these earthquakes.

## *9.1. Focal mechanisms of volcano-tectonic earthquakes: double couple and non-double couple models*

The complete tensor of an earthquake source may be described as a sum of its isotropic and deviatoric parts. The deviatoric part, which is responsible for the production of fracture, may be decomposed into two source types: a double couple (DC) and a compensated linear vector dipole (CLVD) (Fig. 9.1). The dominant type may be discriminated by the relationship between the smallest and largest eigenvalues of the seismic moment tensor $\varepsilon$; it is equal to 0 for a DC source and to 0.5 for a CLVD source. The application of these two source types to describe the fault nature of volcano-tectonic earthquakes is discussed.

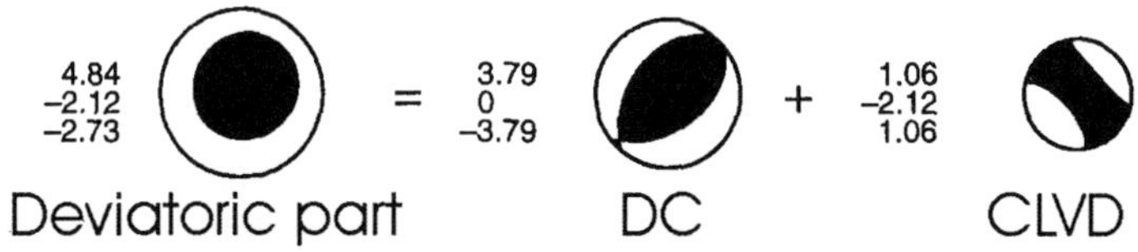


Fig. 9.1. A model for decomposing the deviatoric part of a general moment tensor. DC, the double couple source; CLVD, the compensated linear vector dipole. Numbers to the left of the diagrams give the principal moments for each mechanism, in units of $10^{17}$ N-m. From Julian et al. (1998).

### 9.1.1. Double couple model

Seismological practice and theoretical studies showed the double couple source (Fig. 9.2) as the most adequate for shear faulting that produces volcano-tectonic earthquakes. This model predicts the quadrant distribution of P-wave first motions using stereographic projection and allows the discrimination of the three main types of faulting in earthquake foci: strike-slip, thrust or normal fault. The values of $\varepsilon$ that were obtained for large volcano-tectonic earthquakes related to the eruptions of Mt. St. Helens (1980), Myake-sima (1983), Oshima (1986) and Teishi Knoll (1989) from the Harvard CMT solutions (Zobin, 1992) varied from 0.10 to 0.18. The focal mechanism solution may be obtained from the direct plots of the P-wave first motions on a stereographic projection or by modeling with synthetic seismograms; it has a form shown for DC in Fig. 9.1 (Lay and Wallace, 1995).

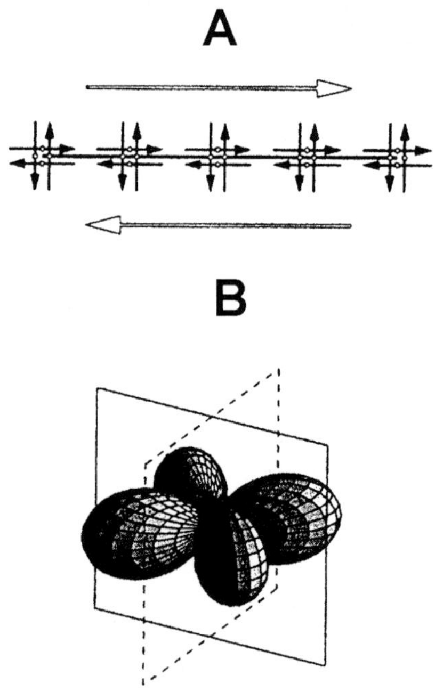


Fig. 9.2. The double-couple earthquake source mechanism. A. Plane view of a vertical strike-slip shear fault in an isotropic medium, showing the direction of slip (open arrows), and the equivalent distribution of double-couple force systems (solid arrow, with forces applied at white dots). B. Radiation patterns of compressional P-waves, showing fault plane (solid lines), and auxiliary plane (dashed lines). Adjacent lobes have opposite polarity. From Julian et al. (1998).

In 1962-1964, the global seismic networks were installed. It allowed collecting enough P-wave first motions to construct the focal mechanism solution on the stereographic projection for volcanic earthquakes of magnitude about 5 or greater. The first P-wave motion focal mechanism solutions for volcano-tectonic earthquakes (Fig. 9.3) were constructed for two magnitude 5 earthquakes preceding the November 11, 1964 directed-blast eruption at Sheveluch volcano, Kamchatka (Zobin, 1970). They indicated a strike-slip faulting for these earthquakes. Later Zobin (1972; 1979a, b) constructed the focal mechanism solutions for the magnitude 5 to 6 volcano-tectonic earthquakes related to the eruptions at Miyake-sima, Tori-shima, Beerenberg, and new Tolbachik volcanoes. All of them were of strike-slip faults. It was shown that one of two nodal planes of the focal mechanism solution of large volcano-tectonic earthquake coincides with the strike of the fissure, where new volcanic centers form (Fig. 9.4). Zobin (1972) showed the similarity between focal faulting of these volcano-tectonic earthquakes and that of tectonic earthquakes surrounding the volcanic regions.

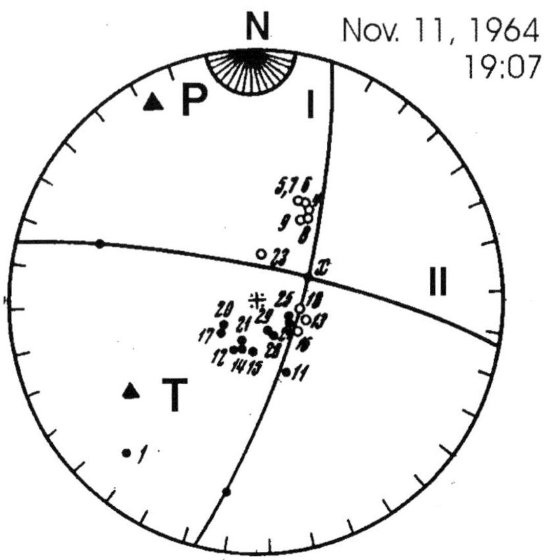


Fig. 9.3. Focal mechanism solution for the volcano-tectonic earthquake that coincided with the November 11, 1964 directed blast explosion at Sheveluch volcano, Kamchatka. The focal mechanism diagram is an upper hemisphere plot. I, II are possible fault planes; P is compression; T is tension; solid circles, compressions; open circles, dilatations. Numbers indicate the seismic stations. From Zobin (1970).

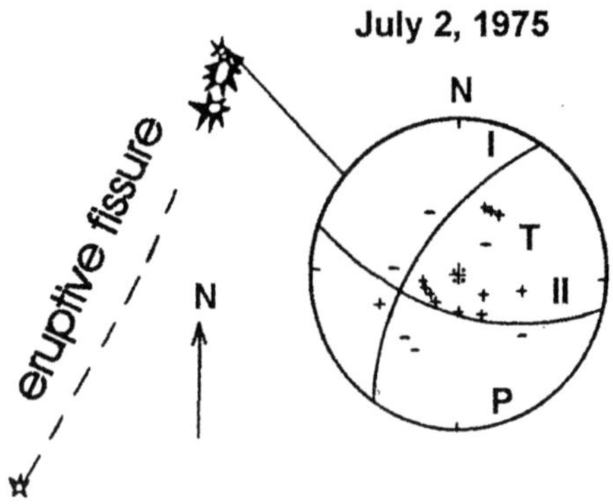


Fig. 9.4. Focal mechanism solution for the July 2 (04:07) volcano-tectonic earthquake preceding the July 6, 1975 New Tolbachik fissure eruption, Kamchatka. It is seen that the fault plane I of the solution coincides with the direction of the 12-km length fissure where the chain of new cones formed. The focal mechanism diagram is a lower hemisphere plot. I, II are the possible fault planes; P is compression; T is tension; + shows compressions; - shows dilatations. From Gorelchik et al. (1990).

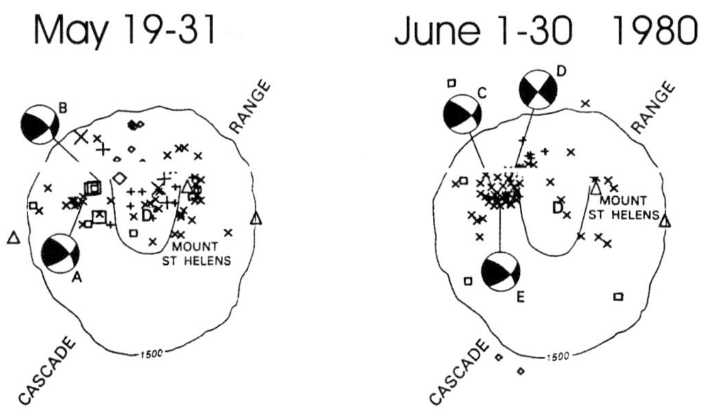


Fig. 9.5. Focal mechanism solutions of volcano-tectonic earthquakes associated with the magmatic activity of Mount St. Helens volcano, Cascades and the epicenters of these earthquakes. Epicenters of earthquakes are shown as small symbols (magnitudes less than 2.8) and large symbols (magnitudes greater than 2.8). Depths are indicated as + (0-5 km), x (5-10 km), square (10-15 km) and diamonds (greater than 15 km). Post-eruption 1500-m contour is shown. The location of lava dome is shown as D. The focal mechanism diagrams are upper hemisphere plots, with darkened quadrants indicating compression and white quadrants indicating dilatations; solid circles, compressions; open circles, dilatations. Mechanism B is a composite solution; all other solutions are single-event solutions. From Weaver et al. (1981).

These first results were supported by studies of focal mechanisms, that were developed for volcano-tectonic earthquakes related to different eruptions in the world. Fig. 9.5 shows the focal mechanism solutions that were constructed for some volcano-tectonic earthquakes originating beneath the dacitic Mount St. Helens volcano, Cascades (Weaver et al., 1981) after the directed-blast eruption of May 18, 1980 during the stage of magmatic eruptions (See Chapter 7). These earthquake focal mechanisms indicated the thrust (A-C, E) and strike-slip faulting at the volcano-tectonic earthquakes foci. All focal mechanisms have steeply dipping fault planes that strike northeast.

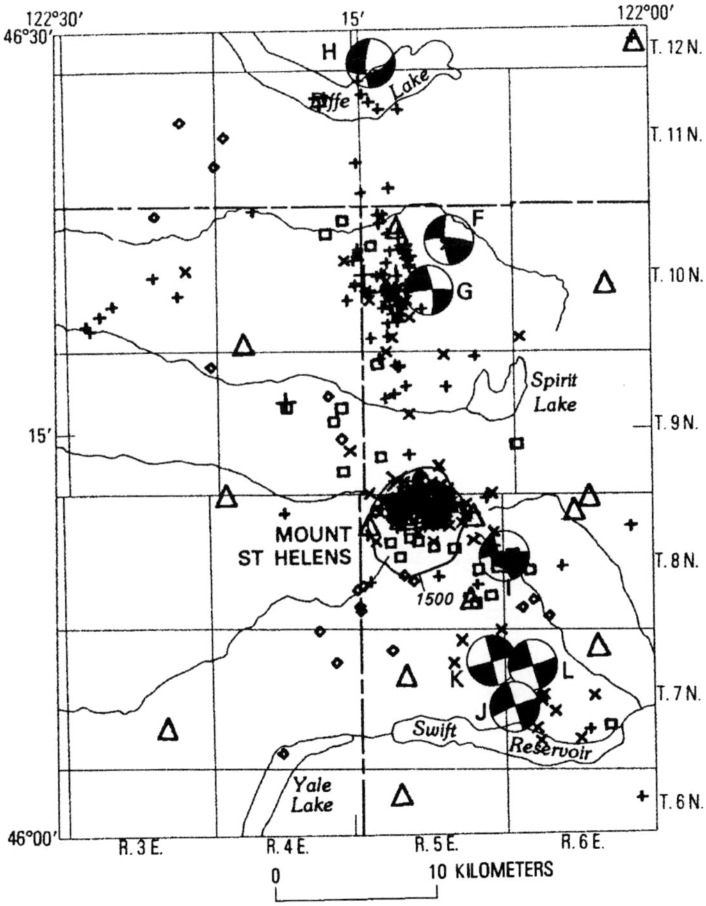


Fig. 9.6. Seismicity and focal mechanism diagrams for some events that occurred within the tectonic environment of Mount St. Helens between May 18 and August 31. Legend is the same as for Fig. 9.5. From Weaver et al. (1981).

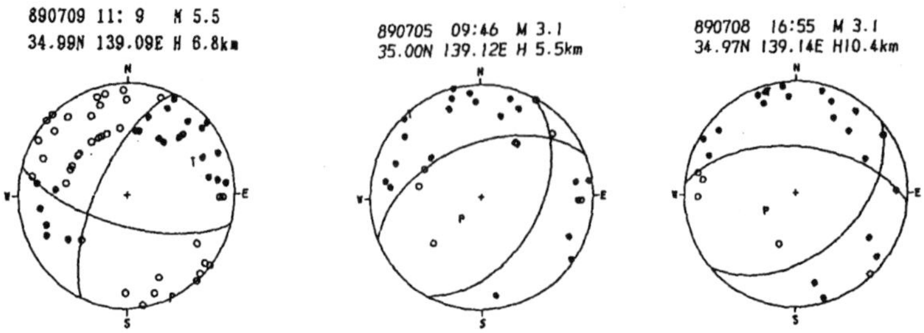


Fig. 9.7. Focal mechanism solutions (lower hemisphere projection) for three volcano-tectonic earthquakes preceding the July 13, 1989 submarine Teishi Knoll eruption, Izu Islands. P is compression; T is tension; solid circles, compressions; open circles, dilatations. From Matsumura et al. (1991).

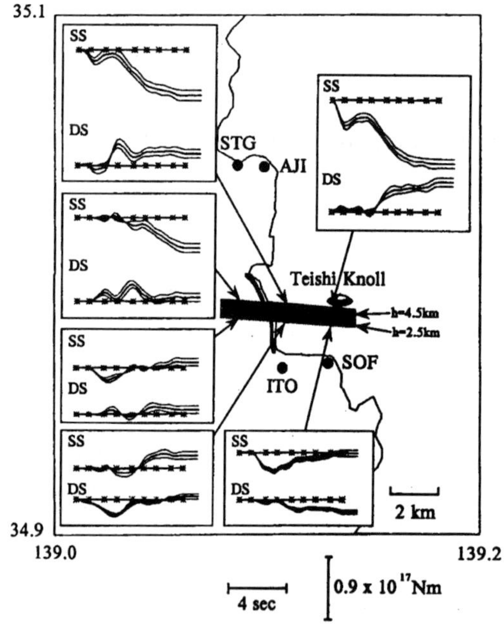


Fig. 9.8. Spatial variation of source time functions with strike-slip (SS) and dip-slip (DS) components for the magnitude 5.5 July 9, 1989 earthquake preceding the July 13, 1989 submarine Teishi Knoll eruption, Izu Islands. The seismic stations whose records were used in the inversion analyses are shown by filled circles. The arrows show the location of six point sources, which represents a 2 km x 2 km subfault. From Takeo (1992).

The study of the focal mechanisms of earthquakes, that occurred during the same year (1980) within the tectonic structures to the north and south of the volcano (Fig. 9.6), shows the similarity in focal mechanism of the earthquakes, originated beneath the volcano at the active eruptive stage, and the earthquakes, originating in tectonic ranges far from the volcano. They were the same thrusts (indexes F, I in Fig. 9.6) and strike-slips (G, J, K, L), as well as normal faults (H).

Focal mechanism solutions for volcano-tectonic earthquakes preceding the 1989 Teishi Knoll submarine basaltic eruption were obtained by two methods: from P-wave polarity distribution (Matsumura et al., 1991) and from the inversion method (Takeo, 1992). Fig. 9.7 shows the P-wave first motion focal mechanism solutions for three earthquakes (Matsumura et al., 1991). The largest earthquake of magnitude JMA 5.5 revealed a strike-slip nature with a normal fault component; two magnitude 3.1 earthquakes indicated thrust faulting. The solution obtained by the inversion method (Takeo, 1992) for the same magnitude 5.5 earthquake is consistent with the P-wave first motion solution. Fig. 9.8 shows the source time functions of the strike-slip and dip-slip components on each subfault of the earthquake fault model marked by arrows. It is seen that a right-lateral strike-slip (SS) motion dominated during the rupturing.

The focal mechanism solutions that were determined by NEIC for seven magnitude mb 4.9 to 5.7 earthquakes, which occurred during the paroxysmal stage of the eruption during June 12-15, 1991 of the dacitic volcano Pinatubo, Philippines, revealed the predominantly strike-slip movement with a minor vertical (normal or reverse) component; only one earthquake was from normal faulting with a large strike-slip component. These solutions suggest that the largest volcano-tectonic earthquakes at Pinatubo were consistent with the east-west maximum regional stress (Bautista et al., 1996).

Okada et al. (1981) determined the composed focal mechanism solutions for small volcano-tectonic earthquakes, which occurred at Usu volcano, Hokkaido during the different stages of dacitic dome activity during October 1977 – June 1979. Dip-slip type of faulting with slight variations within the summit zone was dominant. The strike of the steeply dipping fault planes was roughly parallel to the U-shaped major fault zone within the crater.

During the 1986-1987 activity of the basaltic volcano Oshima, Izu islands, focal mechanism solutions were determined for earthquakes that occurred before and after the fissure eruption. A zone to the north of the volcano was characterized by right-lateral strike-slip faultings; a zone to the south of the volcano was characterized by normal faults (Fig. 9.9). The strike-slip fault motions within the first zone are consistent with the common tectonic stress field around Oshima island;

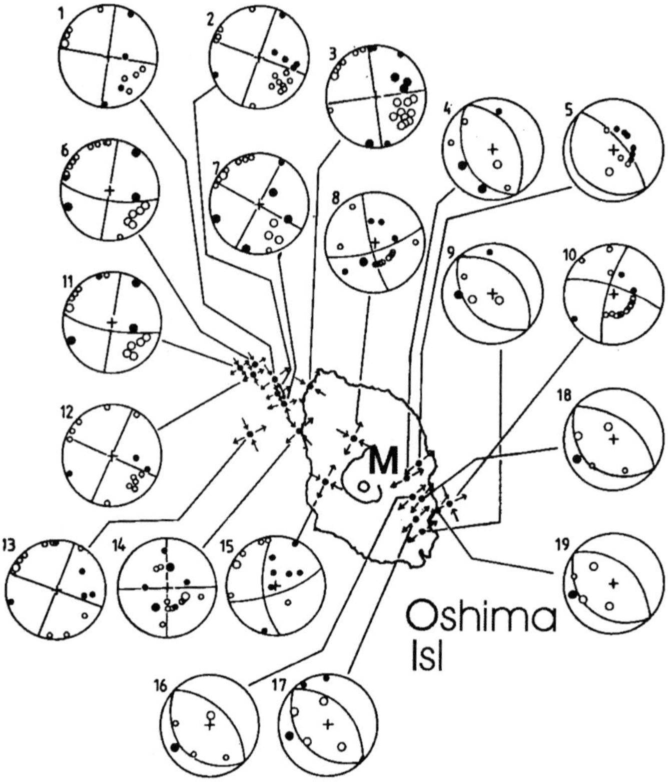


Fig. 9.9. Focal mechanism solutions of volcano-tectonic earthquakes that occurred before and after the 1986 fissure eruption of Oshima, Izu volcano. M marks the summit crater Mihara-yama; solid circles, compressions; open circles, dilatations. Arrows show the pressure and tension axis. From Aramaki (1988).

the peculiar normal fault motions at the second zone may reflect triggering by the fissure eruptions in the tensional stress field (Yamaoka et al., 1988; Aramaki, 1988).

Therefore, the fault nature of volcano-tectonic earthquakes may be various: thrust, normal fault or strike-slip. It may be constrained by the regional or local tectonic stress systems. The focal mechanisms of relatively large earthquakes (magnitude about 5-6), related to fissure eruptions, revealed the coincidence of its one fault plane with the strike of the eruptive fissure.

### 9.1.2. Non-double couple model

The non-double couple sources, or the sources with values 0.2 ≤ ε ≤ 0.5, were determined for a few volcano-tectonic earthquakes located at ring fault structures in Iceland (Ekström, 1994; Nettles and Ekström, 1998) and at Etna and Vesuvius, Italy (Panza and Saraó, 2000; Saraó et al., 2001)). Fig. 9.10 shows the example of focal mechanism determination for the 29 September Mw 5.6 main shock of earthquake sequence preceding the September 30, 1996 eruption of the subglacial volcano Grimsvøtn, Iceland, using broadband teleseismic P waveforms inversion. The ε value for this moment tensor solution was equal to 0.30. Similar solutions were obtained for nine earthquakes that occurred from 1976 to 1994 around the Bárdarbunga caldera (Fig. 9.11). The ε values for these earthquakes vary from 0.24 to 0.41, the magnitudes Mw range from 5.1 to 5.4. They occurred within the range of depths from 3.3 to 6.7 km. Nettles and Ekström (1998) consider that these non-double couple solutions may reflect the faulting on an outward dipping cone-shaped ring fault, beneath the Bárdarbunga caldera. The model of rupture propagation on a cone-shaped fault is shown in Fig. 9.12 and explains the rupturing as a series of subevents propagating around the surface of the cone.

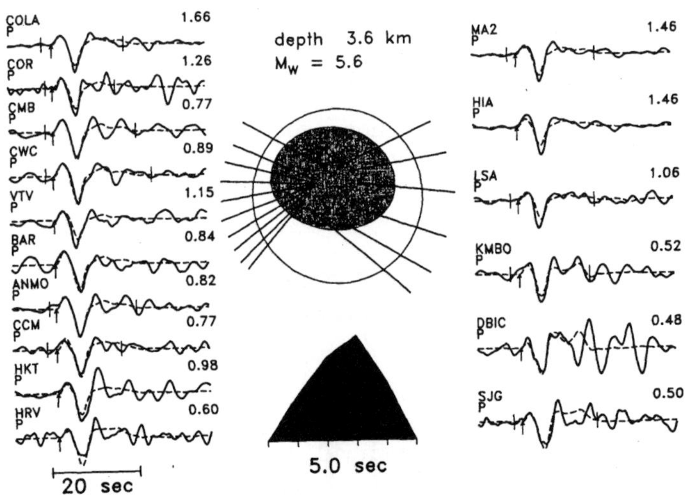


Fig. 9.10. Comparison of the calculated synthetic seismograms (dashed) with the broadband teleseismic P waveforms (solid) for the September 29, 1996 volcano-tectonic earthquake preceding the September 30 flank eruption of Grimsvøtn volcano, Iceland. The focal mechanism corresponds to the full moment tensor solution, and the source time function is determined by the inversion. Maximum amplitudes (in microns) are shown to the right of the seismograms; arrows show picked first arrivals, and the thin vertical bars show the time window included in the inversion. From Nettles and Ekström (1998).

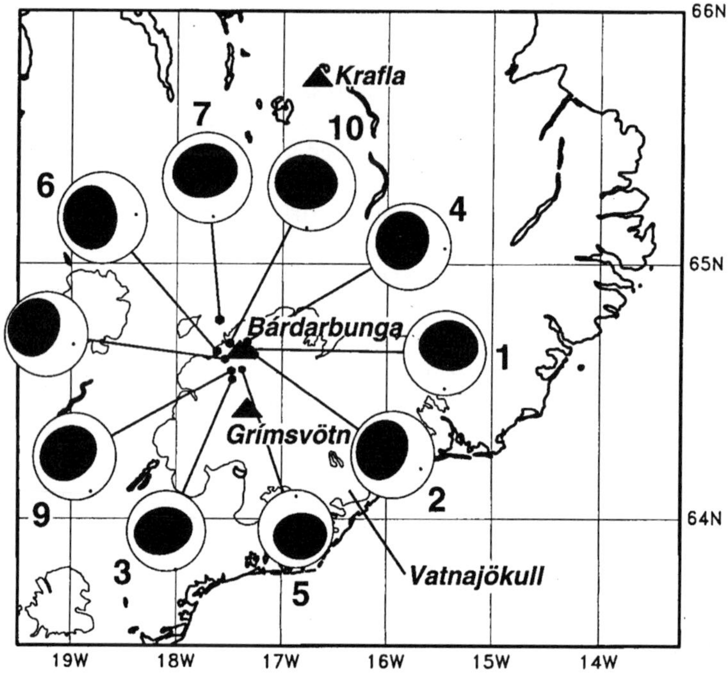


Fig. 9.11. Map showing CMT solutions for the ten earthquakes near Bárdarbunga volcano, Iceland. The earthquakes are highly non-double couple, and have vertical T axes, indicative of horizontal compression. From Nettles and Ekström (1998).

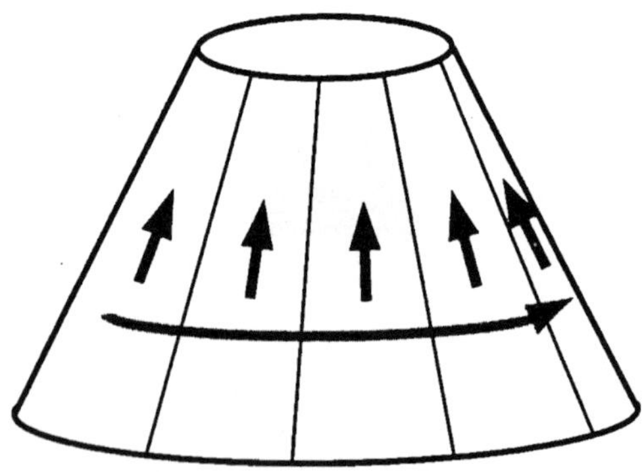

Fig. 9.12. A model of rupture propagation on a cone-shaped fault: a sub-source with reverse-faulting geometry (short arrows) occurs on each fault segment in succession (long arrow), around a specified portion of the cone. From Nettles and Ekström (1998).

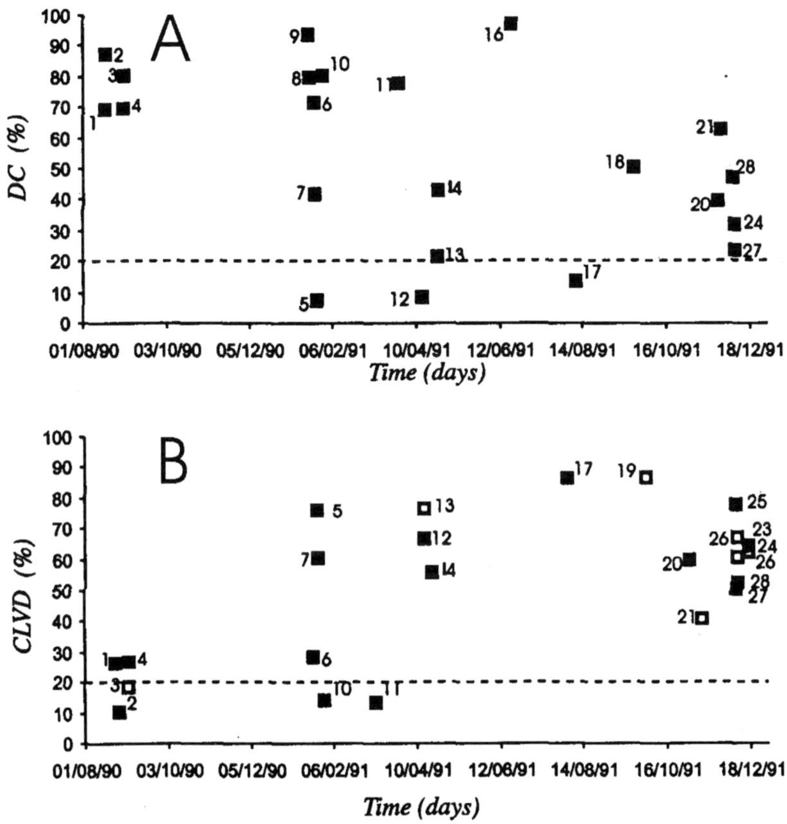


Fig. 9.13. Variation with time of the percentage of DC (A) and CLVD (B) components of the earthquakes preceding the December 15, 1991 flank eruption at Etna, Sicily. The dashed lines represent the limits below which the solutions may be due to numerical noise. Numbering corresponds to temporal appearance of earthquakes. From Saraó et al. (2001).

Saraó et al. (2001) studied the variations in percentage of DC and CLVD components in the sources of volcano-tectonic earthquakes preceding the 1991 eruption at Etna, Sicily (See Chapter 5). The percentage of CLVD was given by $200\varepsilon$, while the percentage DC was given by $(100 - 200\varepsilon)$. Fig. 9.13 shows that between August 1990 and December 9, 1991, the percentage of non-double couple components of earthquakes, statistically significant at the 95 per cent confidence level, increased before the beginning of the 15 December 1991 flank eruption. Panza and Saraó (2000) consider that this fact can be related to fluid movements.

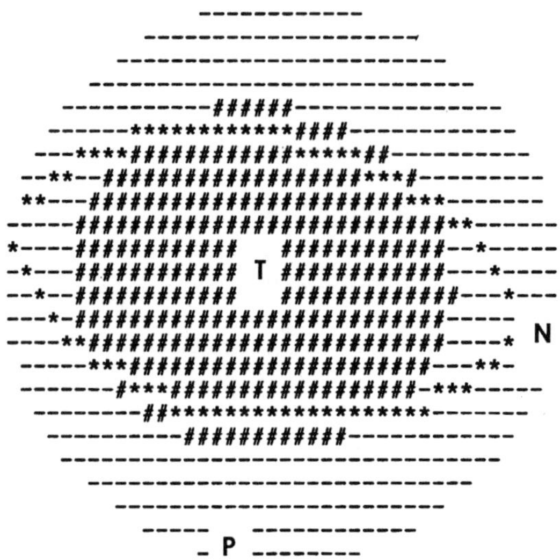


Fig. 9.14. A standard Harvard CMT solution showing two possible interpretations of the solution for the September 29, 1996 earthquake: a double-couple source (marked by stars) or a non-double couple source (marked by filled ellipse area.) T is tension; P is pressure; N is neutral axes for the best double couple.

Therefore, the non-double couple nature of the volcano-tectonic earthquake source may reflect the ring (or other complex) structure of the fault generating the earthquake (Nettles and Ekström, 1998). At the same time, every complex fault may be approximated by a linear structure, and for the majority of the events with the non-double couple solution we have an adequate double couple solution (Fig. 9.14).

The presence of non-double couple components can also depict a process in which a depressurizing magma body and crack formation leads to magma injection towards the shallower portions of the volcano (Panza and Saraó, 2000).

### *9.2. Spectral characteristics of volcano-tectonic earthquakes*

The spectra of volcanic earthquakes are used generally in two types of applications: the total spectra of seismic records may be used for classification and monitoring of seismic activity and the body-wave (or surface wave) spectra may be used for study of source parameters of earthquakes.

## 9.2.2. Spectra of total records of volcano-tectonic earthquakes

Fig. 9.15 and 9.16 show the seismograms, spectrograms, and velocity spectra of records for the four main types of volcano-tectonic earthquakes recorded during the active period at Redoubt, Alaska during 1989-1990 (Lahr et al., 1994). Each seismogram was of 40.6 s duration; the spectra were normalized for better comparison. The long-period events are characterized by a peak frequency near 1.5 Hz corresponding to S and surface waves. The short-period events, that occurred at depths greater than 5 km below the crater floor, are characterized by a peak between 6 and 8 Hz corresponding to the body waves record and a low-frequency maximum between 1 and 3 Hz generated by short seismic coda. Shallower short-period events have two peaks: one, more intensive, between 1 and 3 Hz, and the second, lower, between 7 and 9 Hz. The first peak is produced by surface waves; the second, by body waves. A spectrum of a hybrid earthquake is characterized by a peak between 1 and 2 Hz.

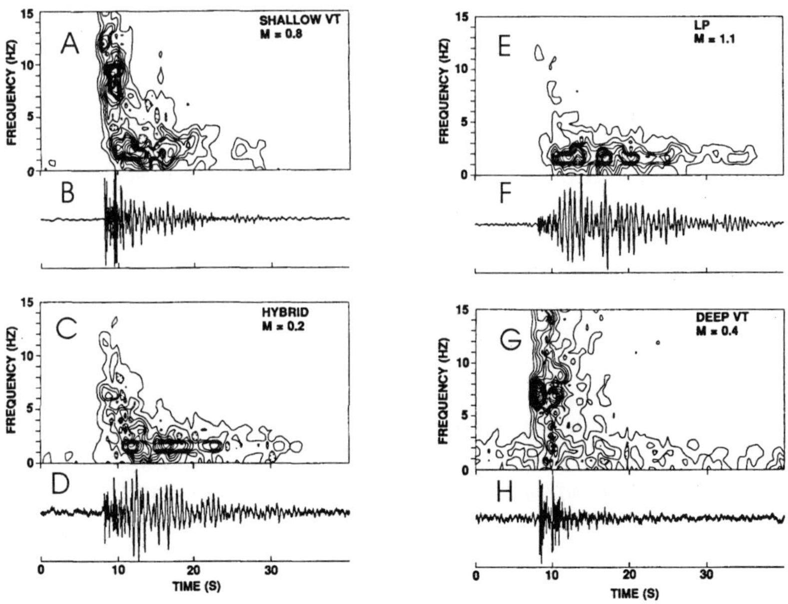


Fig. 9.15. Spectrograms (A, C, E, and G) and seismograms (B, D, F, and H) for the typical volcano-tectonic earthquakes of Redoubt volcano, Alaska. LP are the long-period earthquakes; deep VT are the short-period events that occurred at depths greater than 5 km below the crater floor; shallow VT are the short-period events that occurred at depths less than 5 km below the crater floor; and Hybrid are the hybrid earthquakes. From Lahr et al. (1994).

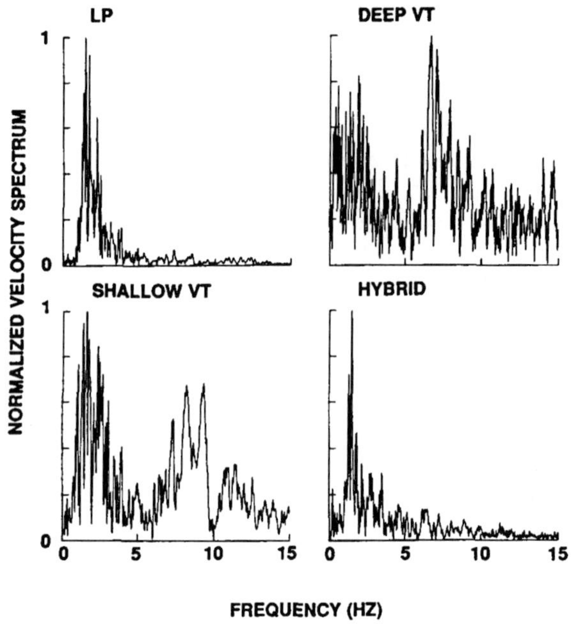


Fig. 9.16. Normalized velocity spectra of the typical volcano-tectonic earthquakes shown in Fig. 9.15. From Lahr et al. (1994).

These differences in spectral content allow a visual discriminating of different types of volcano-tectonic earthquakes for monitoring of volcano activity.

9.2.2. Spectral source characteristics of volcano-tectonic earthquakes and their relationship with the type of volcanic activity

The spectra of body and surface waves of volcano-tectonic earthquakes, as well as synthetic seismogram modeling, may be used for estimating of earthquake source parameters, such as the seismic moment Mo, stress drop $\Delta\sigma$ and spectral corner frequency *fo*. These parameters describe the variations of physical properties within the seismic volume below the volcano and can serve for discriminating obetween the seismic events generated during the different types of volcanic activity. In this review, I use the estimations of source parameters from the Harvard centroid-moment tensor (CMT; Harvard data base) and from body-wave spectra (Brune's model; Brune, 1970). The typical spectrum of P-wave produced by volcano-tectonic earthquake and the Brune's model characteristics are shown in Fig. 9.17.

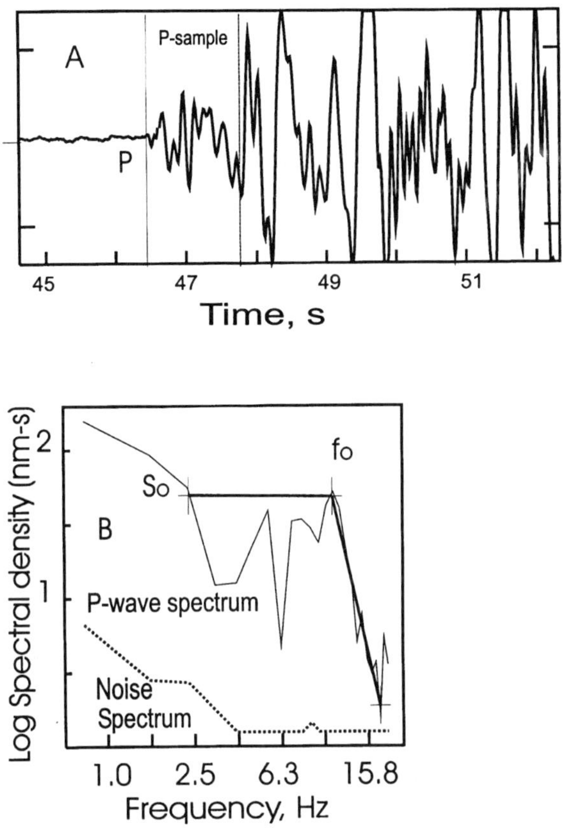


Fig. 9.17. A P-wave sample (A) and its Fourier spectrum (B) for volcano-tectonic earthquake of Volcán de Colima, México. The Fourier spectrum for the noise is also shown. *So* and *fo* are the spectral density of the low-frequency level and the corner frequency, respectively. From Zobin et al. (2002b).

Fig. 9.18 shows a "seismic moment – magnitude" dependence for the volcano-tectonic earthquakes related to central and lateral eruptions. In Fig. 9.18A, the parameters were taken from Harvard CMT database for large 1980-1996 volcano-tectonic earthquakes of magnitude Ms ≥ 5.0. It is seen that the seismic moments of volcano-tectonic earthquakes related to the central eruptions of Mt. St. Helens and Pinatubo volcano are larger at the same magnitudes than the seismic moments of volcano-tectonic earthquakes related to the flank eruptions of Oshima, Miyake-sima, Teishi Knoll, and Grimsvøtn volcanoes. The same tendency was observed at the basaltic volcano Alaid, Kurile islands (Fig. 9.18B) for smaller volcano-tectonic earthquakes related to its 1972 flank and 1981 central eruptions (Zobin, 1988).

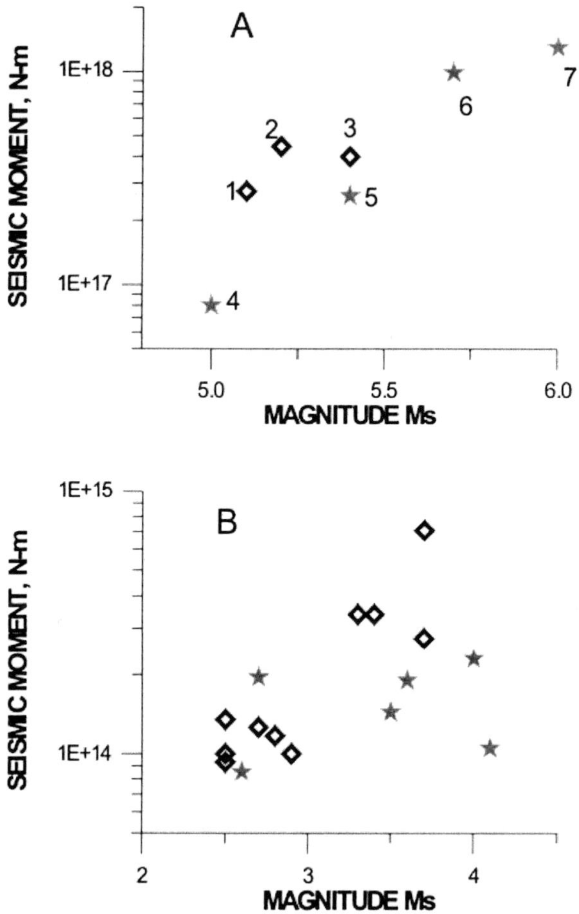


Fig. 9.18. Seismic moment – magnitude relations (A) for large (M≥5) volcano-tectonic earthquakes related to the central eruptions of volcanoes Pinatubo (1, 2) and Mt. St. Helens (3) and the flank eruptions of Teishi Knoll (4), Grimsvøtn (5), Oshima (6), and Miyake-sima (7) volcanoes and (B) for the 1972 flank and 1981 central eruptions of Alaid volcano. The data for central eruptions are marked by filled circles and for flank eruptions, by crosses. Source parameters for Fig. A are taken from the Harvard CMT database; for Fig. B, from Zobin (1988).

This greater amount of long-period radiation expressed in larger values of seismic moments may attribute to the difference in formation of the earthquake source for the two types of eruptions. Apparently, during a central eruption the magma opens rather "smooth" fractures in a medium, which is probably already broken and possibly also hotter from repetitive intrusion of magma along the same path. During a flank eruption, a new eruptive fracture representing a branch of the main central channel is formed every time before the eruption, which causes the generation of asperities or barriers on the rupture. These may cause a higher amount of short-period waves (Zobin, 1988).

### 9.2.3. Temporal variations of spectral source characteristics of volcano-tectonic earthquakes in the course of volcanic activity

The studies of temporal variations in the P-wave spectral parameters of volcano-tectonic earthquakes, that were located before and during the 1975 fissure Tolbachik eruption, Kamchatka, showed that the mean values of stress drops were not stable at different stages of volcanic activity. Stress drops became smaller just before the beginning of eruption. This change in stress drops occurred simultaneously with a change in focal mechanisms (Zobin, 1979b; Zobin and Gorelchik, 1982).

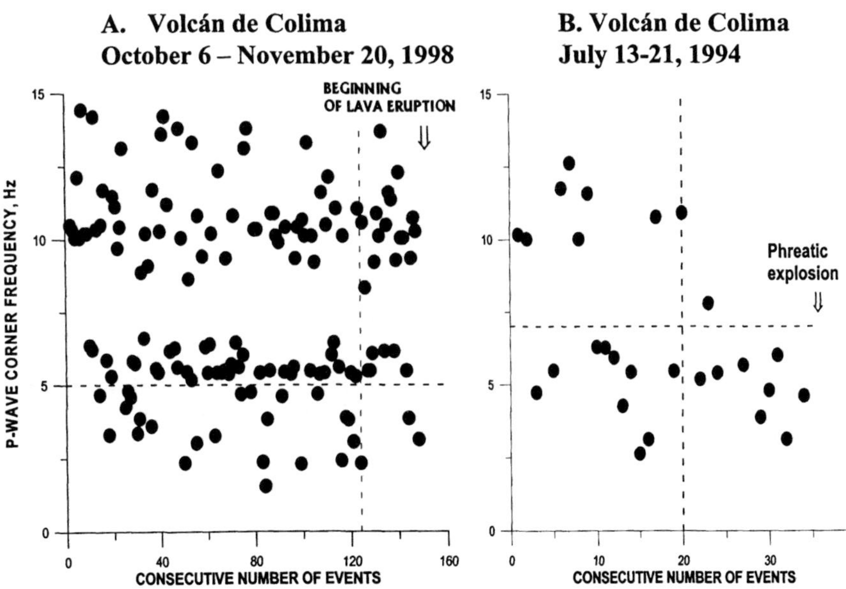


Fig. 9.19. Temporal variations of P-wave corner frequencies for volcano-tectonic earthquakes of Volcán de Colima, México prior to the 1998 lava extrusion (A) and the 1994 phreatomagmatic explosion (B).

A more detailed study of the body wave spectra of volcano-tectonic earthquakes was carried out prior the November 1998 andesitic lava extrusion at Volcán de Colima, México (Domínguez et al., 2001; Zobin et al., 2002a). Five earthquake swarms occurred during one year of seismic activity. For P-wave spectra, two groups of events were observed: low-frequency events (Lfo) with values of *fo* from 2 to 7 Hz and high-frequency events (Hfo) with values of *fo* from 8 to 15 Hz. They were observed simultaneously during all five swarms. Fig. 9.19A shows this effect for the last 5th swarm. These two groups (Fig. 9.20B) are separated at the significance level 99%. The same two groups of Lfo and Hfo events were observed also for S-wave spectra during the June-July 1988 swarm 4 when a three-component seismic station was in operation.

The relative abundances of Lfo and Hfo events changed from swarm to swarm (Fig. 9.20A). Hfo events dominated swarms 1 and 2, whereas Lfo events were twice as abundant as Hfo events in swarm 4. The numbers of Lfo and Hfo events were practically equal for swarm 5, as they were for swarm 3. Very low-frequency (VLfo) events with *fo* from 2 to 4 Hz and very high-frequency (VHfo) events with *fo* from 12 to 15 Hz are also apparent in Fig. 9.19A and 9.20. The VLfo events were practically absent in the November-December 1997 swarm but were recorded during the June-July 1998 and October-November 1998 swarms. The VHfo events were recorded during the November-December 1997, May 1998, and October-November 1998 swarms but they were absent during the June-July 1998 swarm. Two days before the eruption began, the VLfo events with corner frequencies less than 5 Hz practically disappeared (Fig. 9.20A). This effect was tested by the Fisher-Yates 2x2 contingency table and it was found to be significant at the 99% confidence level.

The two groups of Hfo and Lfo events were observed also for volcano-tectonic earthquakes related to the 1994 phreato-magmatic eruption of Volcán de Colima (Fig. 9.19B), the 1975-1976 Tolbachik (Fig. 21A) and the 1991-1993 Etna (Fig. 21B) eruptions. For the 1994 Colima eruption, the Hfo events disappeared one day before the phreatic explosion. For New Tolbachik volcanoes, the Hfo events were observed only before the opening of the fissure when the first cone was formed, six days before the beginning of the eruption. During the two swarms that occurred prior to the eruptions of the third and eighth cones, only the Lfo events were observed.

For the December 1991 Etna events (See Chapter 5), the two groups of Lfo and Hfo events occurred simultaneously during the sharp increase in seismic activity at the initial stage of swarm activity on December 14 (Fig. 9.21B). With the end of peak activity and the opening of the eruptive fissure, the Lfo events disappeared for two hours. Then both groups appeared again up to the beginning of the lava flow on December 15.

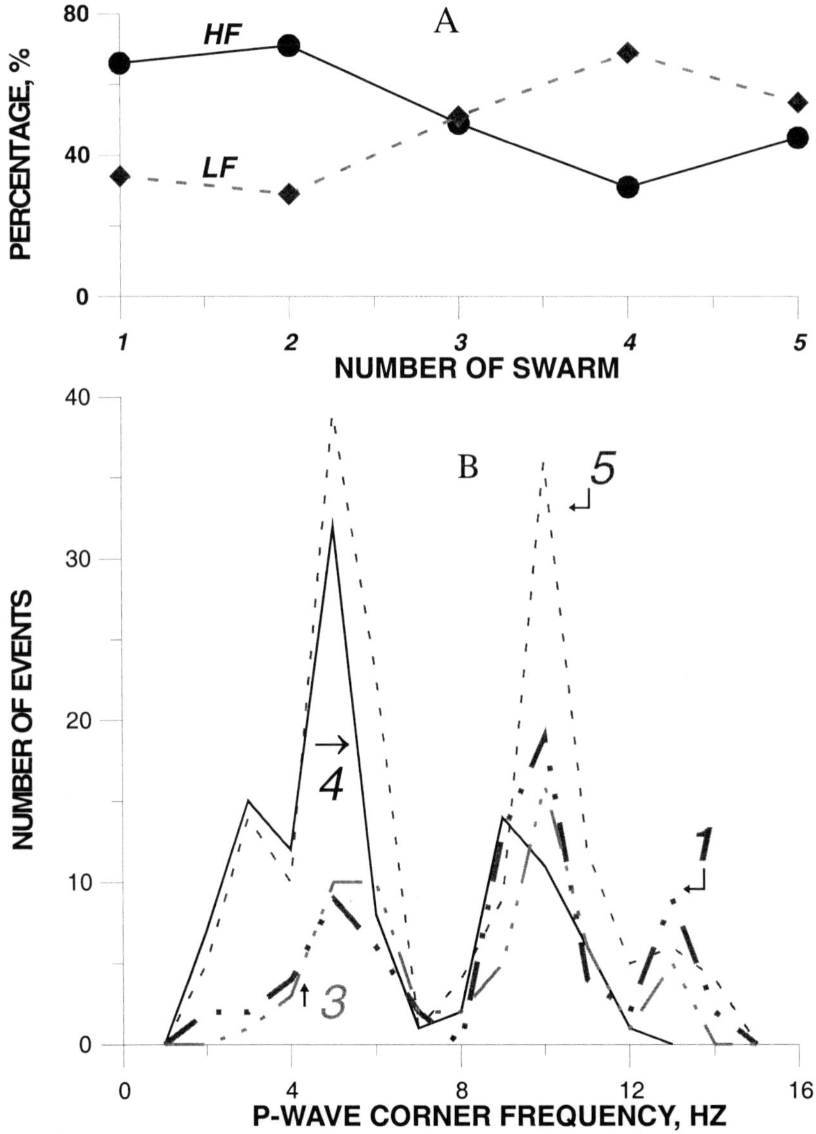


Fig. 9.20. Variations in abundance of Lfo and Hfo events from swarm to swarm (A) and distribution of events with different P-wave corner frequencies during swarms 1, 3, 4 and 5(B). Arrows show the eruptive events. From Zobin et al. (2002b).

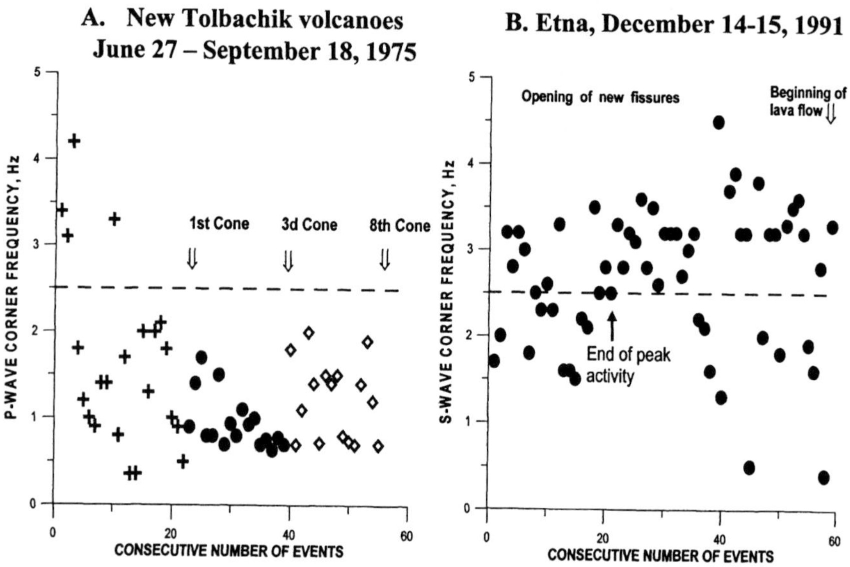


Fig. 9.21. Temporal variations of P-wave corner frequencies for volcano-tectonic earthquakes associated with the 1975 fissure eruption of New Tolbachik volcanoes, Kamchatka (A) and of S-wave corner frequencies for volcano-tectonic earthquakes associated with the 1991 Etna flank eruption (B). Arrows show the eruptive events. Data from Zobin and Chirkova (1984) and Patané et al. (1995).

The volcano-tectonic earthquake sequence model proposed in Zobin et al. (2002b) suggests that a moving magmatic dike could produce Hfo seismic events at distances less than 70 m from the dike tip along existing fractures and trigger the fracturing of local faults at distances greater than 70-100 m from the dike tip as Lfo events. Therefore, the occurring of Hfo events may indicate the moving of magma from depths, and the Lfo events reflect the process of passageway constructing.

From this point of view, a sharp decrease of Hfo events one day before the 1994 Colima phreatic explosion and a sharp decrease of Lfo events two days before the Colima 1998 lava eruption indicate the difference in preparation of the two eruptions: gas movement along the local faults before the 1994 explosion and a magma uplift before the 1998 lava eruption along the prepared passageway.

For New Tolbachik volcanoes, magma came close to the surface on July 30, 1975; during the following six days, a passageway was prepared along the new fissure, then the eruption at the first cone began. During the forming of the third and eighth cones, magma did not rise from depth; it moved laterally from the site of the first cone and from Plosky Tolbachik (See Chapter 5). Therefore, only Lfo events, that ruptured the fissure for new cones, were recorded.

At Etna, during the first 2 hours of seismic activity from 01:55 h on December 14, 1991, the uplift of magma went simultaneously with the formation of a new fissure; then magma went to the surface along new fissure for three hours. From 07:00 h, the formation of a new fissure continued; magma continued its way to the surface. Then the lava flow began.

Therefore, the knowledge of temporal variations of the spectral source parameters may allow the description of the main events during eruptive process.

### *9.3. Seismo-tectonic deformations in volcanic region*

Knowledge of the focal mechanisms and scalar seismic moments of volcano-tectonic earthquakes, associated with an eruption, allows the determination of seismo-tectonic deformations (STD) within the volcanic edifice and surroundings. It is produced by the process of magma deep migration before, during and after the eruption. Seismo-tectonic deformation $\varepsilon_{ik}$ of the volume V that contained the earthquake foci can be estimated (Riznichenko, 1977) using the equation

$$\varepsilon_{ik} = \Sigma \text{ Mo } (\theta_{ik}) \,/2\mu V \tag{9.1}$$

where $\mu$ is accepted to be 3.2 x $10^{11}$ dyne-cm$^{-2}$; V is equal to 30 km$^3$; $\theta_{ik}$ is the unit tensor of seismic moment Mo; and $\Sigma$ is the symbol of sum.

The components $\theta_{xx}$, $\theta_{yy}$ and $\theta_{zz}$ of the unit tensor of seismic moment are estimated using focal mechanism parameters through formulae

$$\theta_{xx} = \sin^2\sigma_T \sin^2\alpha_T - \sin^2\sigma_P \sin^2\alpha_P \tag{9.2}$$
$$\theta_{yy} = \sin^2\sigma_T \cos^2\alpha_T - \sin^2\sigma_P \cos^2\alpha_P \tag{9.3}$$
$$\theta_{zz} = \cos^2\sigma_T - \cos^2\alpha_P \tag{9.4}$$

where $\sigma_P$ and $\sigma_T$ are the angles of P and T axes with the vertical; $\alpha_P$ and $\alpha_T$ are the azimuthal angles of the P and T axes. The X-axis is oriented to the east, the Y-axis to the north and the Z-axis upwards. The STD were calculated for two volcanic zones: for the 1975-1978 volcano-tectonic activity in the region of Tolbachik fissure eruption, Kamchatka (Zobin, 1990) and for the 1997-1998 volcanic pre-lava unrest at Volcán de Colima, México (Zobin et al., 2002a). Fig. 9.22 and 9.23 illustrate the results.

The 1975-1978 seismo-volcanic activity in the region of Tolbachik fissure eruption was characterized by strong 1975-1976 volcanic activity along the new fissure and by the following post-eruptive seismic activity within the surroundings (See Chapter 5). The magnitude 4-5 seismic events were recorded at the site of the cones just before the beginning of the eruption and at the distance of 3 to 50 km from the active cones just after the ceasing of the eruption. Fig. 9.22 shows the spatial distribution of the three components $\varepsilon_{xx}$, $\varepsilon_{yy}$ and $\varepsilon_{zz}$ of extensional deformations that were calculated for all recorded earthquakes of magnitude mb > 3.7 that occurred within the upper crust layer from 0 to 20 km. In the eruptive zone I, the largest compressional deformations dominated with a value of $-5.4 \times 10^{-5}$; while the largest extensional deformations were observed in the latitudinal direction with the maximum value of $3.6 \times 10^{-5}$. The vertical deformations reached to $1.6 \times 10^{-5}$. In the periphery, the maximum deformations were observed in zone IV situated 40 km to the north from the eruptive sites, where the compression was dominant reaching a value of $-1.8 \times 10^{-6}$ whilst the vertical deformation was of $3.4 \times 10^{-6}$.

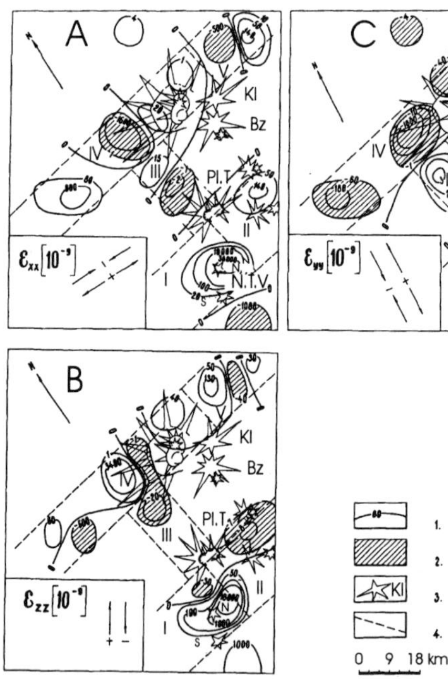


Fig. 9.22. Distribution of extensional deformations in the region of the 1975-1976 Tolbachik fissure eruption, Kamchatka in latitudinal (east-west) (A), meridional (north-sourth) (B) and vertical (C) directions. Kl is Klyuchevskoy, Bz is Bezymianny, Pl.T is Plosky Tolbachik and N.T.V is New Tolbachik volcanoes. N and S indicated the north and south groups of new cones of New Tolbachik volcanoes. 1, the isolines of deformations; 2, the regions of negative deformations; 3, the volcanoes; and 4, a boundary of zones I to V. From Zobin (1990).

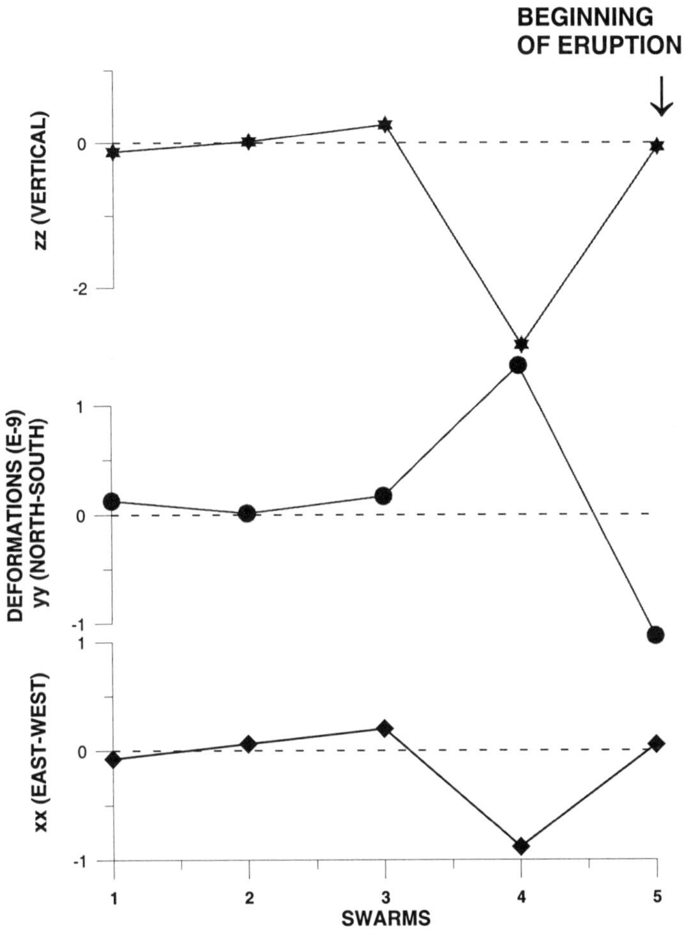


Fig. 9.23. Variations of principal components $\varepsilon_{xx}$, $\varepsilon_{yy}$ and $\varepsilon_{zz}$ of STD from swarm to swarm one year before the November 1998 lava eruption. 1-5, the numbers of swarms according to their temporal appearance. Zero-line is shown by the dashed line. From Zobin et al. (2002a).

Generally, the comparison of seismic activity development (Chapter 5) and the scheme of Fig. 9.22 allow the reconstruction of the process of total crustal deformation within the seismic volume in the region of the fissure eruption. The seismic activity started in zone I, the eruption zone. This had led to considerable bulging of the earth's surface producing extensional deformations in latitudinal and compressional in meridional directions. Later, after the termination of the eruption the seismic activity spread radially from Plosky Tolbachik (zones II and III in Fig. 9.22). Zone II extends along the Upper Pleistocene-Holocene extrusive domes to the east of Plosky Tolbachik. Zone III extends along the western border of distribution of the Upper Pleistocene-Holocene cinder and lava cones. Zones II and III were notable for subsidence of different amplitudes that can be caused by discharge of hydraulically connected magma chambers (Fedotov, 1984b). If this assumption is correct, then the outlines of the subsidence zone in Fig. 9.22C can serve as the boundary of the zone of the near-surface magma chambers that feed the Klyuchevskaya group of volcanoes. Collapses of the near-surface rocks along the margin of the magma chambers could obviously produce compensational bulging in zones IV and V that resulted in the upsurge of seismic activity in 1977-1978 (Zobin, 1990).

The 1997-1998 pre-lava eruption unrest of Volcán de Colima, México (See Chapter 6) was more local and produced lesser deformations, mainly within the volcanic edifice from sea level up to the crater at the altitude of 3,860 m (Zobin et al., 2002a). Fig. 9.23 shows the temporal variations of three components of extensional deformations one year before the eruption. The calculated deformations vary from $1.3 \times 10^{-11}$ to $2.7 \times 10^{-9}$. The maximum deformation was observed in June-July 1998 (swarm 4), and the minimum deformation was observed in March 1998 (swarm 2). The process of deformation was rather complicated. The first earthquake swarm of November-December 1997 resulted in a N-S horizontal elongation that is twice greater than the E-W horizontal shortening. The volume was also subject to vertical shortening. In March and May 1998 (swarms 2 and 3), uniform horizontal N-S and E-W elongations, accompanied by vertical elongation of the volume were observed. In June-July 1998 ($4^{th}$ swarm), the situation of November-December 1997 ($1^{st}$ swarm) was repeated with a N-S horizontal elongation greater than the EW horizontal shortening of the volume and with an intensive vertical shortening. During the last swarm of October-November 1998 (5), the slight E-W elongation of the seismic volume was accompanied by strong N-S shortening and a very slight vertical shortening.

Quantification of the deformation process within the volume for seismic foci recorded prior to the November 1998 eruption of Volcán de Colima enables interpretation of the process, thus giving volcanological meaning to the deformation terms. The seismic volume can be considered to be the volcanic edifice. The process of horizontal elongation is interpreted as the process opening a magma passageway; vertical shortening in the seismic volume is interpreted to reflect magma ascent. The N-S component of deformation coincides with the orientation of the Colima rift and the E-W component of deformation can be associated with the adjusting NE-SW Tamazula fault system (Zobin et al., 2002a).

Based on these assumptions, in November-December 1997 magma began to rise and to form a passageway along the Tamazula fault system. In March and May 1998 the vertical movement of magma practically stopped, and magma began to fill existing fractures in both the Colima rift and the Tamazula fault system. In June-July 1998, a new strong phase of magma ascent began and the passageway was formed along the Tamazula fault system. During the last swarm in October-November 1998 the final passageway was formed, with a N-S orientation within the Colima rift system, and final ascent of magma to the surface was completed (Zobin et al., 2002a).

The study of STD gives us the possibility for a quantitative comparison of two eruptions. The larger magnitude of events recorded during the seismic activity preceding the Tolbachik eruption yielded larger values of STD that reached $10^{-5}$. This difference in four orders of magnitude with the Colima deformations agrees well with the difference in the magmatic volume of erupted products (2.2 $km^3$ for the Tolbachik eruption *versus* 0.04 $km^3$ for the 1998-1999 Colima eruption) (Fedotov, 1984a; Zobin et al., 2002c).

# Chapter 10
# Significant Volcano-tectonic Earthquakes and Their Role in Volcanic Processes

Omori (1912) wrote that a volcanic earthquake could never be equal to a large tectonic shock. He considered that the zones of active volcanoes "may be regarded as a sort of safety valves to, or lines of least resistance in, the earth's crust, preventing an accumulation in their localities of underground stress sufficient to result in a gigantic earthquake. Hence it is that extremely violent or very extensive earthquakes never originate from under volcanoes".

Long-term observations showed that volcanic earthquakes are characterized by significantly lower Mmax and a lower frequency of large earthquake appearance than tectonic earthquakes. The majority of volcanic earthquakes are small; their magnitudes do not exceed 2 or 3. At the same time, significant (moderate-size and large) volcanic earthquakes may occur and produce a local effect of high intensity (Abe, 1979; Zobin, 2000; Yokoyama, 2001).

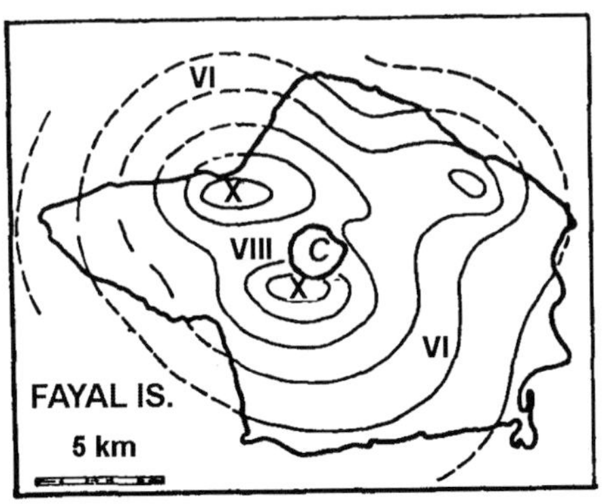


Fig. 10.1. Macroseismic effect of the volcano-tectonic earthquakes of Fayal volcano during the May 13-14, 1958 seismic crisis associated with the flank eruption of the volcano. Numbers VI to X are the Modified Mercalli scale intensities; C indicates a caldera. From Machado et al. (1962).

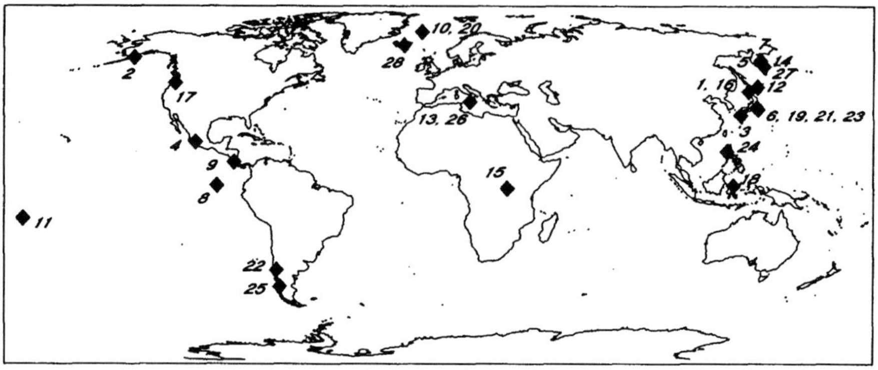


Fig. 10.2. The map of volcanoes under study. The volcanoes are shown by black diamonds with numbers according to Table 10.1. From Zobin (2001).

Machado et al. (1962) described the macroseismic effects that were produced by the May 13-14, 1958 volcano-tectonic earthquake sequence at Fayal island during the flank eruption of Fayal volcano. About 450 earthquakes were felt on Fayal; many of them had a Modified Mercalli scale intensity of X (Fig. 10.1). Of course, these effects were very local, covering an area of about 2-3 $km^2$, and they were strongly multiplied by a surface layer of soft volcanic rocks on the flanks of the volcano. Nevertheless, Blong (1984) wrote about at least 29 deaths associated with the secondary effects (landslides, building collapse) of volcanogenic earthquakes during the eruptions of Sakurajima, Kyushu (1914) and Api Siau, Indonesia (1976). Therefore, a seismic hazard of volcano-tectonic earthquakes for the settlements on the slope of volcano may be a reality.

### *10.1. Selection of significant volcano-tectonic earthquakes that occurred in the XXth century*

Zobin (2001) listed the significant ($M \geq 4.5$) earthquakes, which occurred near volcanoes during the XXth century (Table 10.1). They are very rare, rarer than magnitude 7.5 to 8.5 tectonic earthquakes. Only two magnitude Ms 7.0 volcanic earthquakes were recorded during the century (Abe, 1979; 1992) while the annual average number of $Mw >$ 7.0 tectonic events is about 15 (Lay and Wallace, 1995).

The catalog of Table 10.1 contains 28 events. Three of them are of magnitude $Mw \geq 7.0$, two of magnitude from 6.0 to 6.9, and16 events of magnitude from 5.0 to 5.9. This catalog was prepared for significant seismic events associated with 28 eruptions of 25 volcanoes (Fig. 10.2). The earthquakes were associated with the eruptions of different types

**Table 10.1**
**List of Significant (Moderate-size and Large) Volcanic Earthquakes, 1900-1999**

| # | yymmdd | hhmm | Lat. (deg) | Long. (deg) | Dep. (km) | Mb | Ms | Mw | Volcano |
|---|---|---|---|---|---|---|---|---|---|
| 1 | 100624 | 0749 | | | | | 5.1[1] | 5.4* | Usu |
| 2 | 120608 | 0736 | | | | 6.8[2] | 7.0[2] | 7.0* | Katmai |
| 3 | 140112 | 1028 | | | | | 7.0[1] | 7.0* | Sakurajima |
| 4 | 430215 | 0139 | | | | | 4.5[3] | 4.9* | Paricutin |
| 5 | 560330 | 0611 | | | | | 5.0[4] | 5.3* | Bezymianny |
| 6 | 620826 | 0648 | | | | | | 6.1[5] | Miyakejima |
| 7 | 641111 | 1906 | | | | 5.5[7] | | 5.2[6] | Sheveluch |
| 8 | 680613 | 0733 | | | | 5.4[7] | | 5.4[6] | Fernandina |
| 9 | 680717 | 0623 | | | | 5.1[7] | 4.8** | 5.1* | Arenal |
| 10 | 700918 | 0206 | | | | 5.1[7] | 5.3[12] | 5.7[5] | Beerenberg |
| 11 | 730711 | 2247 | | | | 5.3[7] | 5.0[12] | 5.9[8] | Curacoa Reef |
| 12 | 730714 | 0115 | | | | 4.2[7] | | 4.6[5] | Tiatia |
| 13 | 740129 | 1334 | [9]37.77 | 15.08 | 15 | 4.5[13] | 4.3[9] | 4.7* | Etna |
| 14 | 750702 | 0734 | [15]55.41 | 160.15 | 15 | 4.7[7] | | 4.9[6] | Tolbachik |
| 15 | 770106 | 1833 | | | | 5.2[7] | | 5.1[10] | Nyiragongo |
| 16 | 771004 | 1339 ^ | | | | | 4.3[1] | 4.7* | Usu |
| 17 | 800518 | 1532 | [16]46.21 | –122.19 | 3 | 4.6[7] | 5.2[7] | 5.4[6] | St. Helens |
| 18 | 830719 | 1044 | | | | 5.2[7] | 4.5[12] | 4.9* | Una Una |
| 19 | 831003 | 1333 | [17]34.00 | 139.52 | 15 | 5.8[7] | 6.2[7] | 6.0[10] | Miyakejima |
| 20 | 850107 | 2153 | [7]71.13 | - 7.52 | 10 | 4.9[7] | 4.3[7] | 5.0[10] | Beerenberg |
| 21 | 861122 | 0041 | [17]34.53 | 139.36 | 15 | 5.8[7] | 5.9[7] | 5.9[10] | Oshima |
| 22 | 890224 | 1236 | | | | 5.2[7] | 4.8[7] | 5.2[10] | Lonquimay |
| 23 | 890709 | 0209 | [17]34.99 | 139.11 | 3 | 5.0[7] | 5.2[7] | 5.2[10] | Teisi Knoll |
| 24 | 910615 | 1115 | [12]15.12 | 120.35 | 10 | 5.7[7] | 5.4[7] | 5.6[10] | Pinatubo |
| 25 | 910815 | 1216 | | | | 5.3[7] | 5.6[7] | 5.8* | Hudson |
| 26 | 911215 | 2000 | [18]37.74 | 14.97 | 2 | 4.3[7] | 4.3[13] | 4.7* | Etna |
| 27 | 960101 | 0957 | [14]53.90 | 159.43 | 0 | 5.8[7] | 6.7[7] | 7.1[14] | Akademia Nauk |
| 28 | 960929 | 1048 | [19]64.68 | -17.49 | 4 | 5.2[7] | 5.4[7] | 5.6[10] | Grimsvøtn |

*References:* 1, Abe, 1979; 2, Abe, 1992; 3, Yokoyama and de la Cruz-Reyna, 1990; 4, Okada, 1983; 5, Zobin, 1979a; 6, Zobin, 1988; 7, ISC; 8, Gibowicz et al., 1974; 9, Bottari et al., 1975; 10, Harvard CMT; 11, Wano and Okada, 1980; 12, NEIS; 13, MOS; 14, Zobin and Levina, 1998; 15, Zobin, 1979b; 16, University of Washington, Data Base; 17, JMA; 18, Ferrucci and Patane, 1993; 19, Icelandic Meteorological Office.

*Notes:* (*) Mw was calculated from magnitude Ms using the relation Mw = 0.86 Ms + 1.00; (**) Ms was calculated from the magnitude mb using the relation Ms = 2.5 mb – 8.0 (Zobin, 1999); (^) The first earthquake of magnitude Ms 4.3 Usu earthquakes from a sequence of 16 events of the same magnitude which occurred from Oct. 04, 1977 to Aug. 22, 1978.

and they occurred at different stages during eruptions. It is unlikely that it is the complete catalog for the century but it includes all available published data. Unfortunately, the large amount of information about the seismic events of the large eruptions before 1960s was out of publications.

Only the largest earthquakes for each eruption are listed in Table 10.1. These selected earthquakes had different magnitudes, such as Ms, mb, MLH, Mw, Ks, etc. It was necessary to prepare the homogeneous catalog according to a single magnitude. The magnitude mb was not representative, since it became saturated at the value of magnitude mb = 5.8 (Fig. 10.3). From two other magnitudes, Ms and Mw, the magnitude Mw was used as a basic. The moment magnitudes Mw estimated by Harvard CMT (Centroid-Moment Tensor) group and by the author (Zobin, 1979a; 1988; Zobin and Levina, 1998) and the ISC (International Seismological Centre) magnitudes Ms for 10 events were used for the construction of relationship between Mw and Ms for volcanic earthquakes (Fig. 10.4). It was obtained from the equation

$$Mw = 0.86\ Ms + 1.00 \quad (10.1)$$

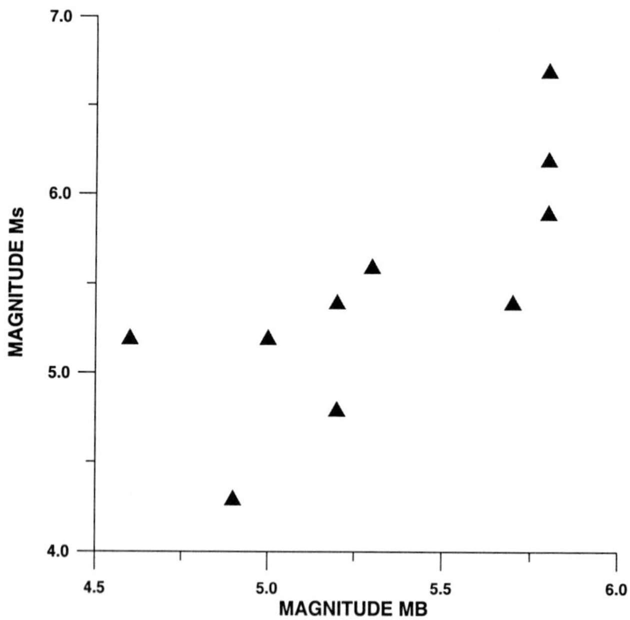


Fig. 10.3. Relationship between the magnitudes Ms and mb for volcanic earthquakes. The values of Ms and mb are taken from *Bulletin of ISC*. It is seen the saturation of magnitude mb at value mb = 5.8. From Zobin (2001).

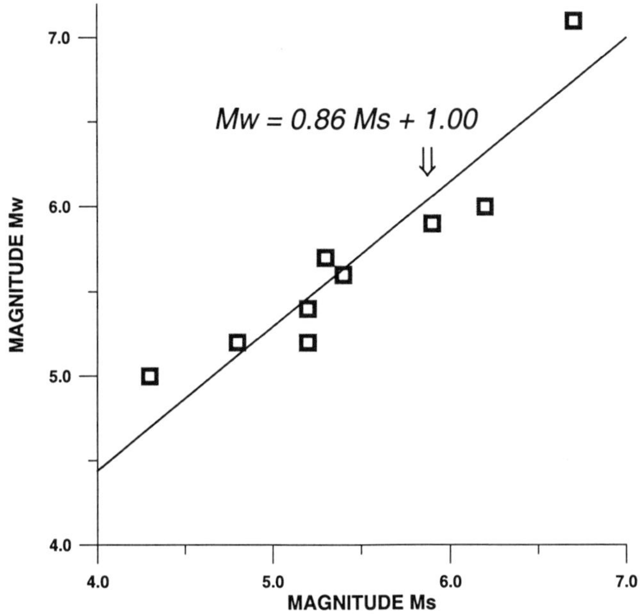


Fig. 10.4. Relationship between the magnitudes Ms and Mw for volcanic earthquakes. The line of regression is shown. The coefficient of regression is 0.927. From Zobin (2001).

which was used then for the classification of earthquakes with an unknown seismic moment. The final magnitude limit for the catalog is Mw ≥ 4.5.

*10.2. Focal rupturing of significant volcano-tectonic earthquakes and its role in volcanic processes*

The rupture process of three significant volcano-tectonic earthquakes (No 23, 27 and 28 of Table 10.1) was investigated by Takeo (1992), Zobin and Levina (1998) and Zobin (1999). These reconstructions of rupture process were realized by the inversion of the near-field strong-motion records for event No 23 (Teishi Knoll volcano) and the broadband teleseismic P waveforms for events No 27 (Akademia Nauk volcano) and No 28 (Grimsvøtn volcano). They were interpreted in terms of barriers or asperities destruction (Ruff, 1983) along the fault plane. The barriers and asperities may be considered as the ancient magmatic bodies that could be destructed by forming the magma intrusion passageway.

### 10.2.1. Rupturing of the magnitude Mw 5.2 earthquake preceding the 1989 Teishi Knoll submarine eruption

This earthquake occurred at the final stage of an earthquake swarm four days before the eruption (See Chapter 5). Its focal mechanism shows that one of its nodal planes coincides with the lineal distribution of epicenters of earthquakes following this event (Fig. 10.5). This nodal plane striking in the direction of N86°W was taken as the fault plane. Takeo (1992) studied the rupture process of the source of the Mw 5.2 earthquake by the inversion of digital 3-component accelerograms recorded at near seismic stations (Fig. 10.5). Fig. 10.6 shows the comparison of synthetics with records.

Takeo (1992) showed that the Mw 5.2 shock (No 23) was followed 45 s later by the M 4.7 earthquake; he studied these two events together as event 1 and event 2 (Fig. 10.7). The rupture of event 1 initiated in the eastern side subfault as a right-lateral strike-slip motion. Then the rupture stopped due to a barrier; Takeo guesses that this barrier could be an old dyke formation. The excitation of long-period seismic waves stopped for 1.5 s after the first stage, but the high-frequency seismic waves excited during the second stage are larger than during other stages.

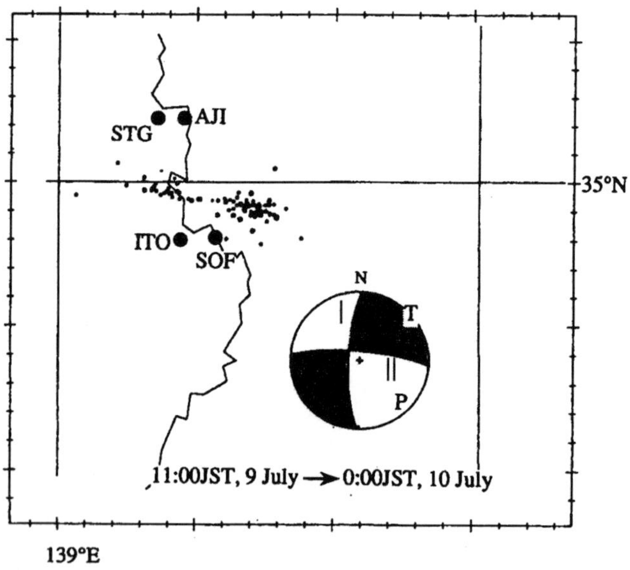


Fig. 10.5. Focal mechanism solution for event 23 and the epicenters of earthquakes following this event for 13 hours. Large circles show four seismic stations whose records were used in the inversion. The nodal plane II was chosen as the fault plane. From Takeo (1992).

Therefore, it can be suggested that only small-scale fractures took place in the barrier during the second stage. Because the rupture in the third stage involved a little reverse component with the dominant right-lateral strike-slip motion, the barrier seems to have been subjected to a vertical shear stress due to the magma intrusion beneath the source zone. The rupture extended westward and the source area of event 1 reached 4 km in length and 2 km in width. The third stage rupture was prevented by the barrier, which consisted of the old dyke formation existing around the western end of event 1. This barrier ended the rupture process of event 1, but a high stress was concentrated in the barrier. A new fracture nucleated 45 s later (event 2) and spread out westward.

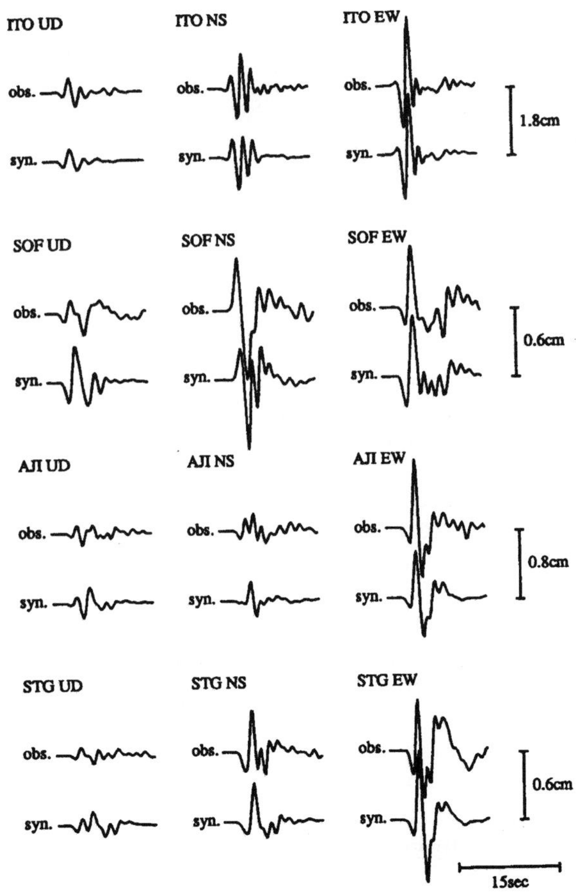


Fig. 10.6. Comparison of observed ground displacements of the Mw 5.2 Teishi Knoll earthquake inferred from accelerograms by integrating twice with synthetics. From Takeo (1992).

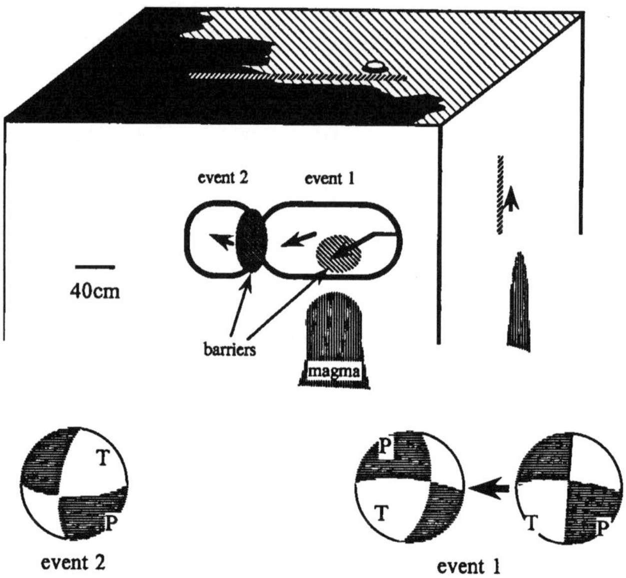


Fig. 10.7. Schematic rupture process of the Mw 5.2 earthquake (event 1) and the following M 4.7 earthquake (event 2) that preceded the 1989 submarine eruption of Teishi Knoll, Izu islands. From Takeo (1992).

Therefore, the rupturing, produced by the Mw 5.2 earthquake, destroyed the old dyke formation and opened the passageway for the magma intrusion. The large stress drop on the east side subfault of event 1 indicates a high stress concentration before its nucleation, due to the magma intrusion. Takeo (1992) guesses that the large stress drop just beneath the Teishi Knoll would assist the magma ascent and may have triggered the eruption.

### 10.2.2. Rupturing of the magnitude Mw 7.1 earthquake preceding the 1996 Akademia Nauk volcano subaqual eruption

The explosive eruption in caldera lake of Akademia Nauk volcano, Kamchatka, occurred after 28, 000 years of dormancy and was preceded and accompanied by an intensive sequence of earthquakes at depths from 0 km to 50 km, beginning on January 1 (Fig. 10.8). This sequence spread from Akademia Nauk volcano to the southwest. The largest earthquake of magnitude Mw = 7.1 (No 27 of Table 10.1) was recorded on January 1, about 4 hrs after the beginning of the seismic sequence and about 14 hrs before the beginning of the eruption.

This earthquake had a strike-slip nature, with the N14°E fault plane, oriented along the regional fault connecting the epicenter with the new eruptive site in Karymskoe lake (Zobin and Levina, 1998). This direction of the fault plane fitted well the distribution of the epicenters of events that occurred before the large shock.

A finite-fault, broad-band teleseismic P waveform inversion (Fig. 10.9) allowed the reconstruction of a dislocation model of the faulting process. The rupture model is presented in terms of a contour map of the final slip of the fault (Fig. 10.10). Two main asperities, outlined by the slip contour of 125 cm, were destroyed; one at depths between 0 km and 15 km and the second at depths between 25 km and 35 km. The majority of small earthquake foci were located away from the maximum asperities and filled a space between two zones of asperities at depths from 12 to 20 km. Probably, this large seismic event represented two shocks, one shallow and the second at a depth of about 30 km (Zobin and Levina, 1998).

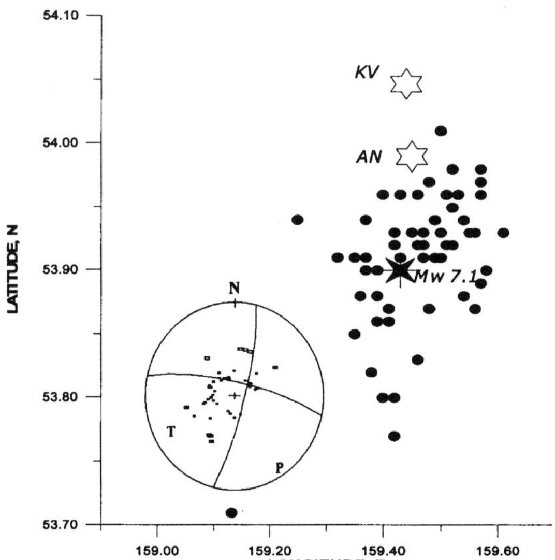


Fig. 10.8. Seismicity preceding the Mw = 7.1 earthquake related to the eruption of Akademia Nauk Volcano (No 27). The foreshocks (filled circles) that were recorded from 06:20 to 09:57 of January 1, 1996 are shown. The volcanoes Karymsky (KV) and Akademia Nauk (AN) are shown as open stars. The epicenter of the Mw = 7.1 earthquake is shown by the filled star. The focal mechanism solution for this event is shown in the left corner. Data from the Kamchatka Regional Earthquake Catalog. The focal mechanism was published in (Zobin and Levina, 1998).

The great amount of juvenile calc-alkaline basalt, representing about 95% of the ejected material (Belousov and Belousova, 2000) supports the inference of the deeper seismic source. It seems possible that during the large earthquake an ancient deep-seated magmatic column was destroyed and a new system of faults was formed that would act as a magma passageway.

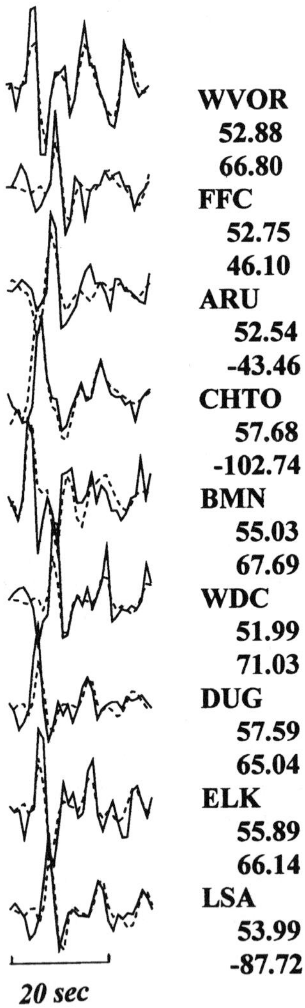


Fig. 10.9. The comparison of observed (solid line) and synthetic (dashed line) P waveforms of the Mw 7.1 earthquake of January 1, 1996 preceding the subaqual Akademia Nauk volcano eruption. The epicentral distance (in degrees) and azimuth from epicenter (in degrees) are indicated for each seismogram pair. From Zobin and Levina (1998).

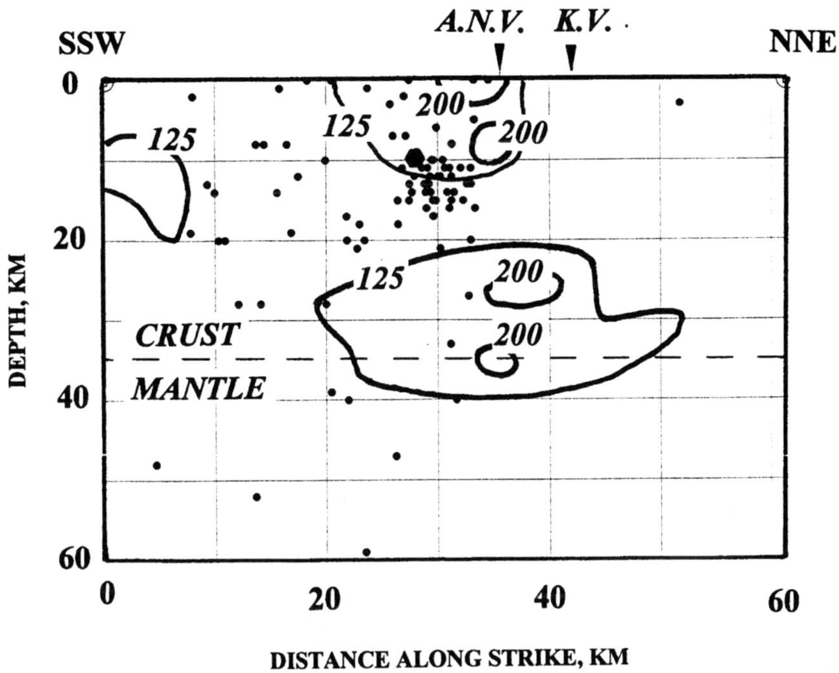


Fig. 10.10. The comparison of the dislocation model, obtained for the Mw 7.1 earthquake preceding the January 1996 Akademia Nauk volcano eruption, with seismic and volcanic characteristics. Solid circles denote the hypocenters of earthquakes of magnitude M>3.0 recorded before the beginning of the eruption. The hypocenter of the mainshock is indicated by a solid hexagon. Cumulative slip is contoured at 75-cm intervals beginning from 125 cm. A.N.V. is the Akademia Nauk volcano; K.V. is the Karymsky volcano. The border between the crust and mantle is shown by the dashed line. From Zobin and Levina (1998).

10.2.3. Rupturing of the magnitude Mw 5.6 earthquake preceding the 1996 Grimsvøtn volcano subglacial eruption

This earthquake (No 28 from Table 10.1) occurred as a mainshock of the preceding earthquake sequence. It occurred 36 hours before the eruption (See Chapter 2). The earthquake had an inverse-type nature, with the submeridional fault planes (Harvard..., 1977-1999). This direction of the fault planes fitted well the distribution of the epicenters of events that occurred just after the mainshock (Fig. 10.11).

A finite-fault, broad-band teleseismic P waveform inversion allowed the reconstruction of a dislocation model of the faulting process. The rupture model is presented in terms of a contour map of the final slip on the fault (Fig. 10.12).

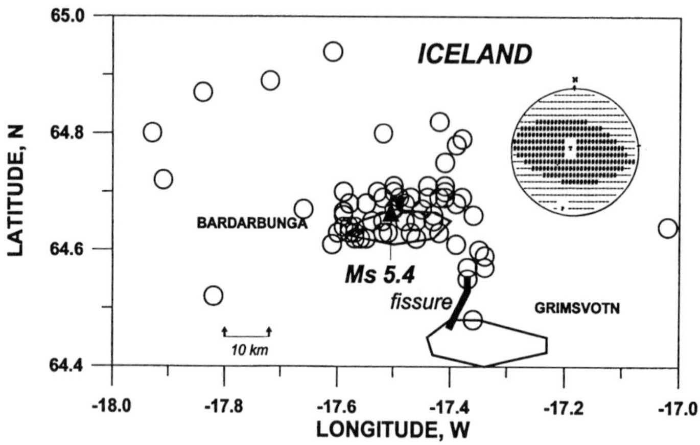


Fig. 10.11. The September 29, 1996 mainshock epicenter (solid triangle), its focal mechanism solution and the epicenters of its aftershocks (open circles), which preceded and accompanied the subglacial eruption of Grimsvøtn volcano, Iceland. The lineal distribution of the aftershocks that occurred during first 19 hours of activity coincides with the directions of two nodal planes of the focal mechanism solution. The black heavy line shows the volcanic fissure, which was formed during the eruption. The volcano calderas are shown by heavy-line ovals. From Zobin (1999).

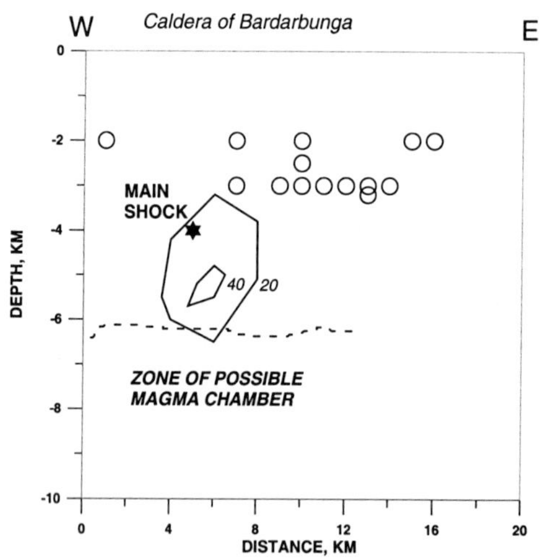


Fig. 10.12. A cross-section below the caldera of Bardarbunga. The circles are hypocenters of the aftershocks ($ML \geq 3.0$); the star is the hypocenter of the main shock. The slip contours are shown for 20 and 40 cm and outline the asperity broken during the rupturing of the main shock fault. A zone of a possible magma chamber is shown below the asperities. From Zobin (1999).

The inversion showed that the rupture had developed downdip from the mainshock hypocenter and the main asperity was ruptured at depths between 4 and 6 km. A comparison with the deep distribution of aftershocks showed that their hypocenters were situated above the destroyed asperities at depths between 0 and 4 km. These space-time seismic patterns and the results of the finite-fault inversion may be interpreted in the following way. There was a magma chamber under Bárdarbunga volcano covered by the strong rocks, or asperities. The main shock of September 29 broke the asperity and formed a path for magma transport. Then a fissure formed connecting the opened magma chamber with the feeding structures of Grimsvøtn volcano, and magma was injected laterally to the site of eruption. The continuous seismic activity during the third period was related to the collapse of rocks around the released parts of the magma chamber. The role of the Mw 5.6 earthquake in this sequence was decisive.

### *10.3. The magnitude 7 volcano-tectonic earthquakes in volcanic processes*

For more ancient large volcano-tectonic earthquakes we had no rupture reconstructions. At the same time, their participation in eruptive processes is of interest. Here I discuss two of them, the largest events with magnitude Mw = 7.0 (No 2 and 3 from Table 10.1).

#### 10.3.1. Event No 2, Katmai, Alaska

The 1912 Katmai eruption was the most voluminous eruption of the twentieth century, producing about 30 $km^3$ of products. The formerly cone-shaped summit of Mount Katmai vanished, leaving in its place a caldera 3 km in diameter and nearly 1 km deep. The collapse was apparently caused by withdrawal of supporting magma from beneath Mount Katmai toward Novarupta, the actual vent of eruption, 10 km west of Mount Katmai (Abe, 1992). There was no information about previous activity of Katmai in historical time.

The seismic activity was recorded beginning about 5 days before the eruption, but the strongest earthquakes were observed during the eruptive activity. The eruption began on June 5, 1912, with frequent minor explosions. Then was observed a series of big explosions from 13 hrs of June 6 to July 8. The eruption continued until July 21 (Simkin and Siebert, 1994), but without significant explosions. The sequence of strong earthquakes ($Ms \geq 6.0$) was recorded beginning from the second tremendous explosion of June 6, and the largest event of magnitude Ms = 7.0 occurred at 21:26 (local time) on June 7, at the final part of the violent phase of the eruption, when from 21 to 22 hrs, 42% of the total

**Table 10.2**
**Short Volcanological Information About the Significant Earthquakes (SE) of Table 10.1**

| # (Table 10.1) | Volcano, year | Type of eruption | Mw of SE | Before eruption | Initial stage | Parox. Stage | Final stage | T (years), remarks |
|---|---|---|---|---|---|---|---|---|
| 1 | Usu, 1910 | 3 | 5.4 | *1 | | | | North flank, T = ∞ |
| 2 | Katmai, 1912 | 1a | 7.0 | | *2 | | | T = ∞ |
| 3 | Sakurajima, 1914 | 2,b, 3 | 7.0 | | *3 | | | The west and east flanks, T = ∞ |
| 4 | Paricutin, 1943 | 1, 2a, 4 | 4.9 | *4 | | | | New volcano, T = ∞ |
| 5 | Bezymianny, 1956 | 1 | 5.3 | | | *5 | | Directed blast, T = ∞ |
| 6 | Miyakejima, 1962 | 2a, 3 | 6.1 | | | | *6 | The NE flank eruption with radial fissure, T = ∞ |
| 7 | Sheveluch, 1964 | 1 | 5.2 | | | *7 | | Directed blast, T = 110 [29] |
| 8 | Fernandina, 1968 | 1a | 5.4 | | | *8 | | Caldera collapse, T = ∞ |
| 9 | Arenal, 1968 | 1, 2[a] | 5.1 | *9 | | | | The west flank and summit, T = ∞ |
| 10 | Beerenberg, 1970 | 1, 2a, 4 | 5.7 | *10 | | | | The NE and SW flanks and summit, T = ∞ |
| 11 | Curacoa Reef, 1973 | 5 | 5.9 | *11 | | | | Submarine, T = ∞ |
| 1 2 | Tiatia, 1973 | 2b | 4.6 | *12 | | | | T = ∞ |
| 13 | Etna, 1974 | 3 | 4.7 | *13 | | | | West flank, T = 211 |
| 14 | Tolbachik, 1975 | 1, 2a, 4 | 4.9 | *14 | | | | New volcano, T = ∞ |
| 15 | Nyiragongo, 1977 | 1, 3 | 5.1 | *15 | | | | The simultaneous eruption of N, S and W flanks, T = ∞ |
| 16 | Usu, 1977 | 1 | 4.7 | | | | *16 | T = 124 |
| 17 | St. Helens, 1980 | 1, 2a | 5.4 | | | *17 | | Directed blast, T = ∞ |
| 18 | Una Una, | 1 | 4.9 | | *18 | | | T = 45 |

| | | | | | | | | |
|---|---|---|---|---|---|---|---|---|
| | 1983 | | | | | | | |
| 19 | Miyakejima, 1983 | 2a, 3 | 6.0 | | | | *19 | SW flank, T = 271 |
| 20 | Beerenberg, 1985 | 2a, 4 | 5.0 | | *20 | | | NE flank, T = 12 |
| 21 | Oshima, 1986 | 1, 3 | 5.9 | | | | *21 | The eruption of central crater with formation of fissure in N and NW of caldera, T = ∞ |
| 22 | Lonquimay, 1988 | 2a, 3 | 5.2 | | *22 | | | The flank crater with radial fissure, T = ∞ |
| 23 | Teishi Knoll, 1989 | 5 | 5.2 | *23 | | | | Submarine eruption, T = ∞ |
| 24 | Pinatubo, 1991 | 1, 2a | 5.6 | | | *24 | | T = ∞ |
| 25 | Hudson, 1991 | 1, 3 | 5.8 | | | *24 | | The radial fissure, T = ∞ |
| 26 | Etna, 1991 | 3 | 4.7 | | *26 | | | SE flank, T = 23 |
| 27 | Akademia Nauk, 1996 | 4[30] | 7.1 | *27 | | | | Subaqual caldera eruption, T = ∞ |
| 28 | Grimsvøtn, 1996 | 2[a], 4[28] | 5.6 | *28 | | | | Subglacial eruption., the fissure eruption on the N flank, T = 58[31] |

*Note.* The data about eruptions are taken from Simkin and Siebert (1994), except of any special marked cases.
*Type of eruption.* 1, central eruption; 1a, caldera collapse; 2a, one-side flank eruption; 2b, two-side flank eruption; 3, radial fissure; 4, regional fissure; 5, submarine eruption. T (in years) is the recurrence time.
*References.* 1, Okada, 1982; 2, Abe, 1992; 3, Yokoyama, 1997; 4, Yokoyama and de la Cruz-Reyna, 1990; 5, Gorshkov and Bogoyavlenskaya, 1965; 6, Minakami, 1964; 7, Tokarev, 1981; 8, Filson et al., 1973; 9, Minakami et al., 1969; 10, Siggerud, 1972; 11, Gibowicz et al., 1974; 12, Zobin, 1976; 13, Bottari et al., 1975; 14, Zobin, 1979; 15, Hamaguchi et al., 1992; 16, Wano and Okada, 1980; 17, Lipman and Mullineaux, 1981; 18, SEAN, 08-07, 1983; 19, Tada, T. and Nakamura, K., 1988; 20, Havskov and Atakan, 1991; 21, Yamaoka et al., 1988; 22, Barrientos and Acevedo-Aranguiz, 1992; 23, Matsumura et al., 1991; 24, Harlow et al., 199; 25, Gonzalez-Ferran, 1995; 26, Ferrucci and

Patane, 1993; 27, Zobin and Levina, 1998; 28, BGVN, 21-09, 1996; 29, Melekestsev et al., 1991; 30, M. Belousova, 1999, pers. com.; 31, Gudmundsson, 1996.

seismic energy was released (Abe, 1992).

Abe (1992) considers that a series of substantial collapses of Mount Katmai occurred during this seismic climax. Therefore, I speculate that the Mw = 7.0 seismic event, which was recorded at the seismic climax and before the third last great explosion, was the decisive one in the Mount Katmai caldera collapse, opening the fault as passageway for withdrawal of supporting magma from beneath Mount Katmai toward Novarupta.

### 10.3.2. Event No 3, Sakurajima, Japan

The 1914 Sakurajima eruption was also one of the largest eruptions of the century (See Chapter 5). The eruption began on January 12 at about 08 hrs (local time) with the emissions of white smoke at the South Peak, and 2 hrs later explosions occurred at the western and eastern slopes, where vents had opened. Just after midnight of January 13, lava began to effuse from the craterlets on both the eastern and western slopes (Yokoyama, 1997). It was the first flank eruption from vents situated on the west and east flanks of the volcano for historical time. There were recorded earlier only the eruptions from the summit crater Minami-dake or from the flank vents situated on the NE, SW, or south slopes of the volcano (Simkin and Siebert, 1994).

Seismic activity near the volcano was recorded beginning January 9. Earthquakes increased in number and magnitude, with the maximum activity on January 11. After the opening of vents in the morning of January 12, earthquake activity rapidly decreased. Then, at 18:30 of January 12, the Mw = 7.0 shock occurred. About 26 hrs later the great lava extrusion began (Yokoyama, 1997). I speculate that this large earthquake opened the passageway for magma and was decisive in the development of eruption from rather calm flank explosions to a greater volume lava eruption.

The analysis of the five largest volcanic earthquakes shows that all events were associated with eruptions of quasi-dormant volcanoes (Katmai and Akademia Nauk) or with a new type of activity for historical time (Teishi Knoll, Grimsvøtn and Sakurajima). The fault nature (events No 23, 27 and 28) and the temporal position of the events in the eruption process allow us to speculate that these earthquakes may have formed new passageways for magma movement.

## *10.4. Seismic hazard of significant volcano-tectonic earthquakes*

**Table 10.3**
**Macroseismic data for three volcano-tectonic earthquakes**

| Site | | Event 19 | | | Event 21 | | | Event 23 | |
|---|---|---|---|---|---|---|---|---|---|
| | $I_{JMA}$ | $I_{MM}$ | R, km | $I_{JMA}$ | $I_{MM}$ | R, km | $I_{JMA}$ | $I_{MM}$ | R, km |
| Miyakejima | 5 | 7 | 17 | 4 | 6 | 66 | | | |
| Oshima | 3 | 5 | 85 | | | | 3 | 5 | 36 |
| Hacijojima | 3 | 5 | 102 | | | | | | |
| Tateyama | 3 | 5 | 109 | 4 | 6 | 46 | | | |
| Niijima | 3 | 5 | 45 | 4 | 6 | 43 | | | |
| Ajiro | 2 | 4 | 112 | 3 | 5 | 44 | 4 | 6 | 6 |
| Katsuura | 2 | 4 | 143 | 2 | 4 | 102 | 2 | 4 | 114 |
| Yokohama | 2 | 4 | 152 | 3 | 5 | 85 | 3 | 5 | 72 |
| Tokyo | 2 | 4 | 180 | 3 | 5 | 113 | 3 | 5 | 42 |
| Irozaki | | | | 3 | 5 | 51 | 3 | 5 | 42 |
| Kofu | | | | 3 | 5 | 127 | 2 | 4 | 92 |
| Mishima | | | | 2 | 4 | 61 | 3 | 5 | 23 |
| Shizuoka | | | | 2 | 4 | 94 | 2 | 4 | 71 |
| Chiba | | | | 2 | 4 | 125 | 2 | 4 | 117 |

Note. Numbers of events correspond to Table 10.1. $I_{JMA}$ is the earthquake intensity of Japan Meteorological Agency scale; $I_{MM}$ is the earthquake intensity of Modified Mercalli scale; R is the hypocentral distance.

The traditional estimation of seismic hazard includes (1) the estimation of maximum magnitude Mmax, (2) its recurrence time, and (3) the law of earthquake intensity decrease with distance for a variety of magnitudes (e.g., Basham et al., 1982). We had no possibility to have a long-term sequence of volcano-tectonic earthquakes for a volcano. So we discuss the seismic hazard for a type of eruption. Table 10.2 describes the types of eruptions and the position of the seismic event in the eruptive process. Classification of 26 of the 28 eruptions was taken from the catalog of volcanoes of the world by Simkin and Siebert (1994). The position of the largest seismic event in the eruptive process was estimated from publications about these eruptions.

Significant earthquakes that occurred near volcanoes but without relation to eruptive activity (Kubotera, 1988; Abe, 1979; McNutt, 1996) were not included in this study. Among them, the Ms 6.3 1916 earthquake near Asama volcano and the Ms 6.1 1975 earthquake to the north of Aso caldera. There was a problem with the magnitudes estimated by near stations for rather small events, since as a rule, they were overestimated. For example, the magnitude of the earthquake of July 7, 1990, related to the eruption of Unzen volcano, was determined by near stations as M 4.8 (Nakada et al., 1999). At the same time, ISC gave the magnitude mb 4.4 and no estimation of Ms. These doubtful events were also not included in our catalog. The Ms 7.1 Kalapana, Hawaii, earthquake of 1975 had triggered (?) the very small summit eruption of Kilauea (Klein at al., 1987). This eruption was very insignificant and short; the real relation between the earthquake and eruption is unclear. McNutt (1996) did not include it in his list of large seismic events near volcanoes, but only mentioned it among the large regional earthquakes that could trigger the eruptions. This earthquake was not included in our Table 10.1.

The data for recurrence time were taken from the catalog by Simkin and Siebert (1994). The data for calculation of the law of decreasing of intensity with distance were taken for three Japanese earthquakes from *Bulletin ISC* (Table 10.3).

For all regressions calculated in this study the correlation was significant at the 95% confidence level.

### 10.4.1. Maximum magnitude Mmax

The problem of maximum earthquake magnitude for any region or local zone is very important in the evaluation of seismic hazard. The size of the maximum magnitude is decisive in the prediction of the destructive effects of seismic activity. The maximum magnitude may be estimated (1) by consideration of the maximum fault area that can break in a single event; (2) by the magnitude truncation in the observed seismicity for source zones in which the return period for the maximum magnitude is shorter than the observation period (e.g., Basham et al., 1982); or (3) by statistical study of the catalog (e.g., Kagan, 1997). However, for our case these types of evidence are not available. We work not with a single volcano, but with a group of volcanoes all over the world. Therefore, we estimate the maximum magnitude Mmax for this set of volcanoes combining them in two ways, based on the data of Tables 10.1 and 10.2: for (1) the type of eruption and (2) the stage of eruption.

The seismicity of volcanoes depends strongly on the type of eruption. Central and flank eruptions are characterized by different development of the swarms of earthquakes (Gorelchik et al., 1990; Zobin, 1988). Earthquake swarms, observed in Kamchatka, preceding central eruptions are characterized by continuously increasing numbers and magnitudes of earthquakes up to the beginning of eruption. Earthquake

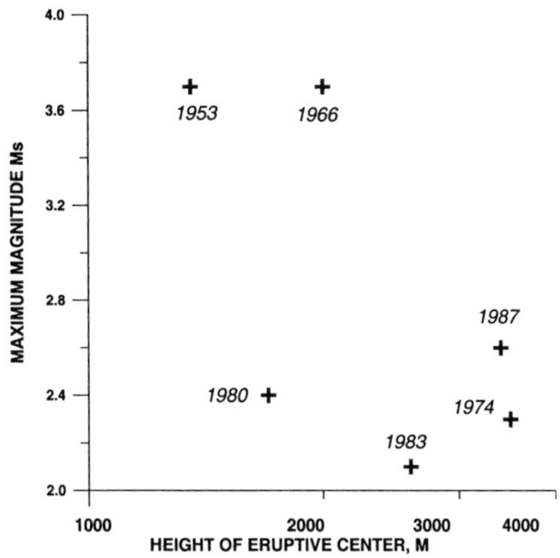


Fig. 10.13. The dependence of maximum magnitude of volcanic earthquakes on the eruption site altitude for Klyuchevskoy volcano. The mean altitudes of flank eruption sites were taken from Khrenov et al. (1991). The magnitudes Ms were calculated from the energy class Ks (Zobin, 1996b). The values of energy class Ks were taken from Tokarev (1981), Zobin (1979a) and V.I. Gorelchik (1989; personal communication, 1999). The numbers near crosses indicate the date of flank eruption. From Zobin (2001).

swarms preceding flank eruptions are characterized by the presence of a maximum number and magnitude of earthquakes at the beginning of the swarm, and a decrease of both just before the eruption.

Gresta et al. (1990) showed that the eastern and western flanks of Mount Etna are characterized by different morphological, structural, volcanic, and seismic data. They recognized the higher "degree" of tectonization on the eastern slope from the morphological and structural data. The eastern slope also has shallower and more intensive seismicity than the western slope. They attribute these differences to higher extensional stresses on the eastern slope.

Fig. 10.13 demonstrates the tendency of the change of the maximum magnitude in volcanic earthquake swarms, preceding the flank eruptions with the altitude of the new-born cones, of the 4750 m height Klyuchevskoy volcano, Kamchatka, in 1953-1987. The values of Mmax of swarms preceding the flank eruptions at lower altitudes (1360-1900 m) were generally greater than the Mmax of swarms preceding the forming of new cones at higher altitudes (2300-3340 m). This tendency may depend on at least two factors: (1) the distance between the new flank cone and the central conduit of the volcano and (2) the mechanical quality of the rocks where earthquake rupture was formed.

Both factors could constrain the length of the earthquake rupture and the size of Mmax. The first factor, that is the decrease of the distance between the central conduit and the new eruptive site with the increase of the altitude of the site, would be the decisive in the length of rupture for producing the earthquake with Mmax. At the same time, the decreasing of mechanical properties of the rocks at high altitudes could prevent large ruptures compared to those at lower altitudes. This example shows that the conditions of forming an eruptive site could be an important factor in the size of maximum seismic event related to the eruption.

Therefore, we divided our data into five groups: earthquakes related to (1) central eruption, (2) caldera collapse, (3) central eruption with flank or fissure eruption, (4) flank and (or) fissure eruption without central eruption, and (5) submarine eruptions.

Fig. 10.14 shows the magnitude distribution for largest volcanic earthquakes associated with these five groups of eruptions. The Mmax was taken as the maximum magnitude of earthquake recorded in the century for each type of eruption. Due to this distribution, central eruptions are characterized by the appearance of earthquakes of Mmax of 5.4. It is the smallest Mmax value among all types of eruptions. The central eruptions following by the appearance of flank and (or) fissure eruptive centers are characterized by slightly higher Mmax of earthquake (5.9).

The earthquakes with the largest Mmax were recorded for the caldera collapses (7.0) and the flank and (or) fissure eruptions without central activity (7.1). The submarine eruptions are characterized by the same order of Mmax for earthquakes as the central eruptions with the appearance of flank and (or) fissure eruptive centers (5.9).

The next set of Mmax values was obtained for the earthquakes occurring at different stages of eruptions (Fig. 10.15). We can see that the largest volcanic earthquakes occur at the preliminary or initial stages of eruption (Mmax 7.1 and 7.0, respectively). The paroxysmal stage is characterized by relatively low magnitude earthquakes (5.6), and at the final stage of eruption rather strong earthquakes can be observed (Mmax 6.1).

The problem of the maximum magnitude of volcanic earthquakes was discussed also by Okada (1983). He examined the earthquakes related to “direct blast”, or “sector collapse” eruptions of Unzen (1792), Bandai-san (1888), Bezymianny (1956), Sheveluch (1964) and Mount St. Helens (1980). In addition to the data listed in Table 10.1, Okada (1983) estimated the magnitude between 5 and 5.5 for the earthquake related to the Bandai-san eruption, according to its macroseismic effect and to its record in Tokyo (epicentral distance 220 km), with amplitude of 0.2 mm on the Gray-Milne seismograph. He also considered the

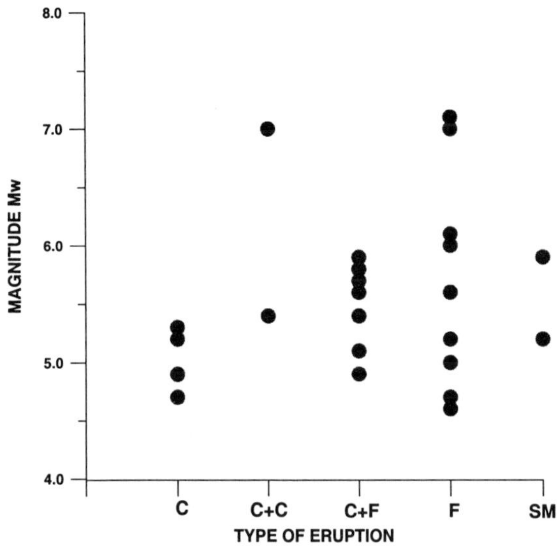


Fig. 10.14. The distribution of magnitudes Mw for 5 types of volcanic eruptions (C, central eruption; C+C, caldera collapse; C+F, the central eruption with flank and (or) fissure eruption; F, the flank and (or) fissure eruption; and SM, the submarine eruption. From Zobin (2001).

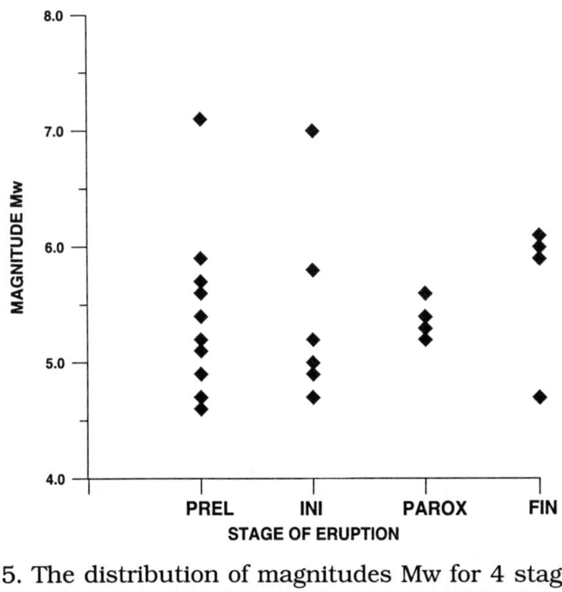


Fig. 10.15. The distribution of magnitudes Mw for 4 stages of eruptions (PREL, the preliminary stage; INI, the initial stage; PAROX, the paroxysmal stage; and FIN, the final stage. From Zobin (2001).

magnitude 5-5.5 for the earthquake that accompanied the explosion of Unzen in 1792, according to macroseismic data. Therefore, he concluded that there tends to exist a maximum limit in magnitude at around 5, except those in which regional tectonics are required for their direct causes.

Benoit (1998) has studied the dependence of the Mmax of volcanic earthquakes upon the type of eruption and the Volcanic Explosivity Index (VEI). He analyzed 180 swarms of earthquakes that occurred near volcanoes, mainly from 1979 to 1989. The Mmax values ranged from 0.5 to 6.2. The relations between Mmax and VEI did not demonstrate any significant correlation. At the same time, Benoit (1998) showed that the eccentric and radial fissure eruptions were preceded by larger shocks than central vent eruptions. This result is in good coincidence with ours.

The reason for the significant difference between the Mmax of earthquakes associated with pure central eruptions and of earthquakes related to more complex eruptions, such as forming of regional or flank fissures, may be the difference in magma transport (Zobin, 1988). It was suggested that large earthquakes formed new fissures for magma discharge in zones of regional faults. The "renovation" of the volcano conduit before and during the central eruption does not require such a large effort. For central eruptions, we suggest that the mechanism of failure is a result of local dyke intrusion within the volcanic edifice, which can involve magmatic driving forces and pore fluid pressurization mechanisms (e.g., Ellsworth and Voight, 1995). This mechanism could trigger moderate-size earthquakes, whose magnitudes would be limited by the size of volcanic edifice and which could be accompanied by a failure of volcanic flanks or sector collapses.

### 10.4.2. Attenuation of intensity with distance for volcanic earthquakes

We have calculated the intensity-distance equation for 31 points shown in Fig. 10.16 in following form (Shebalin, 1968):

$$I = a\,Ms - b\log R - c\,R + d \qquad (10.2)$$

where $R = (\Delta^2 + h^2)^{1/2}$ is the hypocentral distance in km, $\Delta$ is the epicentral distance in km, $h$ is the focal depth in km, the coefficients $b$ and $c$ are the coefficients of geometric spreading of seismic waves and absorption, respectively, and $d$ is a free member. This form of the intensity-distance equation was recommended for macroseismic studies by the New IASPEI Manual of Seismological Observatory Practice (Bormann and Bergman, 2000) and allows the expression of the intensity attenuation as a function of magnitude and distance, as well as the estimation of the inelastic properties of rocks in the region.

We obtain the following law for intensity attenuation:

$$I = 0.66\ Mw - 1.13\ \mathrm{Log}\ R - 0.0072\ R + 3.73 \qquad (10.3)$$

The curve, calculated according to equation (10.3) for the Mw 5.5 earthquake, which occurred at focal depth of 0 km just below the volcanic crater, is shown on Fig. 10.9 and can predict a possible macroseismic effect.

At the same time, the epicenter of a significant volcanic earthquake may not coincide with the volcanic crater position. Therefore, for using equation (10.3) it is necessary to know the possible deviations of the earthquake epicenter from the volcanic crater. For this purpose, the best estimated locations of 11 significant volcanic earthquakes of Table 10.1 were used (Fig. 10.17). The following equation was obtained (Zobin, 2001):

$$\mathrm{Log}\ D\ (\mathrm{km}) = 3.67 - 2.14\ \mathrm{Log}\ H\ (\mathrm{km}) \qquad (10.4)$$

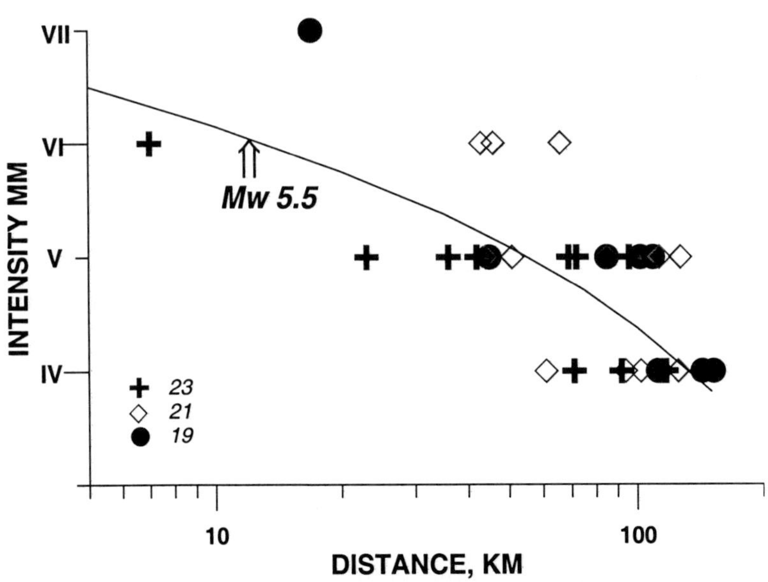


Fig. 10.16. The decrease of earthquake intensity with the distance. The curve calculated from equation (10.3) for the earthquake of magnitude Mw 5.5, focal depth 0 km, is shown. The data are marked by different symbols for three earthquakes, No 19, 21, and 23 from Table 10.1. From Zobin (2001).

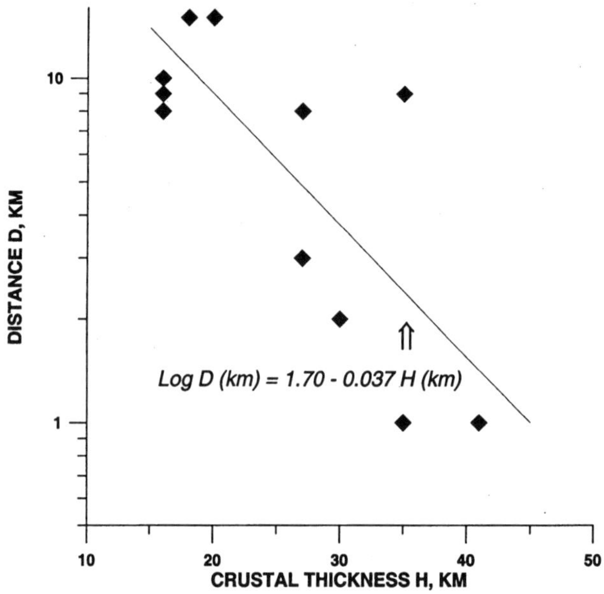


Fig. 10.17. The dependence of distance between the crater and the largest earthquake epicenter and the crust thickness in the volcanic region. The line of regression is shown. The coefficient of regression is –0.746. The numbers near the black diamonds correspond to the numeration of eruptions in Table 10.1. The position of epicenters is shown in Table 10.1. The data about the Earth's crust thickness for events No 13, 14, 17, 19, 20, 21, 23, 26 and 27 are collected in (Zobin, 1996); for the remaining events, they are taken from Mooney et al., 1998 (No 24) and Staples et al., 1997 (No 28).

Equation (10.4) allows correct calculation of possible seismic intensity from equation (10.3) by moving the position of the initial point from the crater to distance D, which is characteristic of the volcano under study. For thick crust, we have larger distance between the crater and earthquake epicenter than for thin crust.

The law of decreasing of earthquake intensity with distance for significant volcanic earthquakes may serve for the prediction of a possible seismic hazard of eruptive activity. This law was calculated based on data from rather deep-seated foci of volcanic earthquakes (focal depth 3 km to 15 km) and therefore it could predict maximum intensity. More shallow seismic sources situated within the volcanic edifice would produce earthquakes of higher attenuation and their intensity would decrease more rapidly.

10.4.3. Recurrence time

I consider the recurrence time T as the recurrence time between similar types of eruption of the same volcano that occurred at the same location. The data for estimation of T were taken mainly from the catalog of Simkin and Siebert (1994). As was shown earlier, the volcano-tectonic activity could be different for different slopes of a volcano (Gresta et al., 1990). Therefore, for the 1914 eruption that occurred on the western and eastern flanks of Sakurajima (event No 3) it was taken at a time T = ∞ but not T = 135 years, counting from the 1779 eruption that occurred on the northern and southern slopes of the volcano.

For the 1962 Miyakejima eruption (event No 6) that occurred on the northeastern flank with the formation of a radial fissure, it was taken T = ∞ because the previous fissure-flank eruptions had all occurred on the other flanks (the 1835 eruption occurred on the western flank and the 1643 eruption occurred on the south-western flank). At the same time, the 1983 eruption of Miyakejima has a recurrence time T = 271 years because the similar 1712 eruption occurred on the same southwestern flank. The information about the magnitude of earthquakes associated with the ancient eruptions is absent; therefore the seismic aspect of the eruption was not taken into account for determination of T.

The majority of information about volcanoes is collected for historical time, which is different for different parts of the world. Nevertheless, it is possible to hope that for the volcanoes of study, listed in Table 10.2, we have consistent information during the historical time for the last 100-150 years. For our 28 eruptions with significant seismic events, the value of T = ∞ (the first event for historical time) is available for 20 events out of 28. What does it mean practically? Two volcanoes (No 4, Paricutin and No 14, Tolbachik) were newborn volcanoes. Two submarine eruptions (No 11, Curacoa Reef and No 23, Teisi Knoll) were born as new volcanoes also. Several volcanoes (No 2, Katmai, No 5, Bezymianny, No 9, Mt. St. Helens, No 24, Pinatubo and No 27, Akademia Nauk) were considered to be dormant volcanoes. For 10 volcanoes, the T = ∞ means that it was the first recorded type of such activity for historical time. For example, for Usu volcano, the flank eruption of 1910 (No 1) was the first mentioned flank eruption after a series of central eruptions beginning from 1663. For Oshima volcano, the eruption of 1986 (No 21) was the first historical eruption with the forming of a fissure on the north flank after a series of central eruptions of 1783-1974 and previous central and flank eruptions.

The remaining 8 eruptions with the T values less than ∞, have these values in the range between 110 yr and 271yr (4 events) and between 12 yr and 58 yr (4 events).

The association between the large recurrence periods for eruptions of the same type at the volcano with significant earthquakes is a very important factor for constraining seismic hazard. Our results show that quasi-dormant volcanoes situated within active volcano-tectonic structures and volcanoes with long-term central crater activity represent higher seismic hazards. They may need to form new passageways for magma ascent. At the same time, the practical application of the recurrence time is problematic. We have the real estimates of recurrence time (not ∞, or less than 200 years) only for 6 volcanoes (Sheveluch, Usu, Una Una, Beerenberg, Etna and Grimsvøtn).

### 10.4.4. Estimation of seismic hazard of volcanic activity of Colima volcano, Mexico

This volcano is one of the most active in Mexico. The history of its volcanic activity (Simkin and Siebert, 1994) shows that it is characterized by central or central + flank eruptions.

The last central + flank eruption occurred in 1869-1875, or 125 yrs ago. We have no information about any significant earthquake related with the eruption and cannot estimate the recurrence time of the event. The Mmax for a possible future significant volcanic earthquake could be taken as Mw = 5.9 for the central + flank eruptions.

The second type of eruption, the central one, was observed here in 1998-1999 without any significant earthquake. Mmax for possible significant earthquakes associated with central eruptions of Colima could be taken as Mw = 5.4.

The thickness of the Earth's crust in the Colima volcanic region is about 36 km (Pacheco et al., 1997). Therefore the position of the seismic event from the central crater, according to equation (10.4), would be about 2 km, or insignificantly far from the crater. We can propose a focal depth of 5 km. The Mmax seismic event could produce, according to equation (10.3), a V-VI (V-VI means that the intensity MM would be 5.5 grades). The following could occur in the case of central + flank eruption: MM intensity in Colima city, situated at 32 km from the volcano, VI MM intensity in the village San Marcos, situated at 14 km from the volcano, and VI-VII MM intensity in the village of La Yerbabuena, situated at 8 km from the volcano. In the case of a central eruption, the intensities MM would be V-VI in Colima city, VI in San Marcos and in La Yerbabuena.

# Chapter 11
# Volcanic Tremor

Volcanic tremor is the very popular seismic signal recorded at volcanoes. McNutt (1992) wrote about 137 volcanic areas where tremors were recorded; he wrote that over 600 (I guess, about 1000 to date) scientific publications have described various features of volcanic tremor. Here I give the general characteristics of volcanic tremor and its relationship with volcanic processes.

## *11.1. Seismograms and spectra*

Volcanic tremor represents a continuous record of monotonic (harmonic) or non-harmonic vibrations that may continue for minutes to months. Kawakatsu et al. (1992) carried out the observations of volcanic tremor at Sakurajima volcano, Japan with a broadband seismometer during the 1991 explosive activity of the volcano at a distance of 4.4 km from the active crater Minamidake. They showed (Fig. 11.1) that for monotonic tremor, its short-period spectrum is characterized by peaks within a narrow frequency range at about 2.5 Hz and its long-period spectrum is characterized by the absence of any predominant peaks. For non-monotonic tremor, the opposite situation is observed: no unique peaks for the short-period spectrum but a narrow frequency range of peaks at about 0.2 Hz for the long-period spectrum.

The volcanic tremor record may be represented by surface waves or body waves (Fig. 11.2 and 11.3).

## *11.2. Location of volcanic tremor*

The sources of volcanic tremor are located at shallow depths beneath the crater floor.

### 11.2.1. Oshima volcano, Izu islands

Yamaoka et al. (1991) estimated the position of the source of the wide-frequency tremor that preceded a small explosive eruption on

November 16, 1987 of Oshima volcano from the cross-correlate analysis of the 0.7 Hz component and the propagation time of the tremor recorded by a dense 13-station seismic network. They obtained a source location at a depth of about 350 m above sea level, which is slightly deeper than the crater bottom before the 1986 eruption (See Chapter 4).

### 11.2.2. Kilauea volcano, Hawaii

Goldstein and Chouet (1994) applied a frequency-slowness analysis to study sources of tremor. They used two dense arrays with respective apertures of 800 and 120 m covering an area of 0.5 km² located 1 km west of Puu Oo as the two seismic antennas. From analyses of particle motions they concluded that most tremor sources at Puu Oo must be located within or in the vicinity of the superficial layers. Synthetics were calculated for two tremor models representing impulsive explosion point sources embedded in the Puu Oo structure and impulsive point sources of pressure exciting the resonance of an embedded fluid-filled crack. Both models required shallow sources in the range of 0 to 100 m (Chouet, 1996a).

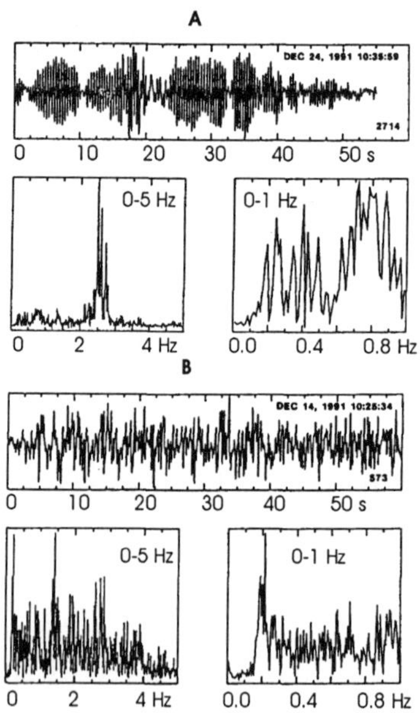


Fig. 11.1. Vertical velocity broadband records of monotonic (A) and non-harmonic (B) tremors observed at Sakurajima volcano, Japan. The initial time of the records and peak-to-peak amplitude (in 10-8 m/s) of each seismogram are given at the upper-right and lower-right corners. Vertical scales of spectra are arbitrary. From Kawakatsu et al. (1992).

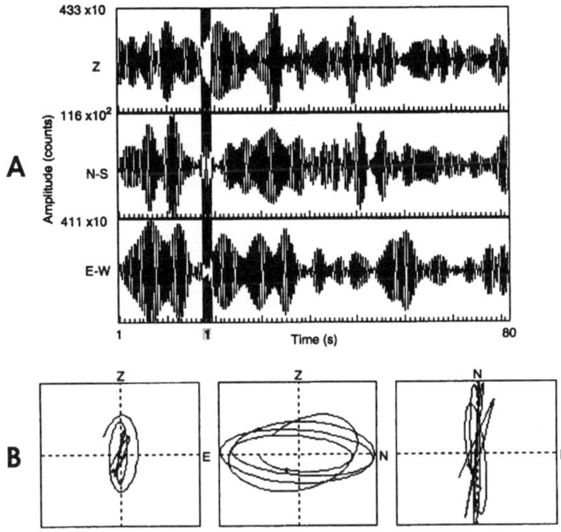


Fig. 11.2. Example of particle-motion analysis on a time series of tremor recorded in 1995 at Stromboli volcano, Italy at a distance of 1.8 km from the crater. A, 3-component short-period seismic records filtered with a Butterworth bandpass between 1.55 and 2 Hz; B, particle motion. The north-south-directed particle motion with subhorizontal incidence indicates a source close to the surface and the P-wave nature of volcanic tremor. From Falsaperla et al. (1998).

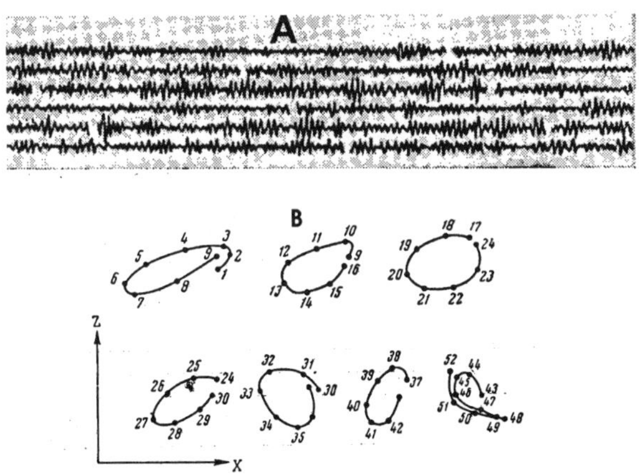


Fig. 11.3. Example of particle-motion analysis on a time series of tremor recorded in 1961 at Klyuchevskoy volcano, Kamchatka at a distance of 9 km from the crater. A, the E-W component of short-period seismic records; B, particle motion. The ellipse-shaped particle motion indicates the Raleigh-wave nature of volcanic tremor. From Tokarev (1966).

*11.3. Volcanic tremors in eruptive process*

Volcanic tremor reflects many important variations in eruptive activity. Below I show some cases of these relationships. For proper comparison between episodes of tremor at different volcanoes, the standardization of tremor amplitudes to a common reference is used. For surface waves, the following formula is used:

$$\text{R.D. (reduced displacement)} = A\sqrt{\lambda r} / 2\sqrt{2M} \quad (11.1)$$

where A is amplitude in cm peak-to-peak, *r* is a distance from source to seismic station in cm, M is seismograph magnifications at the tremor frequency, and $\lambda$ is wavelength in cm (McNutt, 1992).

11.3.1. Etna volcano, Italy

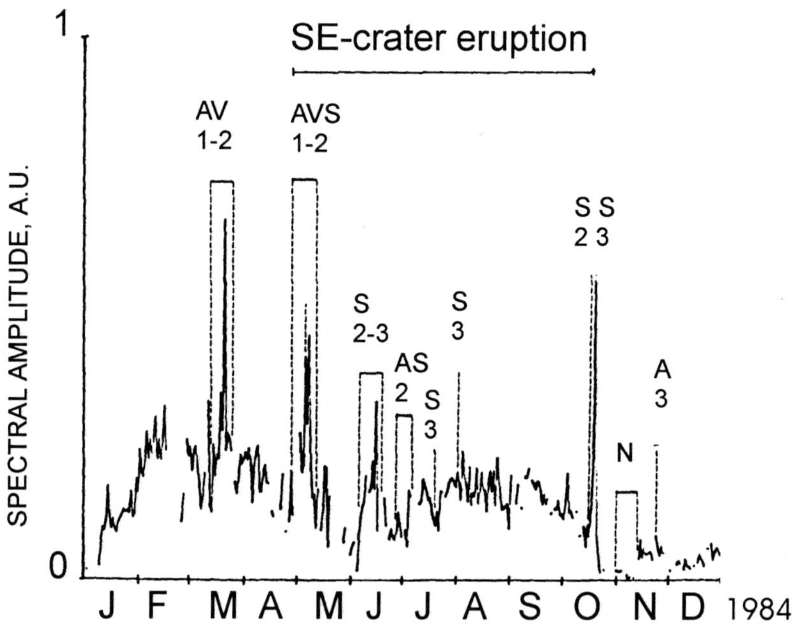


Fig. 11.4. Amplitude variations of spectral peak at 1.2 Hz of volcanic tremors recorded at Etna volcano, Sicily during its activity in 1984. The types of volcanic activity are marked by A (ash ejection), V (emission of vapor and/or gas) and S (Strombolian activity). The intensity of eruptive activity is shown by 1 (moderate), 2 (strong) and 3 (very strong). N corresponds to the absence of any activity. A.U. mean the arbitrary units. From Gresta et al. (1991) with modifications.

Gresta et al. (1991) demonstrated a good relationship between the course of the 1984 activity at Etna volcano, Sicily and the variations in the amplitude of volcanic tremor. They showed that this relationship is clearer for the spectral amplitudes than for overall amplitudes. The most sensitive were the amplitudes of a 1.2 Hz peak (Fig. 11.4); it was the only one that displayed coincidence with the more or less remarkable events of volcanic activity of different types and magnitudes, which occurred at all four summit craters of the volcano. At the same time, it can be seen that an increase in the spectral amplitudes of the tremor does not depend on the type or intensity of volcanic activity; it may only indicate a surface eruptive manifestation. The spectral amplitudes of the peaks of 0.95, 1.45, 1.65, 1.8 and 2.4 Hz, that were also dominant during the eruptive activity, were characterized by a lower sensibility to eruptive manifestations.

11.3.2. Pavlof volcano, Alaska

The eruption was continuous from September 16 to December 4, 1996, declined, and then had two stronger bursts on December 10-12 and December 27-29 (McNutt, 2000b). Fig. 11.5 demonstrates the ability of tremor to reflect the explosive phases of the eruption by an increase in amplitude. The two largest peaks in amplitude correspond to the stronger December bursts.

11.3.3. Kilauea volcano, Hawaii

The 1984 east-rift eruption that produced about $11 \times 10^6$ $m^3$ of lava

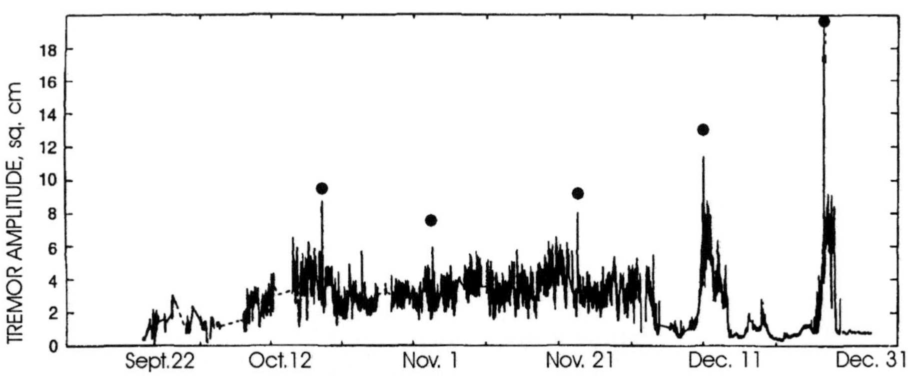


Fig. 11.5. Variation of normalized amplitude of volcanic tremor during the 3.5 months of the 1996 Pavlof eruptions. Stronger tremor accompanied more explosive phases (shown by filled points) on October 18, November 4 and 23, and December 10 and 27. From McNutt (2000b).

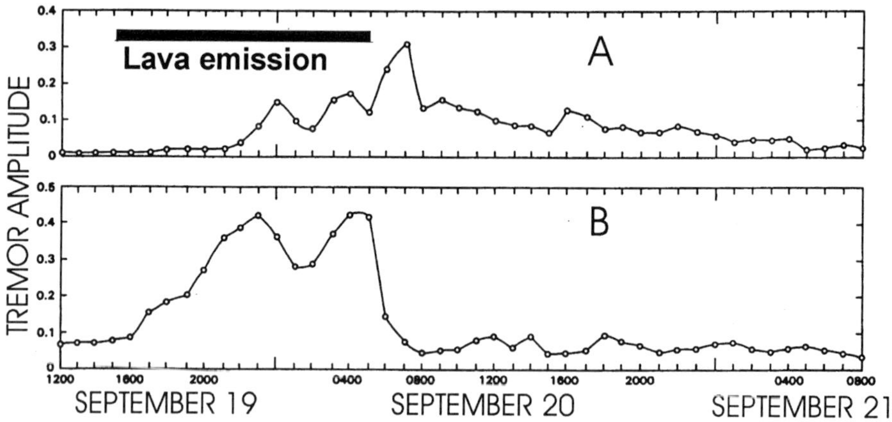


Fig. 11.6. Variation in the tremor peak-to-peak amplitudes (in micrometers) during and after the September 1984 east-rift lava eruption at Kilauea volcano, Hawaii at the volcanic summit (A) and at a distance of 6 km west from the site of lava emission in the East Rift zone (B). The period of lava emission is shown. From Koyanagi et al. (1987).

began on September 19 at 16 h and continued for 13.5 hrs. This eruption was accompanied by harmonic volcanic tremor recorded at a distance of 6 km west from the site of the eruption and at the summit of Kilauea volcano. A volcano-tectonic earthquake swarm did not precede the eruption. Fig. 11.6 shows the variations of the tremor amplitude recorded at volcanic summit and in the East Rift zone, near the site of the eruption. The tremor near the site of the eruption was more intensive than at the summit of Kilauea. The tremor began with the start of the eruption, gradually reached its maximum value over seven hours, and stayed at this level up to the end of the eruption. A rapid decrease of tremor accompanied the end of sustained high fountaining.

The summit tremor began in six hours after the start of the eruption and reached its maximum amplitude two hours after the cease of the lava emission (Koyanagi et al., 1987).

### *11.4. Relationships between the intensity of seismic signals and volcanic events*

Fig. 11.7 and 11.8 show a good correlation between the amplitudes of volcanic tremors and the height of lava fountains and pyroclastic material. The dependence of the amplitude of volcanic tremor upon the height of the lava fountain is shown in Fig. 11.7. The data obtained for two volcanoes (Kilauea, Hawaii and Pavlof, Alaska) overlap, suggesting similar relationships for both volcanoes. McNutt et al. (1991) noted that

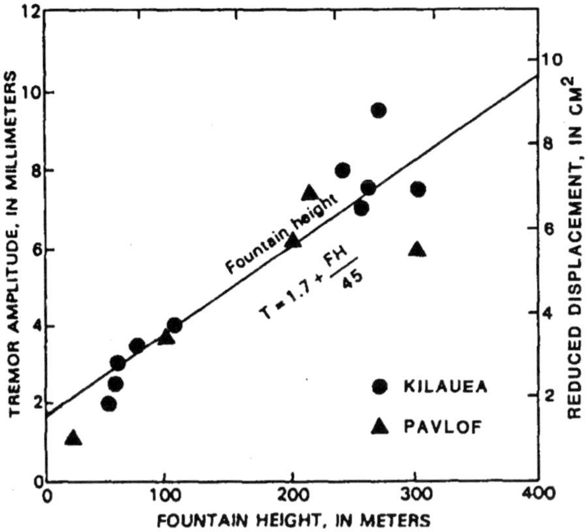


Fig. 11.7. Relationship between the fountain height at basaltic volcanoes Kilauea, Hawaii (round dots) and Pavlof volcano, Alaska (triangles) and tremor amplitude. Both data sets were normalized to units of reduced displacement. The line represents a lineal regression fit to Kilauea data. From McNutt et al. (1991).

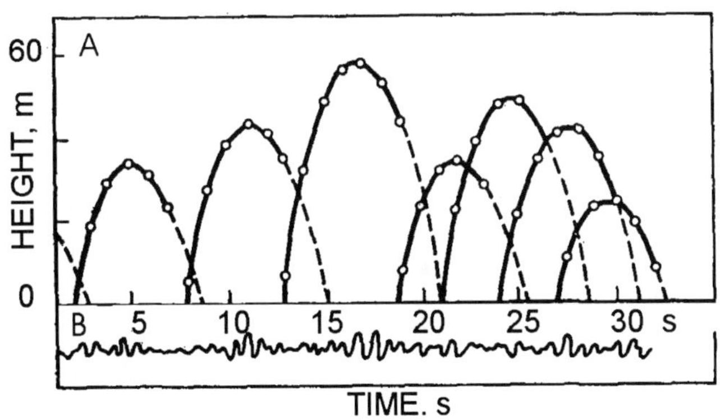


Fig. 11.8. Dependence of the amplitude of volcanic tremor upon the height of pyroclastic material rising out during the explosions at New Tolbachik volcanoes, Kamchatka). The shape of pyroclastic cones (A) and a seismogram (vertical component) of volcanic tremor (B) were recorded simultaneously by filming and by short-period seismometer, respectively at a distance of about 3 km from the site of the eruption. The seismogram is moved to the left for 2.5 s to correct for the seismic wave propagation from the site of eruption to seismic station. From Tokarev et al. (1984b).

the Kilauea data are from the first phase of the 1959 eruption; later data showed stronger tremor for the same fountain heights.

Tokarev et al. (1984b) showed that the time of output of incandescent material coincided with the time of origin of an intensive seismic impulse in the sequence of volcanic tremors recorded during the November 1976 Strombolian explosions at the southern site of New Tolbachik volcanoes in Kamchatka (Fig. 11.8, left side). The amplitude of volcanic tremor was well correlated with the height of pyroclastic material rising out of the volcanic crater. With an increase in the explosion frequency, the tremor impulses interfere with the decreasing of the amplitude (Fig. 11.8, right side).

A more general relationship between the maximum amplitude of volcanic tremor and the volcanic explosivity index (VEI) was obtained by McNutt (1994) based on the data from 21 eruptions at 14 volcanoes (Fig. 11.9). A lineal regression of amplitudes (in R.D. units) plotted versus VEI yielded the following regression:

$$\text{Log R.D. (sq. cm)} = 0.48\,(\text{VEI}) + 0.08 \qquad (11.1)$$

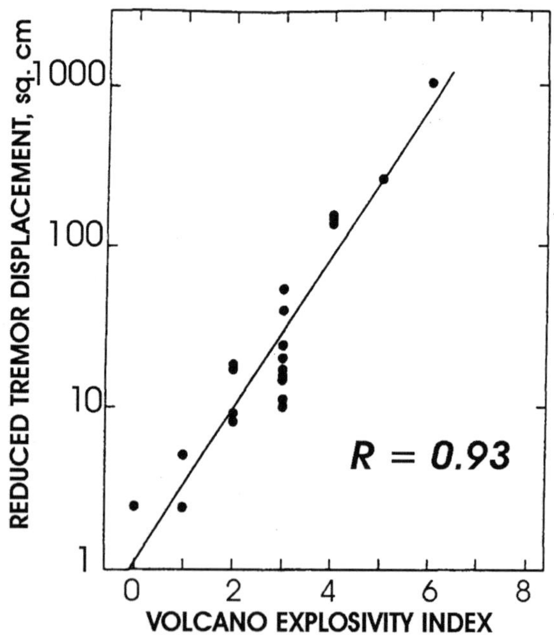


Fig. 11.9. Volcanic tremor reduced displacement versus volcano explosivity index (VEI, Newhall and Self, 1982). *R* is the correlation coefficient. The least square regression line is shown. From McNutt (1994).

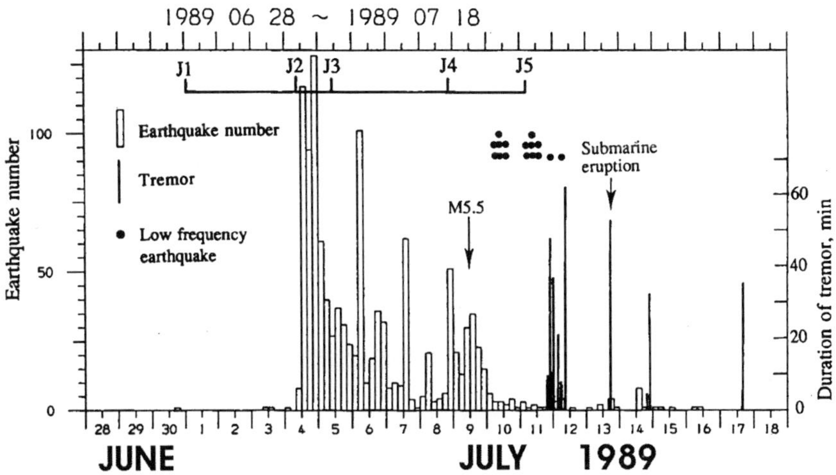


Fig. 11. 10. Sequence of the seismotectonic events associated with the 1989 Teishi Knoll, Izu islands eruption. Earthquake number is shown for every 4 hours; the epochs J1 to J5 reflect the increase of crustal tilt from J2 to J5 that exceeded 20 mrad. From Ukawa (1993).

### *11.5. Special cases of volcanic tremors*

#### 11.5.1. Isolated tremors

Large-amplitude volcanic tremor was observed before the 1989 Teishi Knoll, Izu islands eruption. Fig. 11.10 shows the position of these signals in the total sequence of seismotectonic events preceding the eruption (Ukawa, 1993). The sequence commences with a swarm of volcano-tectonic earthquakes (See Chapter 5) that continued for about seven days from July 4 and was accompanied by a large increase in crustal tilt. After cessation of the volcano-tectonic activity and during the maximum tilt, a sequence of low-frequency earthquakes occurred. At a coda of the sequence of low-frequency earthquakes, the continuous tremor began. It culminated with the large-amplitude tremor that took place for about one hour and conclusion was coincident with the beginning of the submarine eruption.

Figs. 11.11 and 11.12 show the waveforms of the volcanic tremor and their spectra. Here F1 and F2 indicate the episodes of continuous tremor of July 11 and 12, and the large-amplitude tremor of July 13 is shown as the main phase and isolated events. The power spectra of these events demonstrate that every tremor has its spectral peak at around 0.8 Hz. The spectral shape of the main phase is similar to that

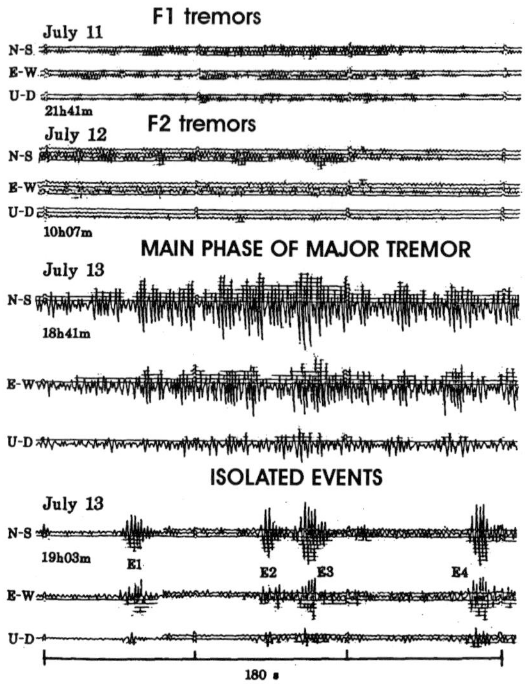


Fig. 11.11. Seismograms of the isolated tremors preceding the 1989 Teishi Knoll, Izu islands eruption. Recorded at a distance of about 8 km from the site of submarine eruption. From Hashimoto et al. (1991).

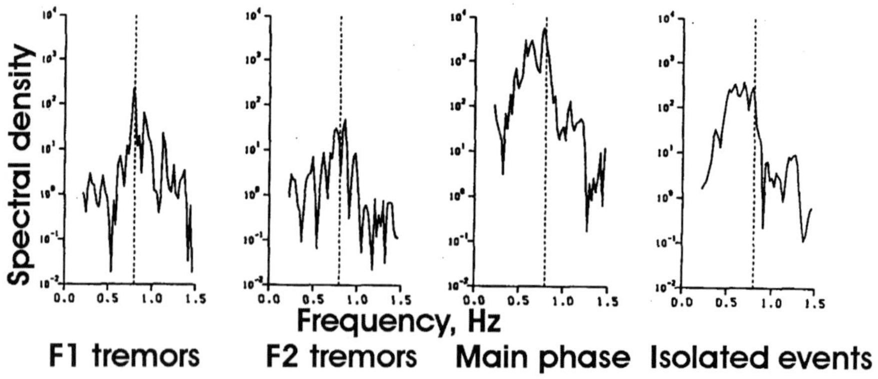


Fig. 11.12. Power spectral density (in $\mu m^2$-s) curves of NS component of isolated tremors shown in Fig. 11.10. Dashed lines indicate the frequency of 0.8 Hz. From Hashimoto et al. (1991).

of the isolated events. The spectra of the continuous tremor F1 and F2 are different in shape from those of the large-amplitude tremors and are characterized by narrower peak bands. These spectral properties indicate that the F1 and F2 tremors were produced by smaller-scale events compared with the large-amplitude tremors (Hashimoto et al., 1991).

Oshima et al. (1991) considered that the continuous tremors F1 and F2 of July 11 were associated with the formation of a new volcanic knoll (cryptodome). The large-amplitude tremor of July 13 was believed to be caused by the final moments of pre-eruptive and eruptive magma movements. The hypocenters of the isolated events are located near the position of the submarine eruption (Goto et al., 1991).

A study of the particle motions for the large-amplitude tremors (Hashimoto et al., 1991) indicates their surface-wave nature and shows that the radial and vertical components of the motions are represented by Rayleigh waves while the transverse components are Love waves (Fig. 11.13).

The newborn New Tolbachik cones in Kamchatka were formed along a freshly opening fault on July 6, 1975 (See Chapter 5). The 16-minute record of low-frequency volcanic tremor was observed just before the beginning of the preceding swarm of volcano-tectonic earthquakes (Fig. 11.14).

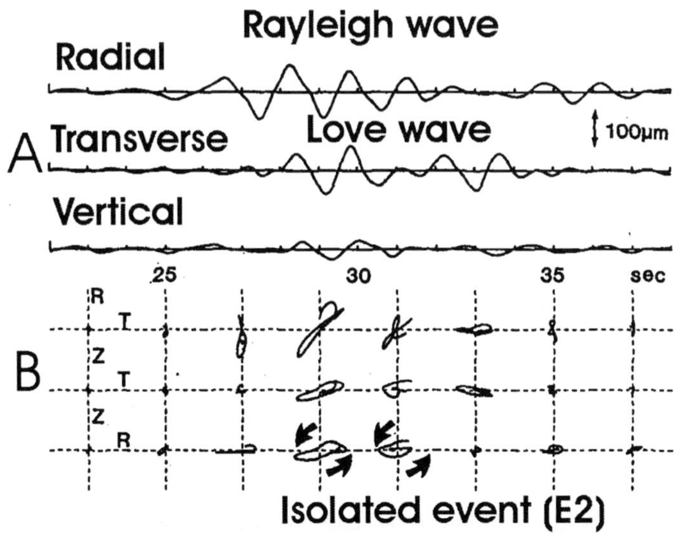


Fig. 11.13. Waveforms of the isolated event E2 recorded at a distance of about 8 km from the site of submarine eruption (A) and the particle motions (B) at 2-s intervals. The arrows on R-Z plane indicate retrograde motions that characterize the Rayleigh-type surface waves. From Hashimoto et al. (1991).

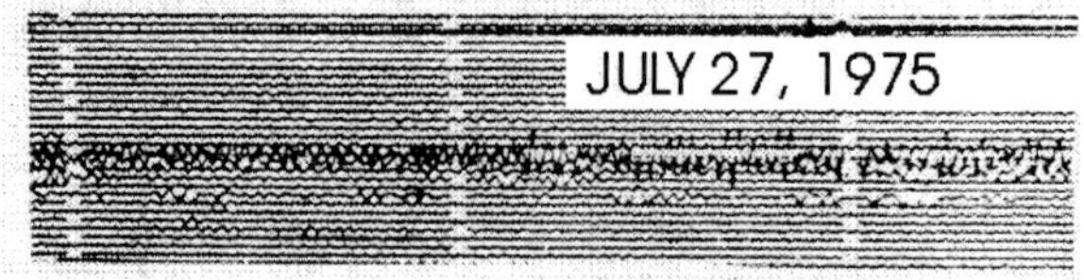


Fig. 11.14. Low-frequency isolated volcanic tremor recorded by a short-period seismic station at a distance of about 50 km from the site of a newborn cone of New Tolbachik volcanoes, Kamchatka in June 1975, 10 days before the beginning of the eruption. From Tokarev et al. (1984b).

It represented continuous vibrations with a gradual increase in the amplitude. Maximum amplitudes were four times larger than the initial amplitudes. After reaching a maximum, the amplitudes began to gradually decrease. With the onset of the volcano-tectonic earthquake swarm, the tremors ceased and were not observed for following two-week seismic activity up to the beginning of the explosive activity of the new-formed cone (Tokarev et al., 1984b).

### 11.5.2. Deep tremor

Deep tremors are distinguished from shallow ones by their amplitude distribution. For Kilauea, Hawaii, deep tremor shows a more uniform amplitude for many stations over a large area as compared to the shallow events, which show the concentration of large amplitudes only a few nearby stations. Deep Hawaiian tremor demonstrated an identical period at almost all stations but it had a tendency to change with time. Fig. 11.15 shows the source locations of 24 deep tremor episodes that were determined using the onset times. The tremor sources were distributed at a distance from 13 to 40 km to the southwest of Kilauea volcano within a depth range of 27 to 75 km (maximum number of events were recorded between the depths from 35 to 45 km) with the mean error in depth estimation of about 6 km (Aki and Koyanagi, 1981).

There exists a dependence of the duration of deep volcanic tremor episodes on their reduced displacements (Fig. 11.16) shows that the duration decreases exponentially with an increase in reduced displacement. Aki and Koyanagi (1981) consider that it may indicate a unique length scale involved in the source process of deep volcanic tremor such as the average size of the conduits.

Aki and Koyanagi (1981) found that the deep magma transport inferred from the deep tremor data is remarkably stationary and relatively unaffected by volcanic activity near the surface.

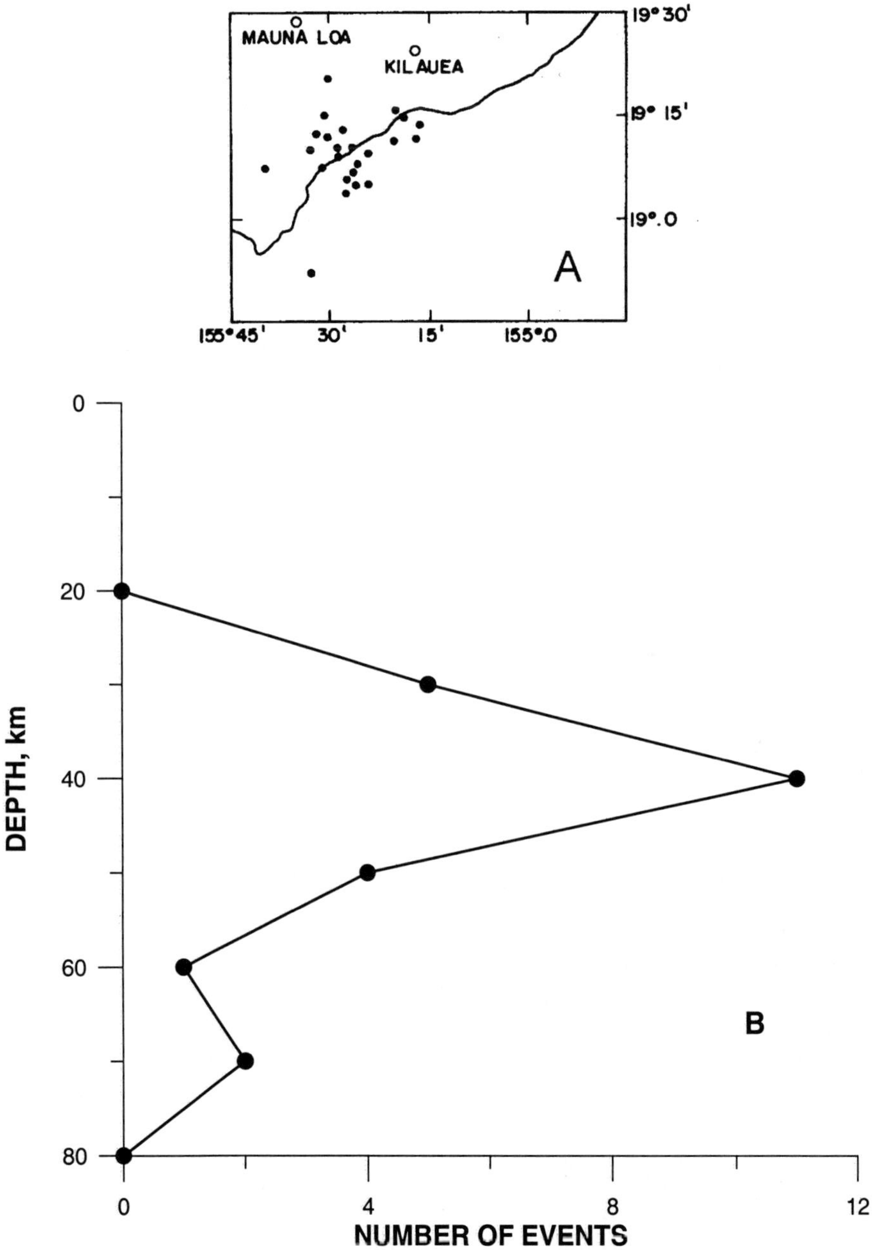


Fig. 11.15. Locations of the 1973-1977 deep tremor sources under Kilauea volcano, Hawaii (A) and their depth distribution (B). A, from Aki and Koyanagi (1981), B, data from Aki and Koyanagi (1981).

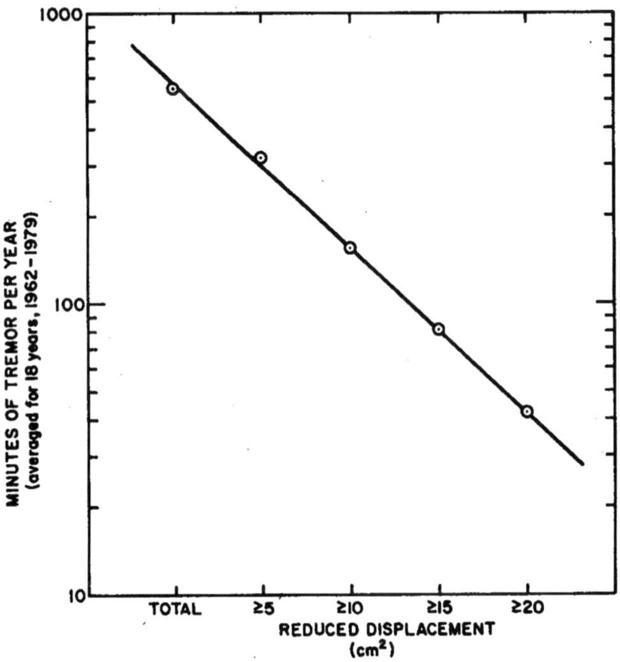


Fig. 11.16. The duration of tremor per year recorded with the reduced displacement for deep tremors under Kilauea volcano, Hawaii during 1962-1979. The duration decreases exponentially with the increase in reduced displacement. From Aki and Koyanagi (1981).

## 11.5.3. Long-period tremor

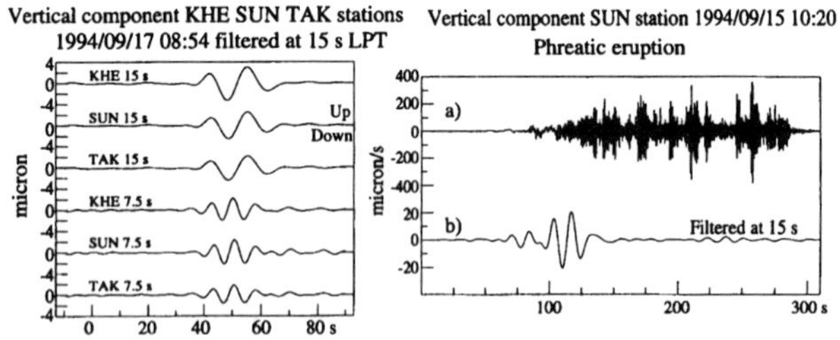


Fig. 11.17. Broadband records of long-period seismic events recorded at Aso volcano, Kyushu without surface activity and during the phreatic explosion. Both types of records filtered at 15 s have comparable waveforms. From Legrand et al. (2000) with modifications.

Long-period (longer than 7 s) signals of rather short duration (several tens of seconds) have been observed at Aso, Kyushu volcano by a broadband network in 1984 (Kawakatsu et al., 2000). These long-period signals occurred at the volcano during the small phreatic explosions or without surface activity (Fig. 11.17). The most energetic period of 15 s is always clearly recorded.

These events were located within the depth range from 1.2 km to 1.6 km above the volcanic crater. Legrand et al. (2000) used a linear inversion to estimate the moment tensor of the events. The inversions showed a similar isotropic mechanism for all 43 long-period tremors and the six phreatic explosions.

# Chapter 12
# Earthquakes Related to Pyroclastic Flows and Rockfalls

Pyroclastic flows and rockfalls are common for explosive and effusive activity at andesitic and dacitic volcanoes. The fall-and-flow of incandescent pieces of lava dome or lava flows as well as the numerous rockfalls generate seismic signals. These signals ("fall-and-flow" earthquakes) may be the main component of seismicity during block lava emission and lava dome destruction.

## *12.1. Occurrences of "fall-and-flow" earthquakes*

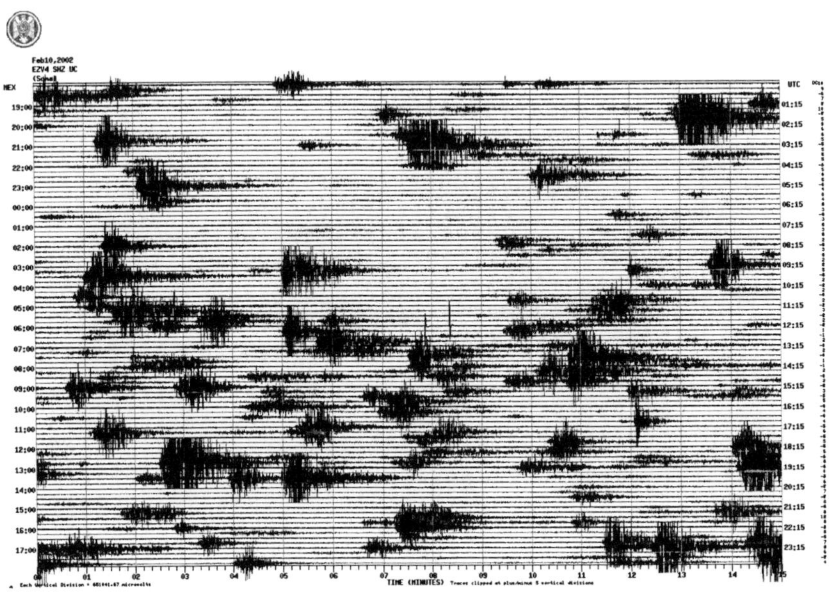


Fig. 12.1. A seismogram of station EZV4 situated at a distance of 1.7 km from the crater of Volcán de Colima, México. The seismic records of February 10, 2002 were produced by rockfalls and pyroclastic flows. Minute marks are shown. Courtesy of Colima Volcano Observatory.

The "fall-and-flow" earthquakes (or avalanche signals, rockfall signals, guguran) occur in dense groups, sometimes completely filling the seismogram (Fig. 12.1). The total number of these events recorded during the 80-days andesitic block-lava emission at Volcán de Colima, México in November 1998 – February 1999 exceeded 6 000. The rate of seismic energy release during the seismic "fall-and-flow" activity was comparable with the rate of seismic energy release during the preceding swarm of volcano-tectonic earthquakes (Fig. 12.2).

The "fall-and-flow" signals replace the volcano-tectonic earthquakes after the beginning of lava emission (Fig. 12.3). Malone (1983) wrote that a typical dome-building eruption episode at Mt. St. Helens, Cascades began with a few volcano-tectonic earthquakes several weeks before the actual eruption. When the eruption actually starts, the volcano-tectonic events are replaced by the avalanche signals and explosion events. Both types of events may last for several days, slowly decreasing in number.

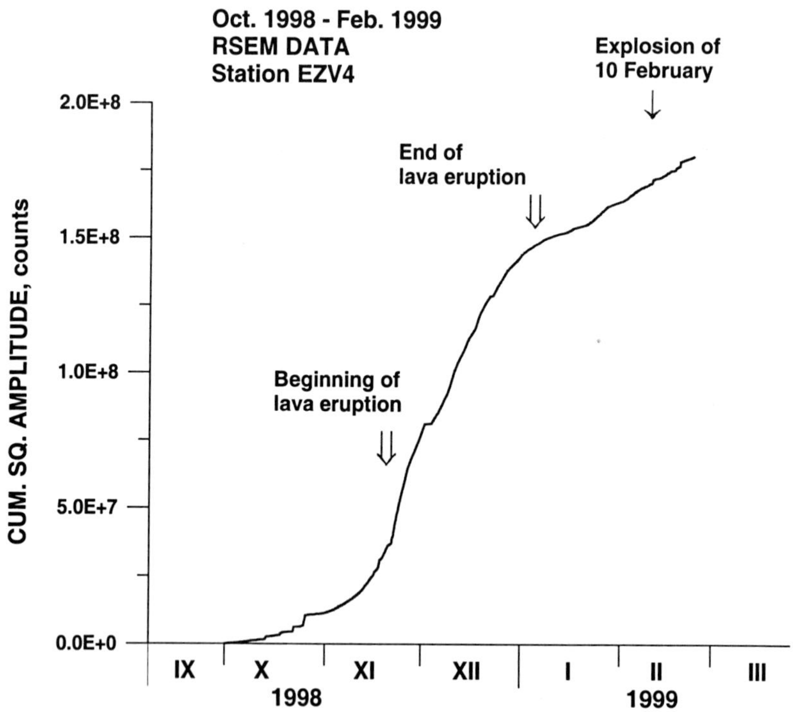


Fig. 12.2. Cumulative RSEM (real-time seismic energy measurement) curve of variations in seismic energy release at Volcán de Colima in October 1998-February 1999. It is seen that the rate of seismic energy release did not change at the beginning of the lava eruption.

### *12.2. Relationship between the "fall-and-flow" earthquakes and seismo-volcanic activity during lava emission*

The relationship between the "fall-and-flow" seismic signals and eruption dynamics has been observed by many authors. Neuberg (2000) wrote that these signals correlated visually with the avalanches of material from the lava dome at Soufriére Hills volcano. Ratdomopurbo and Poupinet (2000) note that the duration of these seismic signals ("guguran" in Indonesian) at Merapi volcano, Java usually lasts several minutes. Their frequencies are higher than that of tremor. Purbawinata et al. (1997) wrote that the occurrences of "guguran" indicate the level of stability of the lava dome. When the dome is unstable, "guguran" occurrences may have a rate of up to more than 500 events/day. Whereas, when the dome is stable, only 3 events per day occur. A rate of more than 50 events/day can usually be used to indicate that the dome is in an unstable state.

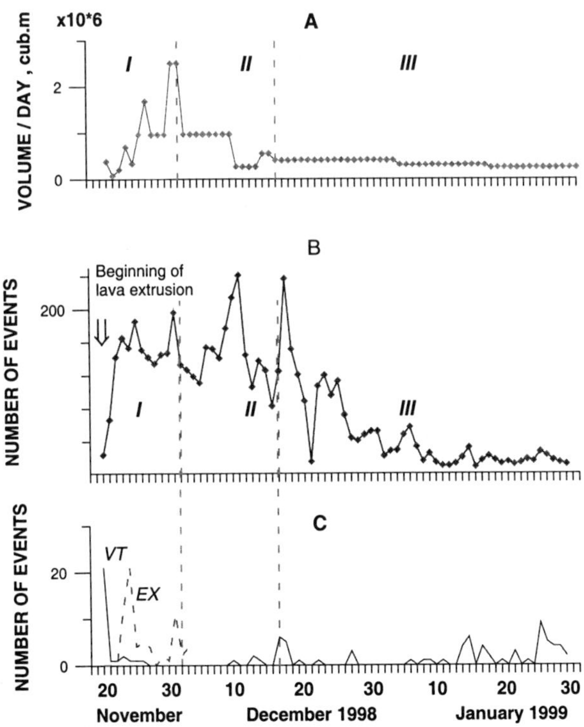


Fig. 12.3. Daily variations in the volume of lava emission (A), number of the "fall-and-flow" earthquakes (B) and volcano-tectonic (VT) and explosion (EX) earthquakes (C) during the 1998-1999 lava eruption at Volcán de Colima. I, II, and III indicate three stages of lava eruption.

A dacitic lava dome at Unzen volcano, Japan, emerged in May 1991, and its subsequent collapse resulted in successive generations of pyroclastic flows. Yamasato (1997) studied the seismic signals produced by the collapses of lava blocks from the dome, the fall of blocks down the slope and the migration of pyroclastic flows on the mountain slope. The comparison of the seismic signal with the video records allowed him to identify seismic signals that were produced during the different stages of the formation of a pyroclastic flow. Small amplitude seismic waves with the predominant frequency of 2-3 Hz were excited almost simultaneously with the collapse of the lava dome. In a few seconds, when lava blocks fell onto the slope, the amplitude of seismic signal become large with the appearance of a low-frequency (about 0.5 Hz) component. Then the fragmented pyroclastics started to flow generating high-frequency (more than 2 Hz) seismic waves.

Fig. 12.3 gives an illustration of seismic activity that was observed during 1998-1999 lava eruptions at Volcán de Colima, México. The three stages may be distinguished in this eruption: (I) from November 20 to December 1, 1998, the stage of intensive lava flow; (II) from December 2 to December 17, 1998, the stage of slow lava flow; and (III) from December 21 to January 4, 1999, the lava flow practically ceased. By early February 1999, the lava flows were no longer being fed from the summit crater, the activity changed to an explosive stage.

Three types of seismic signals were observed during the block-lava effusion in November 1998 – January 1999. The first type was represented by volcano-tectonic earthquakes. They were similar to those recorded before the lava eruption. The second type of signal was produced by explosions and exhalations in the crater. The signals of the third type, the most numerous, were generated by the movement of pyroclastic flows and rockfalls down the slope of the volcanic edifice. These signals were of a spindle form; their durations varied from 30 to 350 s.

These three types of seismic events were distributed unevenly during the lava emission (See Fig. 12.3B and C). With the beginning of lava emission, the number of volcano-tectonic earthquakes sharply decreased; from November 21, 1998 to January 16, 1999 they were practically absent. They appeared again in the second part of January, when the lava emission terminated. Explosion earthquakes were rather numerous on November 23-25 during the period of maximum velocity in lava emission; then they appeared rarely without regularity. The majority of seismic signals was of ""fall-and-flow"" type and was produced by the movement of pyroclastic flows and rockfalls of incandescent fragments of the lava flows. These seismic signals sharply increased in number with the beginning of lava emission. Their number was still high when the lava emission rate decreased but it gradually decreased with the termination of lava emission.

### *12.3. Quantification of "fall-and-flow" earthquakes*

An earthquake signal may be characterized by their envelope area, amplitude and duration. Norris (1994) showed that the envelope area for avalanche seismograms is correlated with the maximum amplitude of the signal; therefore, these two parameters are equivalent.

Fig. 12.1 illustrates the seismic signals recorded by a short-period seismograph at Volcán de Colima, México. Their durations and their amplitudes vary from event to event. The simultaneous observations with the short- and long-period seismographs at Unzen volcano, Japan showed that the long-period records are significantly shorter than the short-period records (Uhira et al., 1994). Uhira et al. (1994) demonstrated that the long-period amplitude may vary by about a factor of ten.

Norris (1994) wrote that the rockfall sequences (Mt. St. Helens, Cascades; Makaopuhi crater, Hawaii) appeared to show a poor correlation between the maximum signal amplitude recorded by a short-period seismometer and the volume of the rockfalls. Mills (1991) used the RSAM (real-time seismic amplitude measurement) system for study of the rockfall activity at Mt. St Helens and reported a similar poor correlation; rockfalls that differed in volume by several orders of magnitude could be distinguished by their seismic amplitude or duration, but those closer in volume could not.

The comparison of seismic data with visual observations of pyroclastic flows at Soufriére Hills volcano, Montserrat showed that short, sharp RSAM peaks represented discrete pulses of large-volume, energetic flows (Calder et al., 2002). Uhira et al. (1994) showed a good correlation between the maximum signal amplitude recorded by long-period seismometer and volume of rocks of pyroclastic flows at Unzen volcano, Japan. As noted by Calder et al. (2002), the relationship between volume of pyroclastic flow and signal amplitude is not simple.

Here I present separately the data of quantitative analysis of seismic signals obtained using short-period and long-period seismometers.

#### 12.3.1.Quantification of "fall-and-flow" earthquakes recorded by short-period instruments at Volcán de Colima, México

*Relationship between the duration and maximum amplitude of "fall-and-flow" earthquakes.* Fig. 12.4 shows the relationship between duration and maximum amplitude of "fall-and-flow" earthquakes measured at the short- period station EZV5 at a distance of 4 km from the crater of Volcán de Colima. Generally, these do not correlate.

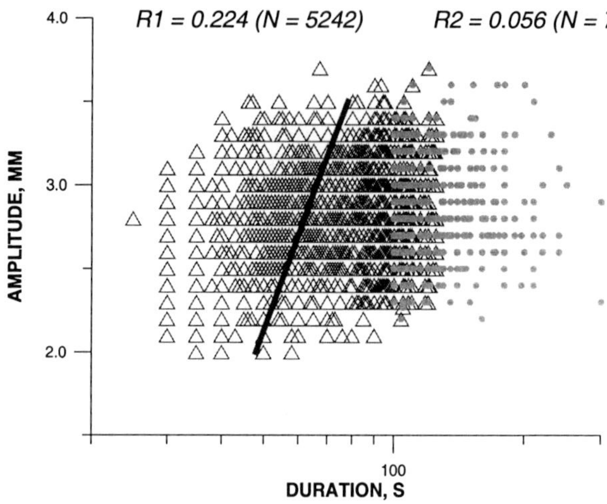


Fig. 12.4. Relation between the amplitude and duration of seismic signals of "fall-and-flow" earthquakes recorded at the short-period station EZV5. Triangles show the data for the records with duration of less than 100 s; the filled points show the data with durations longer than 100 s. The least-square regression line for the first set of data is shown. R1 and R2 are the coefficients of correlation; N is the number of events.

However, for smaller events, with durations less than 100 s, the coefficient of correlation is $R = 0.224$, which is significant at 99% level. For larger events, with durations greater than 100 s, the coefficient of correlation $R = 0.056$ is not significant. Therefore, the duration and maximum amplitude of short-period records of ""fall-and-flow"" earthquakes are not related to one another and may be used for the earthquake quantification.

*Temporal variations in the durations of "fall-and-flow" earthquakes.* Fig. 12.5 shows the variations of maximum daily values of signal durations (B) of "fall-and-flow" earthquakes that were measured at two short-period seismic stations EZV5 and EZV6, situated at a distance of 4 and 7 km, respectively, from the crater of Volcán de Colima. The measurements that were made at the two stations vary similarly. By comparing the variations in seismic characteristics with the variations in lava emission, shown in Fig. 12.5A, was concluded that the variations in the maximum daily durations of "fall-and-flow" earthquakes generally correlated with the variations in the rate of lava emission.

Fig. 12.6 demonstrates that a correlation between the maximum (Fig. 12.6A) and the cumulative (Fig. 12.6B) daily durations of the

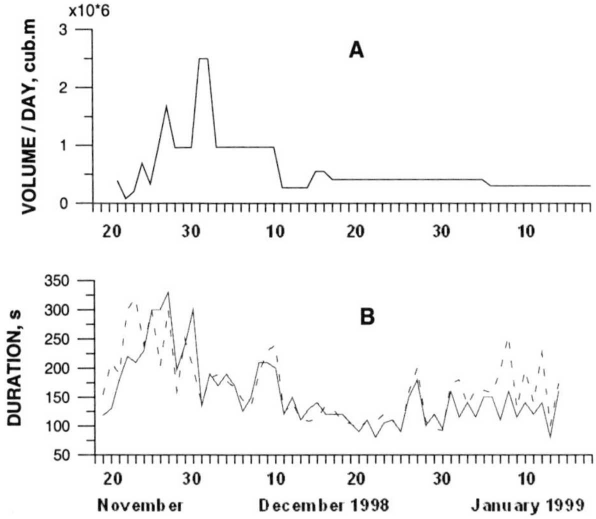


Fig. 12.5. Daily variations in the volume of lava emission (A) and the maximum duration (B) of "fall-and-flow" earthquakes records during the 1998-1999 lava eruption at Volcán de Colima, México. Data of station EZV5 are shown by solid lines; the data of station EZV6 are shown by dashed lines.

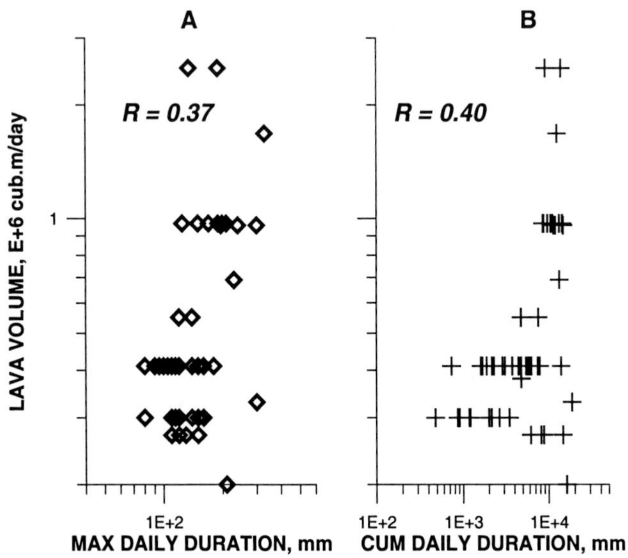


Fig. 12.6. Relationship between the daily maximum (A) and cumulative (B) amplitude of "fall-and-flow" earthquakes recorded at the short-period station EZV5 and the daily output of lava during the 1998-1999 lava eruption at Volcán de Colima, México.

seismic signal at seismic station EVZ5 and the volume of daily lava emission is significant at 99% confidence level.

*Duration-frequency relationship of "fall-and-flow" earthquakes at different stages of lava emission.* Gutenberg and Richter (1956) showed that the relation between the magnitude of seismic events and their frequency might characterize the physical conditions of material fracturing. Supposing that "fall-and-flow" earthquakes are products of the destruction of lava bodies, we construct the magnitude-frequency plots for three stages of lava eruption in Fig. 12.7. We use the durations of seismic signals as a magnitude parameter of these seismic events.

The distribution of events with different signal durations at stage I of the intensive emission of lava is the most representative with durations from 100 to 330 s (durations less than 100 s are not considered here as they are not representative). A log-log plot of the duration intervals *versus* the number of events of this duration gives a good linear dependence between the parameters. It indicates a power-law distribution of seismic signals related to the "fall-and-flow" earthquakes. This relation is typical for fractal sets (Scholz, 1990) and indicates the self-similarity in the dimensions of the pieces that were broken off during the gravitational destruction of the lava body.

During the second and third stages of lava emission, the durations of seismic signals of "fall-and-flow" earthquakes were more limited. At stage II, the maximum duration did not exceed 250 s, and at the third stage, it was not longer than 200 s. This may reflect a decrease of maximum volume of lava body pieces that were broken off during these stages.

*Relationship between the magnitude of "fall-and-flow" earthquakes and the volume of pyroclastic flows.* The results of the above sections suggest that the duration of short-period records of the "fall-and-flow" earthquakes may provide a measure to quantify these earthquakes. The following magnitude scale was proposed according to the durations of "fall-and-flow" earthquakes measured at a short-period station EVZ5 situated at a distance of 7 km from the crater: magnitude 1 for the events with durations of seismic signal less than 100 s; magnitude 2 for events with durations between 100 and 200 s; and magnitude 3 for events with durations greater than 200 s (Fig. 12.8). The events of magnitude 1 can be attributed mainly to rockfalls; the magnitudes 2 and 3 events may be associated with pyroclastic flows and large rockfalls.

Self-similarity in the dimensions of the pieces that were broken off during the gravitational destruction of lava body and the positive dependence between signal durations and lava volume output allow us

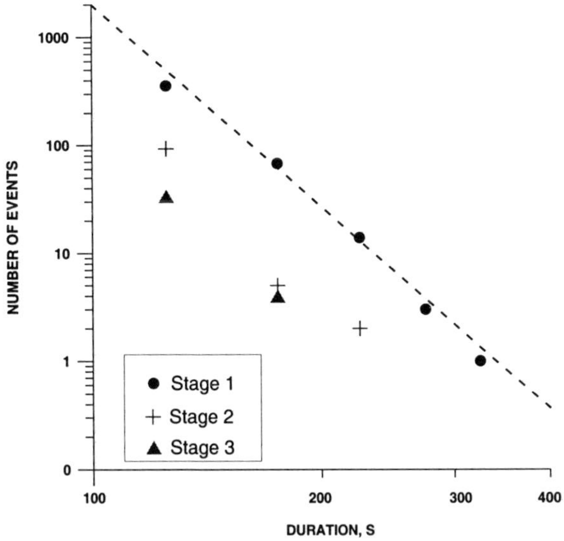


Fig. 12.7. Number of "fall-and-flow" earthquakes *vs* the durations of seismic signal of these earthquakes. The events that were observed at different stages of the eruption are shown by different symbols. The least-square regression line *log N = 15.7 - 6.2 log D (s)* for stage I is shown. Here N is a number of events.

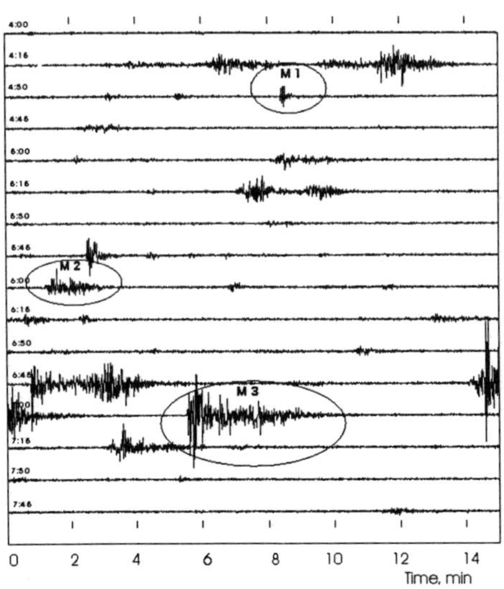


Fig. 12.8. Seismic "fall-and-flow" signals recorded during the 1998-1999 lava eruption at Volcán de Colima, México. The records of the short-period seismic station at a distance of 4 km from the crater are shown. M1, M2 and M3 denote the records of "fall-and-flow" earthquakes of magnitude 1, 2, and 3.

to estimate the mean values of the material that was transported by pyroclastic flows of different magnitude.

The volume of pyroclastic flows that occurred between November 25 and 26, 1998 ($7.2\times10^5$ $m^3$) and of the total volume of pyroclastic flows that occurred during the eruption in November 1998 – January 1999 ($1.6 \times 10^6$ $m^3$) were estimated by Navarro-Ochoa et al. (2002) and by Saucedo et al. (2002), respectively. Thus the approximate mean volume of flows corresponding to the seismic records of different durations was estimated. For 25 and 26 of November, three events of magnitude 3, 123 events of magnitude 2, and 219 events of magnitude 1 were recorded. For the total period of the eruption, four events of magnitude 3, 577 event of magnitude 2, and 4939 events of magnitude 1 were recorded. Field observations showed that the magnitude 1 events correspond to the rockfalls of the mean volume of 50 $m^3$. According to these data, a system of two linear equations was prepared:

$$577V_2\ (m^3) + 4V_3\ (m^3) = 1.6\times10^6\ (m^3) - 50\ (m^3)\times4939 \qquad (12.1)$$
$$123V_2\ (m^3) + 3V_3\ (m^3) = 7.2\times10^5\ (m^3) - 50\ (m^3)\times219 \qquad (12.2)$$

Here $V_2$ and $V_3$ are the mean volumes of material corresponding to the pyroclastic flows of magnitudes 2 and 3, respectively. From these two equations, mean volumes of $1.4\times10^5$ $m^3$ and $1.3\times10^3$ $m^3$ obtained for events of magnitude 3 and a volume of as a mean value for the events of magnitude 2, respectively.

These values are very approximate but may be useful for monitoring an eruption. They are in agreement with the estimates of the volume of pyroclastic flows made at Unzen volcano, Japan ($0.6\times10^4$ to $5.7\times10^4$ $m^3$; Uhira et al., 1994; Yamasato, 1997) and Soufriére Hills volcano, Montserrat (small pyroclastic flows, $10^3$ – $10^4$ $m^3$; medium-sized pyroclastic flows, $10^4$ – $10^6$ $m^3$; Calder et al., 2002).

### 12.3.2. Relationship between the amplitude of the long-period records of "fall-and-flow" earthquakes and the volume of pyroclastic flows

Uhira et al. (1994) showed that the maximum amplitude of long-period seismic records Amax produced by pyroclastic flows is proportional to the volume of the blocks fallen from the dome V (Fig. 12. 9). The records were made at a distance of about 4 km from the lava dome.

I calculated the least-square regression between these two parameters based on Fig. 22 from Uhira et al. (1994):

$$V\ (10^5\ m^3) = 0.22\ \mathrm{Amax}\ (\mathrm{micron}) + 0.06 \qquad (12.3)$$

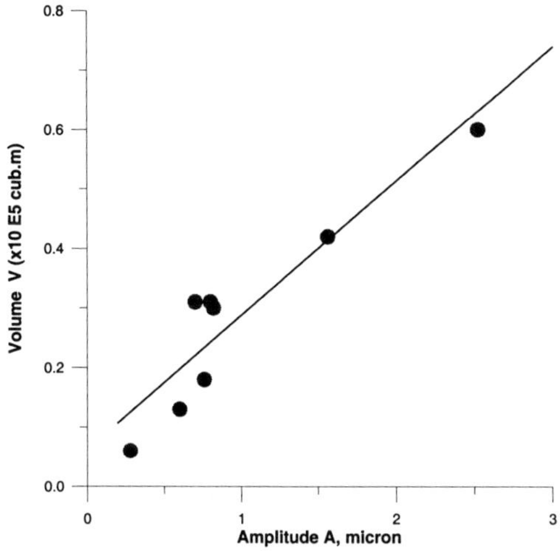


Fig. 12.9. Relationship between the volume of pyroclastic flows and the maximum amplitude of long-period records (vertical component) of the earthquakes produced by pyroclastic flows at Unzen volcano, Japan in 1991. Seismic station was situated at a distance of about 4 km from the crater. From Uhira et al. (1994).

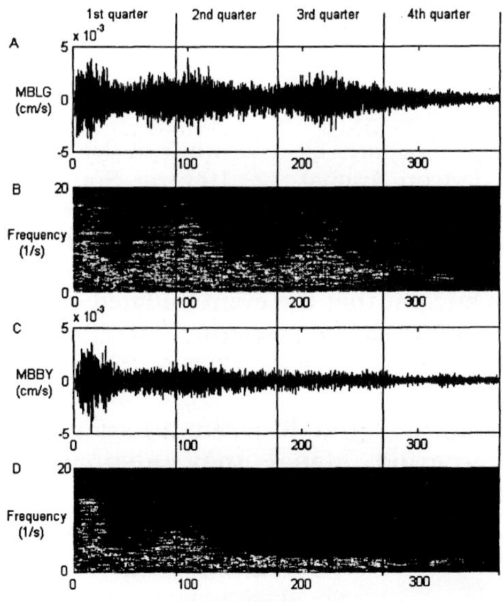


Fig. 12.10. Seismograms (A and C) and spectrograms (B and D) of pyroclastic flow recorded on August 12, 1999 at Soufriére Hills volcano, Montserrat at the distance of about 2 km NE from the dome (station MBLG) and about 2 km SW from the dome (station MBBY). From Jolly et al. (2002).

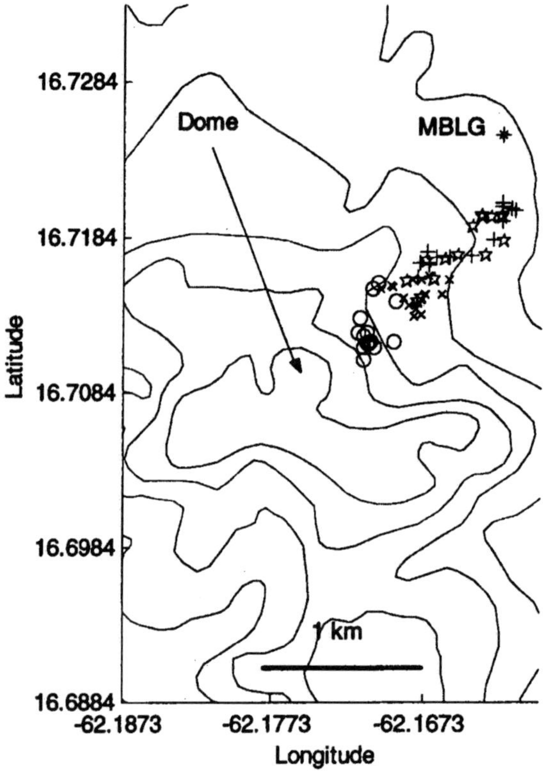


Fig. 12.11. Amplitude-based locations by a network of five seismic stations for the pyroclastic flow recorded on August 12, 1999 at Soufriére Hills volcano, Montserrat. The event is divided into four time intervals to show event location through time (See Fig. 12.10). The first time quarter is designated by circles, the second by crosses, the third by plus signs, and the fourth by stars. The amplitude-based locations suggest that the event initiated on the southern part of the dome and subsequently propagated down the Tar River Valley. From Jolly et al. (2002).

The correlation coefficient is equal to 0.93; therefore, the maximum amplitude of the long-period signal may also be used for the quantification of pyroclastic flows.

### *12.4. Location pyroclastic flows using amplitude signals of "fall-and-flow" earthquakes*

The amplitude signals of "fall-and-flow" earthquakes recorded by a high dynamical range seismograph network may be used for locating of the trajectory of movement of pyroclastic flows and rockfalls (Yamasato,

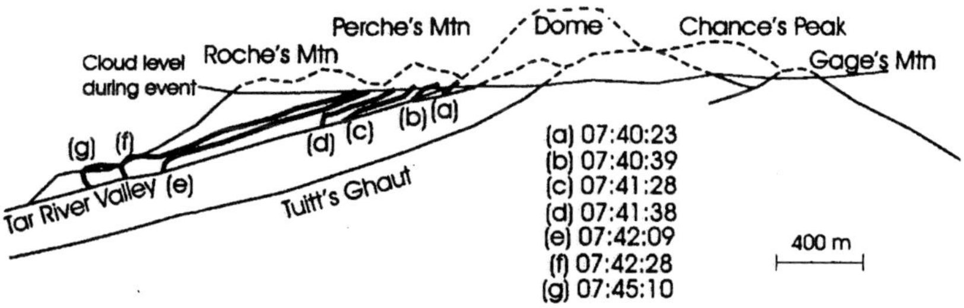


Fig. 12.12. Schematic diagram extracted from video footage recorded on August 12, 1999 at Soufriére Hills volcano, Montserrat, showing pyroclastic flow location through time. The view of the video camera is to the north from the dome. From Jolly et al. (2002).

1997; Jolly et al., 2002). Fig. 12.10 shows the typical seismograms and spectrograms of "fall-and-flow" earthquake recorded at Soufriére Hills volcano, Montserrat during the movement of the pyroclastic flow.

The signal observed at one or more seismic stations records the onset that initiates at a single location, following by a multiple moving sources that might be spread over a large area. The amplitudes of the totality of sources can be exploited to determine an average location for the event. Jolly et al. (2002) used the spectral amplitudes of the signal at frequency 7-9 Hz for the location of moving sources of pyroclastic flow. The seismic record was divided into four quarters (Fig. 12.10), and the locations were made for every quarter of seismic signal correlated from station to station. Fig. 12.11 shows their locations made for the pyroclastic flow of August 12, 1999 at Soufriére Hills volcano. It is seen that the pyroclastic flow propagated from the dome to NE and passed the distance about 1.4 km.

The comparison with the video camera record of the pyroclastic flow movement (Fig. 12.12) strongly supports the seismic amplitude-based reconstruction of the movement of pyroclastic flow.

# Chapter 13
# Explosion Earthquakes

Explosion earthquakes are directly related to volcanic events, they occur with volcanic explosions: Strombolian or Vesuvian. Each explosion produces an earthquake; the seismic record is commonly accompanied by the acoustic signal. Therefore, the monitoring of explosion earthquakes allows the monitoring or reconstruction of explosive volcanic activity.

## *13.1. Waveforms of explosion earthquakes*

Three-component broadband displacement seismograms of the explosion earthquake recorded at Sakurajima volcano, Japan (Kawakatsu et al., 1992) are shown in Fig. 13.1. These records show a simple seismic signal accompanied by an impulse of air wave. The nature of this impulse is problematic within a seismogram, but it is clear when the simultaneous recording of seismic and acoustic waves is possible.

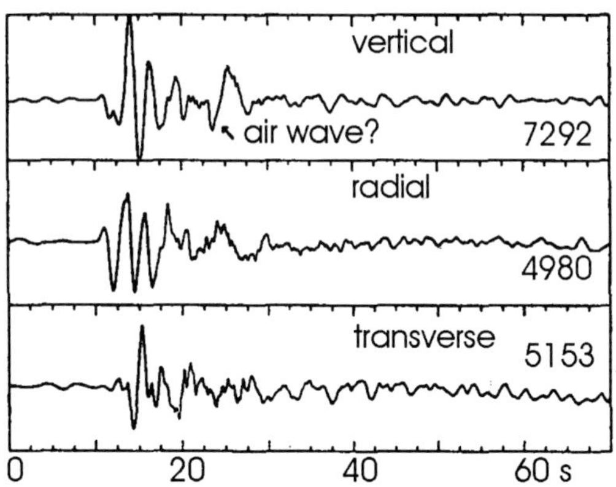


Fig. 13.1. Three-component broadband displacement seismograms of the explosion earthquake recorded at Sakurajima volcano, Japan. Number in the lower-right corner of each box indicates the peak-to-peak amplitude given in a unit of $10^{-8}$ m. The horizontal components are rotated taking the active Minami-dake crater of the volcano as the origin. From Kawakatsu et al. (1992).

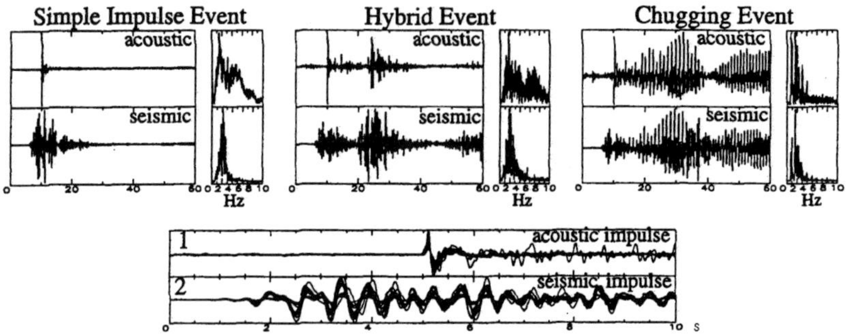


Fig. 13.1 Three types of explosion earthquakes recorded at Karymsky volcano, Kamchatka at a distance of about 2 km from the crater during the 1997 activity: simple impulse event, hybrid event and chugging event. The acoustic and broadband seismic traces and frequency spectra are shown for each event. From Johnson et al. (1998).

Fig. 13.2 shows the three types seismic and acoustic waveforms of explosion earthquakes that were recorded at Karymsky volcano, Kamchatka (Johnson et al., 1998). For one half of the explosions, simple impulse events were recorded. For 10% to 20% of the events, the explosion was characterized by a 1 to 3 minute interval of low-frequency volcanic tremor. These signals were accompanied by audible, steam-engine-like exhalations, and using the terminology introduced by Benouit and McNutt (1997), they were named as "chugging events". Other long-duration explosions, hybrid events, contain a combination of higher frequencies and non-harmonic tremor.

A seismo-acoustic study of the explosion earthquakes that occurred at the andesitic Tokachi-dake volcano, Hokkaido (Okada et al., 1990) showed that the amplitudes of the seismic signals did not correlate with the amplitudes of the air waves (Fig. 13.3). The explosions accompanied by pyroclastic flows were characterized by larger seismic amplitudes with a similar intensity of the air waves.

The seismic records of the explosion earthquakes are very similar (Johnson et al., 1998; Rowe et al., 2000). Fig. 13.4 shows the records of lava lake Strombolian explosions of Erebus volcano, Antarctica. Strong repeatability of the gross waveform characteristics is evident among events up to 20 s into the coda, demonstrating a repetitive source process. A high similarity in the scaled seismic waveforms for initial explosions at the same station was observed at Karymsky volcano. Johnson et al. (1998) believe that the initial source process and location were the same for all Karymsky explosive events. The complication of seismic records with the hybrid and chugging events may reflect the

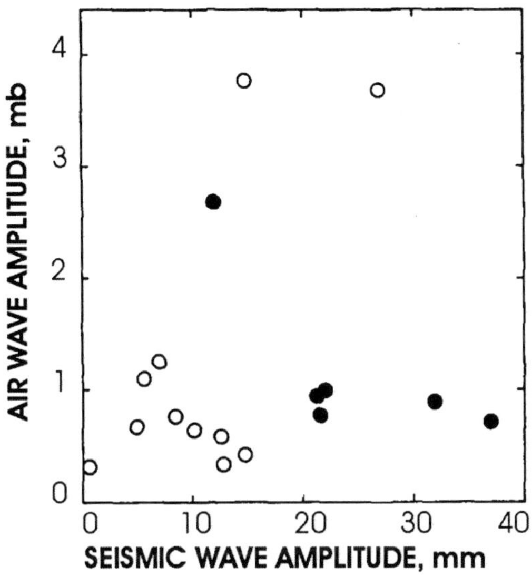


Fig. 13.3. Relationship between the amplitudes of air waves and seismic waves that were produced by explosion earthquakes at Tokachi-dake volcano, Hokkaido in 1988-1989. The amplitude of seismic waves is peak-to-peak maximum amplitude recorded by a short-period seismograph at a distance of about 7 km from the crater; the amplitude of the air wave was measured at a distance of about 3 km from the crater. The filled circles indicate the explosions accompanied by pyroclastic flows. From Okada et al. (1990).

type of extended degassing process during the explosion.

*13.2. Occurrences of explosion earthquakes and their location*

Explosion earthquakes occur just beneath the crater floor. Tokarev (1989) wrote that the explosion earthquakes at Karymsky volcano, Kamchatka occurred at a depth of less than 100 m below the crater. Ishihara (1990) located the explosion earthquakes at Sakurajima volcano, Japan within a tube-like narrow zone extending beneath the summit crater from sea level to a depth of 2 km. The area of the epicenters was elongated in an east-west direction and corresponds to the configuration of the two active vents.

The depth of explosion earthquakes recorded at Stromboli volcano, Italy, was determined using a broadband waveform inversion. The observed waveforms were well reproduced by the inversion and the two source centroids that fit the data were offset 220 and 260 m beneath, and about 100 m north-west of the active vents (Chouet et al., 2002).

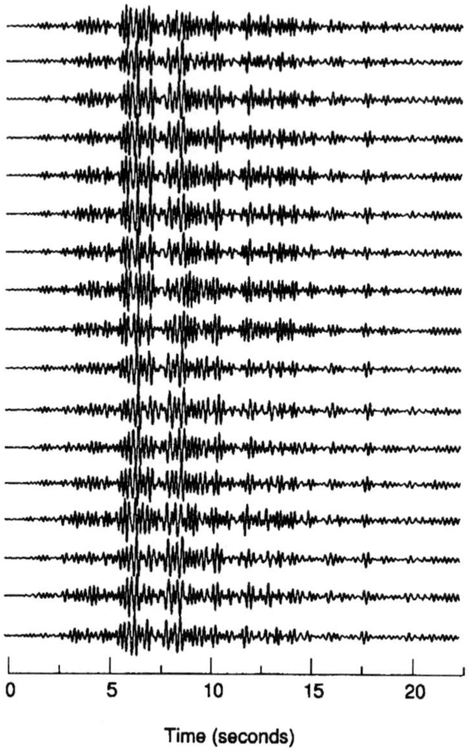


Fig. 13.4. Illustration of the similarity of seismic records for explosion earthquakes. Seventeen lava lake Strombolian explosions of Erebus volcano, Antarctica produced these short-period seismograms of explosion earthquakes that were recorded at a distance of 10 km from the lava lake. Traces have been bandpass filtered from 1 to 7 Hz to reduce incoherent signals. From Rowe et al. (2000).

Explosion earthquakes may occur as a single event or as long sequences. Tokarev and Firstov (1967) studied seismic activity of Karymsky volcano in 1965, when small explosions of Vulcanian type occurred accompanied by a slow emission of dacitic lava. During three weeks of observation, 727 explosion earthquakes were recorded.

Volcanic activity at Mount Erebus, Antarctica is dominated by eruptive activity within a summit lava lake. Common eruption styles range from passive degassing to Strombolian explosions, which typically occur several times daily, and occasionally in swarms of up to 900 per day (Rowe et al., 2000).

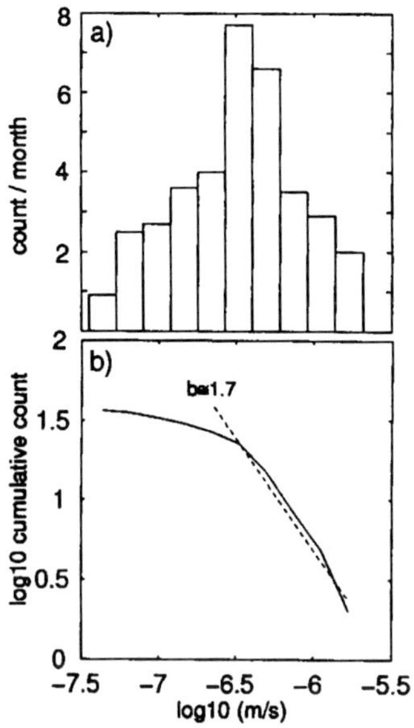


Fig. 13.5. Size (maximum ground velocity) distribution of 364 explosions of Erebus volcano, Antarctica recorded between October 1997 and July 1998 at a distance of 10 km from the lava lake. A, histogram of explosion counts versus $\log_{10}$ of seismic amplitude, representative of explosion seismic magnitude. B, $\log_{10}$ of monthly cumulative counts (for each size bin and higher) versus $\log_{10}$ seismic amplitude. From Rowe et al. (2000).

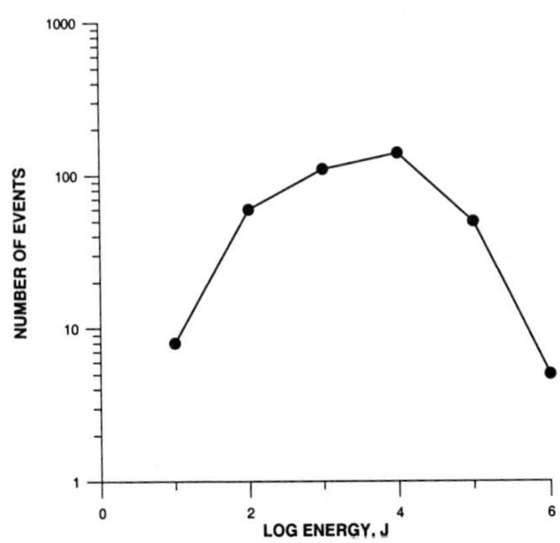


Fig. 13.6. Magnitude (Energy in J) – frequency dependence for explosion earthquakes of Karymsky volcano. Data from Tokarev and Firstov (1967).

Magnitude-frequency relationships for explosion earthquakes differ from the lineal regression common for volcano-tectonic earthquakes. Fig. 13.5 shows the size (maximum ground velocity) distribution of 364 explosions of Erebus volcano, Antarctica (Rowe et al., 2000). Note that sharp bend in the cumulative curve of Fig. 13.5B, corresponding to the peak in the histogram of Fig.13.5A. For 727 Karymsky explosion earthquakes, the magnitude (energy) - frequency relation has a more clear parabolic form (Fig. 13.6). Tokarev and Firstov (1967) consider that the maximum in Fig. 13.6 corresponds to the optimal energy for explosions in the crater, according to the properties of the magmatic agents and conduit dimensions ("critical magma radius"; Nishimura, 1995); all remaining events represent the deviations from this optimal case.

*13.3. Explosion earthquakes in the eruptive process*

Fig. 13.7A shows the 10-year variations in number of explosion earthquakes at Karymsky volcano. The mean rate of lava emission during each eruptive episode is shown in Fig. 13.7B. It is possible to see that the episodes with lava emission are characterized by an increase in the frequency of explosive earthquakes. At the same time, no dependence between the volume of lava emission and the maximum number of seismic events was observed. These earthquakes were observed also without lava emissions (Khrenov et al., 1982).

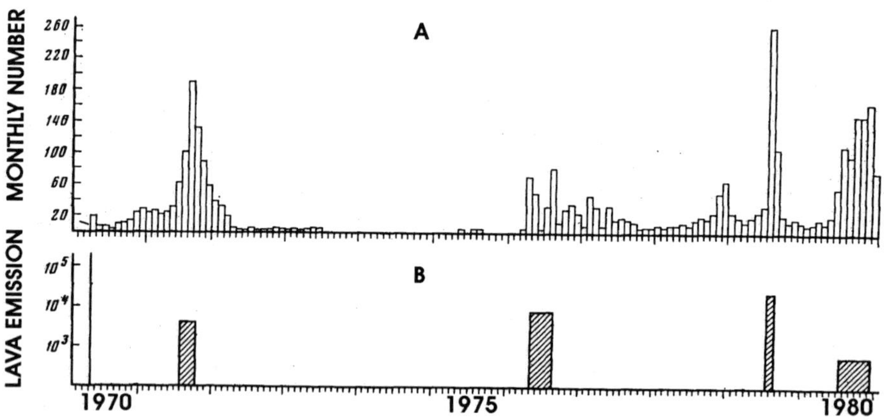


Fig. 13.7. Variations in the monthly number of explosion earthquakes (A) and in the mean rate of lava emission (in kg/s) during eruptive episodes of Karymsky volcano in 1970-1980. From Khrenov et al. (1982).

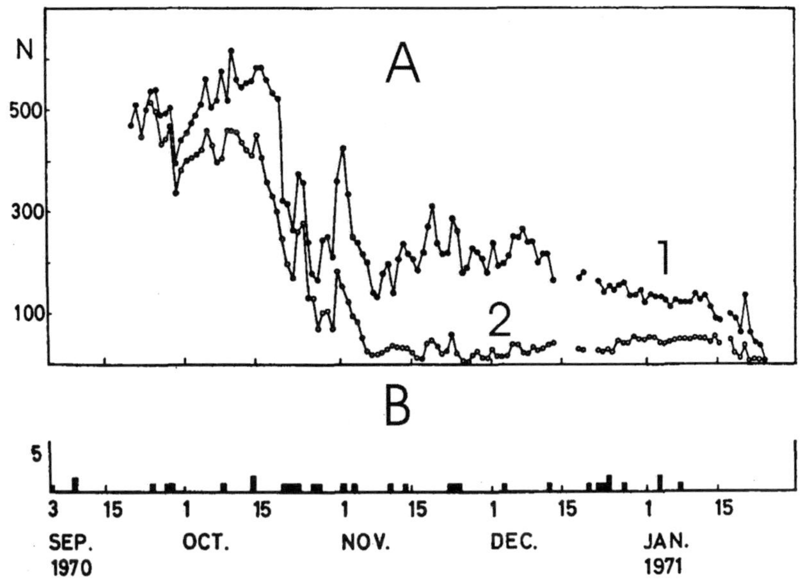


Fig. 13.8. Daily number of explosion earthquakes (A) and volcano-tectonic earthquakes with S-P times of less than 3.0 s (B) recorded during the eruption of Akita-Komaga-take volcano, Japan at the seismic station situated at a distance of about 4 km from the crater. Open circles (curve 2) represent the number of explosion earthquakes with a maximum amplitude larger than 200 μkine. From Tanaka et al. (1972).

Tanaka et al. (1972) studied the explosion earthquakes that occurred during the 1970-1971 eruption of Akita-Komaga-take volcano, Japan. The eruption began with a sequence of Strombolian explosions and basaltic lava emission on September 18, 1970 and continued until January 25, 1971. About 33 000 explosion earthquakes were recorded. They were the main seismic signals for this eruption. Single volcano-tectonic earthquakes (Fig. 13.8B) were very innumerous. Fig. 13.8 shows the variation in the number of explosion earthquakes that were observed from the very beginning of the eruption. These earthquakes were directly connected with volcanic explosive activity and may serve for monitoring of this type of eruption.

### *13.4. Size of explosion earthquakes*

The model of a single force acting at the moment of an explosion

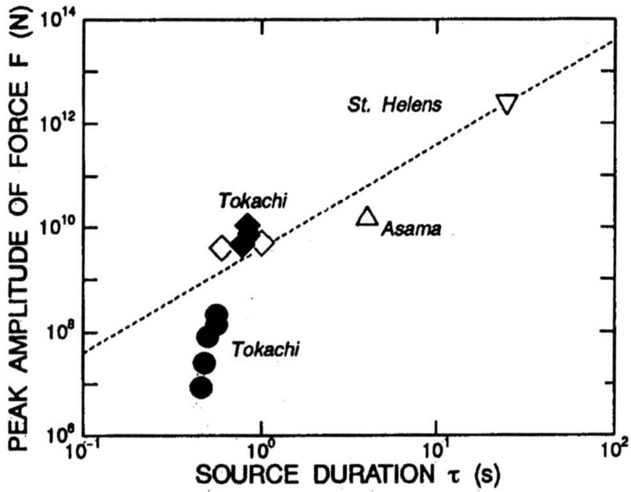


Fig. 13.9. Relationship between the peak amplitude F and the source duration τ of the single force for small explosion earthquakes of Tokachi volcano, Japan (solid circles) and large explosion earthquakes of Tokachi volcano, Asama volcano, Japan, and Mt. St. Helens, Cascades (other symbols). From Nishimura (1995).

eruption that was proposed by Kanamori and Given (1982) and developed by Nishimura and Hamaguchi (1993) lead to the determination of the magnitude of the counter force of eruption (explosion). The values of this force vary from $2.6\times10^{12}$ N for the great May 18, 1980 explosion at Mt. St. Helens, Cascades (Kanamori et al., 1984) to $10^7$ N for small explosions at Mt. Tokachi, Japan (Nishimura, 1995). Fig. 13.9 shows the scaling law for explosion (eruption) earthquakes (Nishimura, 1995). Here the peak amplitude of the single counter force F is plotted versus the source duration τ of the single force. It is interesting to see that the scaling law for large events ($F>10^9$ N) differs from the law for smaller events. For large events, the empirical law

$$\text{Log } F \text{ (N)} = 2 \text{ Log } \tau \text{ (s)} + 9.6 \qquad (13.1)$$

was obtained (Nishimura and Hamaguchi, 1993). For smaller events, the dependence of the peak force upon the source duration is too weak; τ is almost a constant value. Nishimura (1995) considers that this difference is caused by the difference in a "critical magma radius" of the hypothetical spherical gas-rich magma batch that produces the volcanic explosion.

# Chapter 14
# Seismic Monitoring of Volcanic Activity and Prediction of Volcanic Eruptions

Chapters 5 to 8 of this book described and discussed the properties of volcano-tectonic earthquakes, and demonstrated the temporal and spatial regularities in the earthquake sequence development before volcanic eruptions. It allows the forecasting of volcanic eruptions using seismic data. This chapter discusses the special aspects of seismic monitoring of volcanoes and the methodology of prediction of volcanic eruptions.

## *14.1. Methodology of seismic monitoring of volcanic activity*

Seismic monitoring of volcanic activity serves to forecast an eruption and to issue an alert of a volcanic hazard to the public (Shimozuru, 1971; McNutt, 1996; 2000b; McNutt et al., 2000). Therefore, the system of monitoring must be capable of precisely locating seismic events in the region of a volcano and able to quickly process the obtained information. To resolve this problem, a seismic network has to consist of a minimum of three to five seismic stations, with a spacing of a few kilometers and they need to be installed within 1 to 10 km from the active crater. Digital and analog seismic records are usually telemetered to a center of monitoring for processing.

A typical seismic network around an active volcano, Volcán de Colima, México is shown in Fig. 14.1. It consists of five short-period vertical-component seismic stations distributed at a distance from 1.3 km to 7.0 km and covering a broad azimuthal range. The stations are situated within the altitude range from 1700 m to 4000 m; it is possible to locate earthquakes of magnitude 0 to 4, which occur within the edifice of this volcano. The center of telemetrical monitoring is situated in Colima city at a distance of about 30 km from the stations. The analog data are transmitted by radio to the center where they recorded on drums and stored as digital time series (Castellanos and Jiménez, 1995). This seismic network is of minimal capability; the destruction of any seismic stations during an eruption could stop the monitoring. The absence of three-component stations lowers the precision of locations.

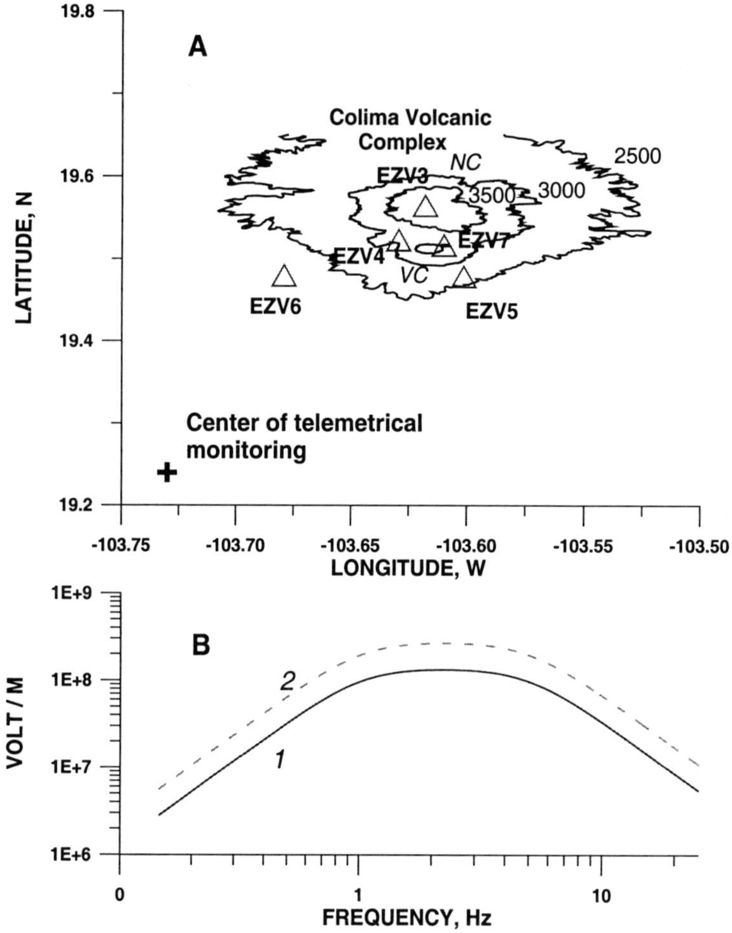


Fig. 14.1. The position of seismic stations (A) and their amplitude responses (B). The stations EZV3 to EZV7 are shown as triangles; the Center of telemetrical monitoring is shown by cross. Shown are the contour lines for 2500 m, 3000 m, and 3500 m. VC is Volcán de Colima; NC is Nevado de Colima. Lines labeled 1 and 2 in B are two types of seismic station responses. From Zobin et al. (2002b).

A good illustration of the importance of three component stations and the use of both P- and S-waves for volcanic earthquake location is shown in Fig. 14.2. The 40 best-constrained earthquakes recorded at Etna volcano, Sicily are relocated in the absence of some essential data (Ferrucci, 1995). The seismic network consists of 13 stations, 11 of them are of three components. Location of events by all stations (Fig.

14.2B) constrains seismicity to within a focal volume of less than 12 $km^2$ and confines it between sea level and a depth of 2 km. In Fig. 14.2C, the seismic cluster is shown after excluding of the two three-component stations nearest to the crater, that were situated above the seismically active volume. It causes a dramatic spreading of the epicenters towards the southeast and an increase in the depth interval. In Fig. 14.2D, all stations are included but only P-wave data have been used; the result being only a weak effect on the epicentral distribution, but it strongly increases the depth interval.

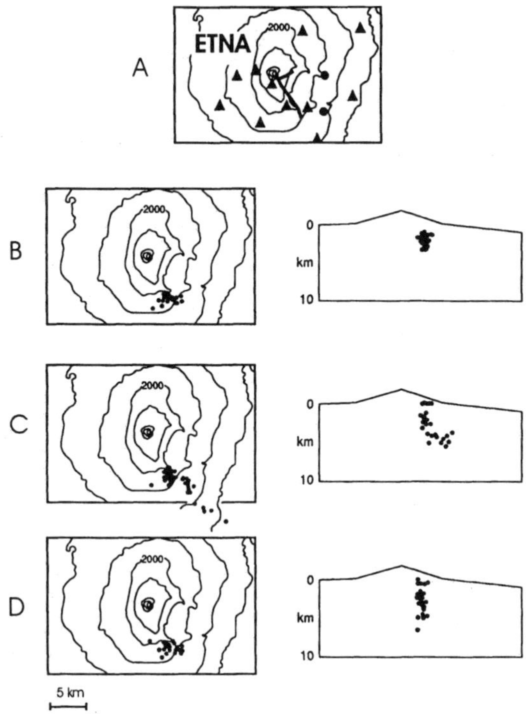


Fig. 14.2. Illustrations of the importance of the use of three-component stations in seismological routine and the adequate positioning of the stations according to the seismic volume. (A) seismic network (triangles, three-component stations; circles, one-component stations); (B) epicenters (left) and hypocenters (right) computed using all the P- and S-wave data; (C) epicenters (left) and hypocenters (right) computed without the two three-component stations nearest to the crater that were situated above the seismically active volume; and (D) epicenters (left) and hypocenters (right) computed using only the P-wave data. From Ferrucci (1995).

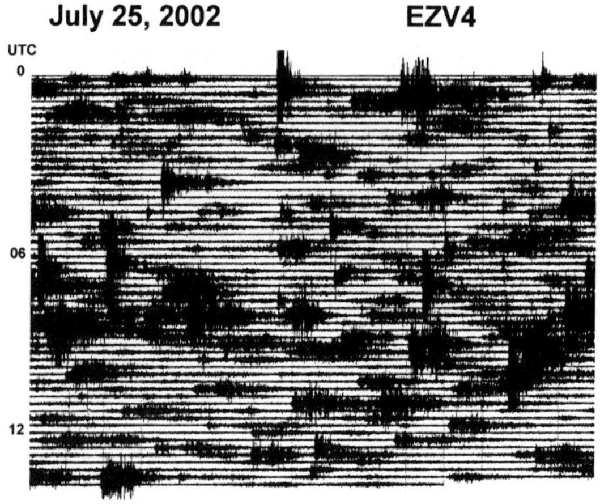


Fig. 14.3. A seismogram recorded at a distance of 1.7 km from the crater of Volcán de Colima, México during the lava eruption of 2002. The numerous seismic signals were produced by rockfalls and pyroclastic flows. Low-amplitude volcanic tremor is also seen. Courtesy of Colima Volcano Observatory.

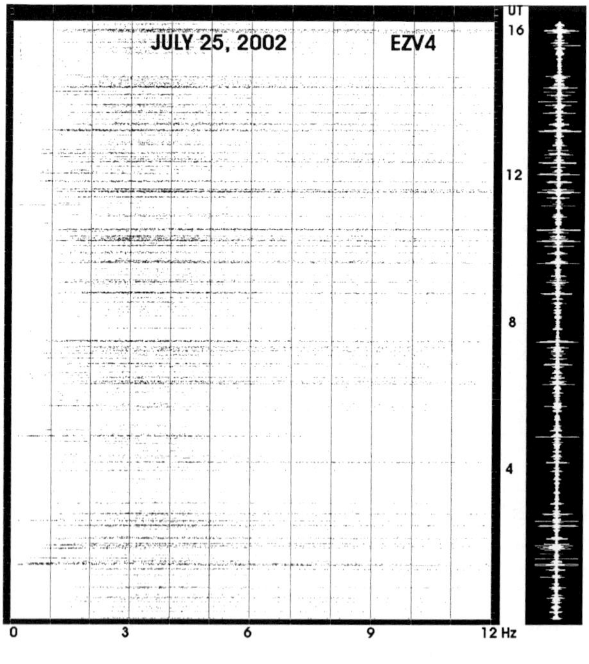


Fig. 14.4. A spectrogram of the same record as in Fig. 14.3. Courtesy of Colima Volcano Observatory.

Recently, many seismic networks at volcanoes began to be equipped with digital broadband seismometers. The first results show a greater potential for these stations for detailed investigations of volcanic earthquakes (Chouet, 1996b; Aster et al., 2000).

The processing of seismic data is realized in real-time and after wards the earthquakes are localized. For real-time processing, the digital and drum-recorded seismograms and spectrograms are used, as well as the compact presentation of seismic information with the RSAM (Endo and Murray, 1991) or RSEM (De la Cruz-Reyna and Reyes, 2001) methodologies. Figs. 14.3 to 14.5 show these three types of presentation of seismic information. A seismogram of volcanic earthquakes in an active period (Fig. 14.3) gives the general impression of the different types of the seismic signals associated with high-frequency or low-frequency volcano-tectonic earthquakes, explosions, tremor or rockfalls. At the same time, it can be overfilled by a sequence of numerous and saturated signals. The spectrogram (Fig. 14.4) helps to separate low- and high-frequency events, to see the temporal variations of these signals, as well as the period of volcanic tremor.

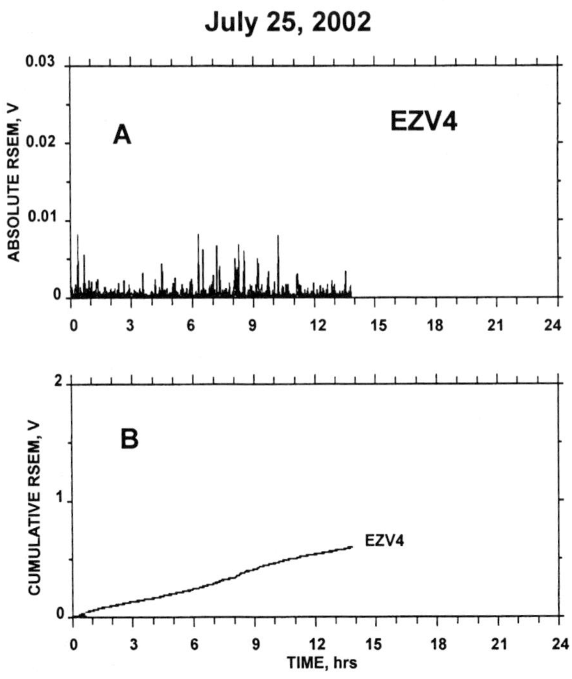


Fig. 14.5. The 1-min average RSEM values of the seismic signals of the same record as in Fig. 14.3. A, the variations of absolute values of RSEM; B, the cumulative curve. Courtesy of Colima Volcano Observatory.

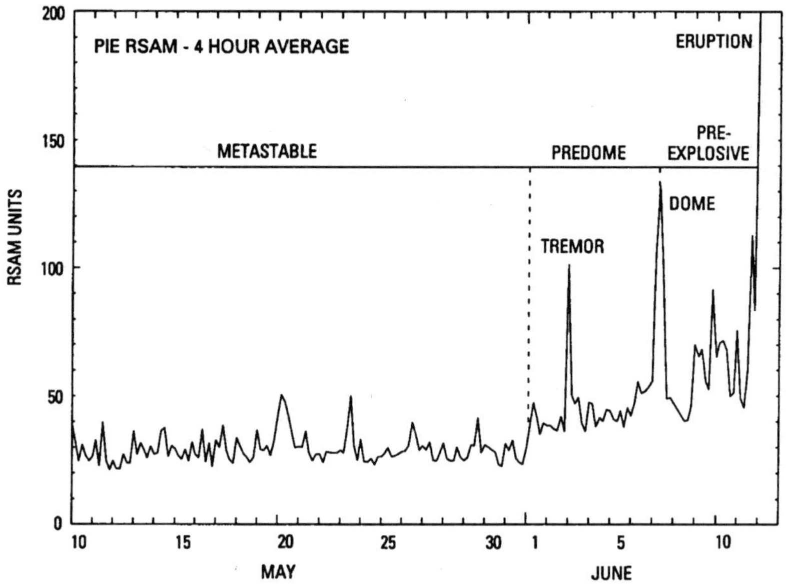


Fig. 14.6. The 4-hour average RSAM values showing the marked division between the metastable and preexplosive buildup seismic phases of the 1991 eruption of Pinatubo volcano, Philippines (See Chapter 7). From Harlow et al. (1996).

The RSEM (real-time energy measurement) (Fig. 14.5) or RSAM (real-time amplitude measurement) (Fig. 14.6) curves give the characteristics of strain release associated with seismic activity. The RSAM (RSEM) continuously samples the absolute amplitude (or square amplitude) of seismic signal; the average of each signal is obtained summing the measurements made in a certain time interval, typically 1-10 min, and dividing by the number of measurements (or unit time). These data can be represented in absolute (Fig. 14.5A) or cumulative (Fig. 14.5B) forms and can be used for recognizing anomalies in the temporal behavior of seismic activity.

For more detailed analysis of the situation, we need the maps and cross-sections of the focal distribution of the earthquakes. Fig. 14.7 illustrates these characteristics of seismic regime for Volcán de Colima in May and June of 2001, during the initial stage of lava dome growth. The events that occurred in May and June are marked by different symbols. This map and cross-section show that the June events became shallower and may reflect the intensification in lava dome growth.

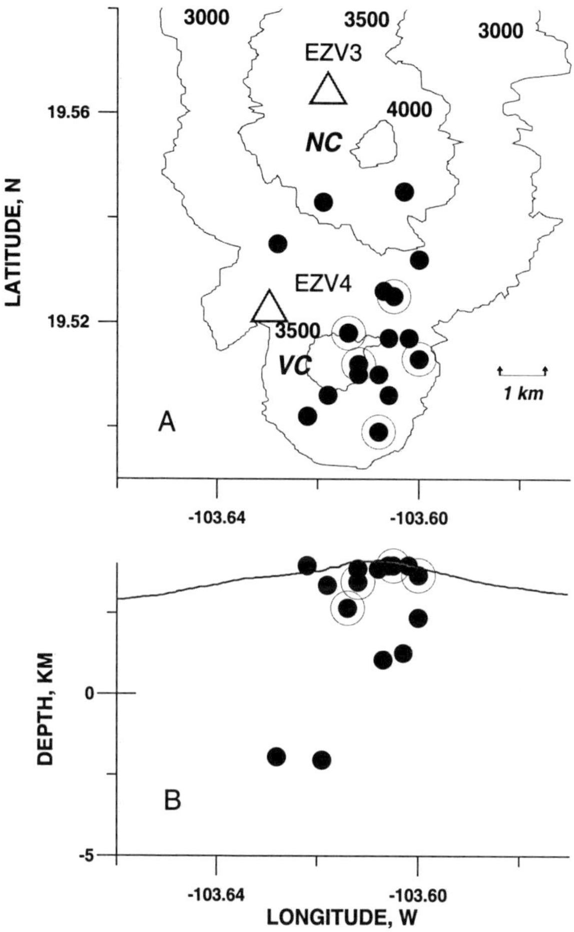


Fig. 14.7. The epicenters (A) and hypocenters (B) of small volcano-tectonic earthquakes located at Volcán de Colima, México in May (filled circles) and June 2001 (stars) during the initial stage of lava dome growth. The stations EZV3 and EZV4 are shown as triangles. The contour lines of 3000 m, 3500 m and 4000 m are shown. VC is Volcán de Colima; NC is Nevado de Colima.

## *14.2. Applications of volcanic seismicity in the prediction of volcanic eruptions*

The case studies of Chapters 4-6 show that volcano-tectonic seismicity occurs before the majority of volcanic eruptions, indicating the site of the forthcoming event. F. Omori had written in May 1911, after the first historical instrumental study of seismic and volcanic activity during the 1910 Usu, Hokkaido eruption: "I believe the problem of prediction of great volcanic eruption is, in some cases, not very difficult" (Omori, 1911-1913). He wrote later (Omori, 1914): "Unlike large destructive earthquakes, which originate along seismic zones, but are not repeated from one and the same center, the eruptions of a given volcano take place usually from the central crater or from new side vents, being anyhow thus localized to the mountain itself. Hence, it may be that the great outbursts from a given volcano at widely different epochs resemble each other more or less, both in the eruptive phenomena and in the precursory events".

The prediction of any event consists of two main elements: place and time of event appearance. Omori pointed out an advantage in predicting of volcanic eruption: we know the place of the event. Therefore, in short, the forecasting of volcanic eruption is based on a simple fact that with a sharp increase of seismic activity near a volcano, we may have an eruption. The most effective forecasting of volcanic eruptions has had no special methodology, more important has been the human factor of experience.

The July 6, 1975 eruption of New Tolbachik volcanoes in Kamchatka was predicted officially, with the publication of the date and time of eruption in the regional newspaper on July 5 (Tokarev, 1978). This prediction was based on unusual seismic activity in an unusual place. The Tolbachik Valley, a region to the SW from Plosky Tolbachik volcano, where the swarm of volcano-tectonic earthquakes was localized, has been calm at least from 1960. Therefore, the appearance of a seismic swarm could nave been a signal of possible eruption. The swarm continued for two weeks giving time for the analysis of the situation and to take the correct decision. As a result of the great experience of Dr. Pavel Tokarev, the forecasting of the forthcoming eruption was made based upon the similarity in seismic activity with that before the flank eruptions of Klyuchevskoy volcano. No people live within the Tolbachik Valley; no evacuation was made. But the scientific importance of this prediction was very high. The first expedition of Institute of Volcanology arrived at the site of the future eruption on July 1; their camp was only 400-500 m from the place where a fissure opened on July 6. A few minutes after the opening the fissure, the eruption was being observed by the volcanologists.

The prediction of the June 15, 1991 large explosions of Mount Pinatubo, Philippines was carried out also according mainly to a continuous increase in seismic activity beneath the volcano, after moderate manifestations of volcanic activity (Harlow et al., 1996). In April-May, seismic activity was recorded mainly within a cluster located about 5 km NW of the volcano, and it was only 50% probable that a large eruption may occur. During the first week of June, the seismic activity beneath the volcano strongly increased, and the lava extrusion occurred on June 7. From June 8 to 12, the majority of earthquakes occurred beneath the new dome; the tremor steadily increased from June 9. This seismic data together with the observations of an increase in $SO_2$ flux from the volcano allowed the forecasting of the possibility of an increase in volcanic activity. This information was communicated to the officials, and the communities close to volcano, as well as nonessential personnel from Clark Air Base, were evacuated on June 10, two days before the first explosive eruption. The continuous increase in seismic activity beneath the volcano, including long-period and shallow volcano-tectonic earthquakes, and volcanic tremor showed that a climactic eruption was possible. During the morning of June 15, at 07:30 the monitoring team and the remaining military personnel abandoned Clark Air Base. The climactic eruption began at 13:42. This prediction of the volcanic eruption was realized only because the USGS volcanological team had a great experience in the study of volcanic activity in other volcanic regions. They had no fixed methodology for this case, but they were prepared to try and understand the situation and take a decision.

Therefore, it is difficult to propose a theory for volcanic eruption prediction. At the same time, some empirical methodologies have been developed. The place of eruption may be marked by earthquake swarm epicenters, but the problem of the timing of the event, the onset of the eruption remains. To resolve this problem, two approaches were developed: a study of the variations in volcano-tectonic earthquake parameters and a study of the variations in complex behavior of all types of volcanic earthquakes. The first approach works better before the unrest of a volcano; the second may be important to predict any events during the volcanic unrest. Of course, all these approaches benefit from the use of additional types of monitoring such as deformation, chemical parameters, etc.

### 14.2.1. Methods based on the statistical variations in the parameters of volcano-tectonic earthquakes

Minakami was a pioneer in the development of the seismic statistics method (Minakami, 1960; 1974b). The studies were based on the monitoring of seismo-volcanic activity at the andesitic Asama volcano, Honshu. His famous diagram showing the relationship between the number of B-type volcano-tectonic earthquakes and the daily frequency of volcanic events observed at this volcano (See Fig. 2.2 in Chapter 2) became classic. It was a typical example, in which premonitory earthquakes increased in number before an eruption. Minakami proposed the empirical laws (1960 and 1966) for predicting the Asama

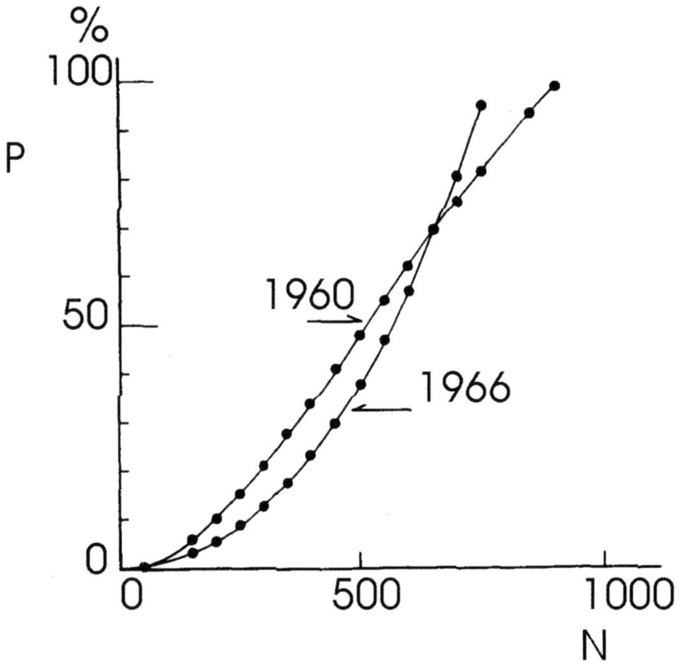


Fig. 14.8. The empirical curves-formulae which were derived in 1960 and 1966 for predicting the Asama eruptions based on the five-day frequencies of B-type earthquakes. P is the probability of eruptions in the next five days; N is the seismic frequency of past five days. From Minakami (1974b).

eruptions based on the five-day frequencies of B-type earthquakes (Fig. 14.8).

These laws predict that with an increasing number of B-type earthquakes, there is an increasing probability of an eruption. The practical application of the 1960 formulae is shown in Fig. 14.9. The problem of forecasting was dealt with using the seismic frequencies of five successive days, as follow:

$$n_1 + n_2 + n_3 + n_4 + n_5 = N_5 \qquad (14.1)$$
$$n_2 + n_3 + n_4 + n_5 + n_6 = N_6$$
$$n_3 + n_4 + n_5 + n_6 + n_7 = N_7$$
$$\cdots\cdots\cdots\cdots\cdots\cdots$$

where $n_1$, $n_2$, $n_3$, etc. indicate seismic daily frequencies $F_2$ of earthquakes and $N_5$, $N_6$, $N_7$, etc. indicate seismic frequencies N of five successive days. So if the value of N on one day is, for example, 800, the curve 1960 of Fig. 14.8 gives the probability P of an eruption within the following 5 days as 0.85, showing a very dangerous state of the volcano (Minakami, 1974b).

Tokarev has developed this methodology for the prediction of volcanic eruptions in Kamchatka using, not only the earthquake frequency, but also seismic strain release (Tokarev, 1963; 1971; 1978; 1983; 1985). For andesitic volcanoes, he proposed the hyperbolic law of cumulative strain release ε:

$$\varepsilon_t = (at + b) / (t - c) \qquad (14.2)$$

where a, b, and c are the coefficients for the given eruption and the time t is expressed in days. Eruptions occur in a period of time Tu ≈ c. Fig. 14.10 illustrates this methodology. Coefficient c that was large at the initial stage of earthquake swarm, strongly diminished with the approach of the eruption. In practice the date of the forthcoming eruption was estimated on the fortieth day of activity, one week before the eruption.

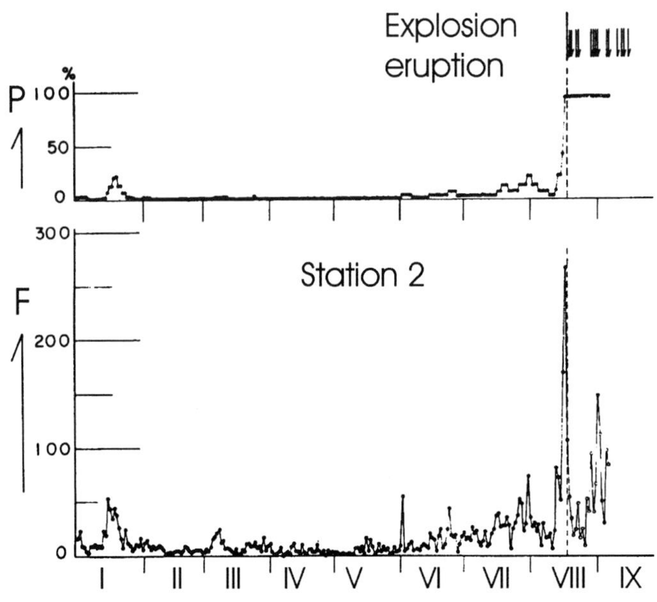


Fig. 14.9. Practical application of the 1960 formula to eruptions that took place in 1961, based on the daily frequency of B-type earthquakes. Arrows indicate the eruptions; P is the probability of eruption given from the 1960 formula; F2 is the daily frequency of B-type earthquakes originating from Asama. From Minakami (1974b).

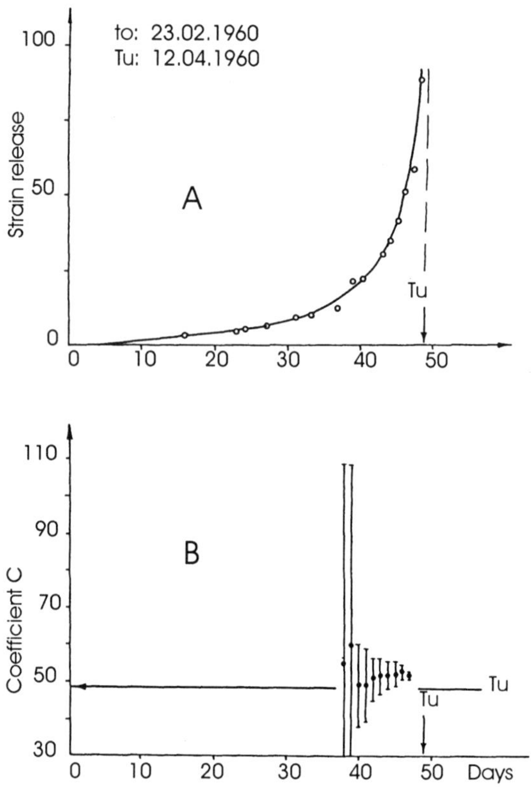


Fig. 14.10. Practical application of Tokarev's method for forecasting eruptions of andesitic volcanoes. The variations of cumulative strain release (A) and coefficient C (B) before the April 12, 1960 Bezymianny volcano, Kamchatka eruption are shown. Here $t_o$ is the time of the first earthquake in the swarm. The strain release is expressed in units of $[10^3]xJ^{1/2}$; c is expressed in days. Arrows show the time of beginning of the eruption Tu. From Tokarev (1983).

Recently, this statistical methodology was strongly improved by including the material failure model (FFM) as a theoretical basis for the forecasting of an eruption (Voight, 1988; Voight and Cornelius, 1991; Cornelius and Voight, 1994; 1995). This FFM method is based on the empirical law describing the failure of materials. It is proposed that the acceleration (the second derivative) of any variable of state $\Omega$,which could be a possible precursor such as seismicity, deformation, or gas emission, is proportional to the rate (the first derivative) of this variable to the order of $\alpha$ (equation 14.3):

$$\Omega'' = A x \Omega'^{\alpha} \tag{14.3}$$

The empirical coefficient of proportionality A and the constant $\alpha$ are dimensionless. For $\alpha > 1$, there is a time of singularity, or the time at which the rate theoretically becomes infinite, and is therefore an upper bound to the time of failure (or the onset of the eruption , in our case).

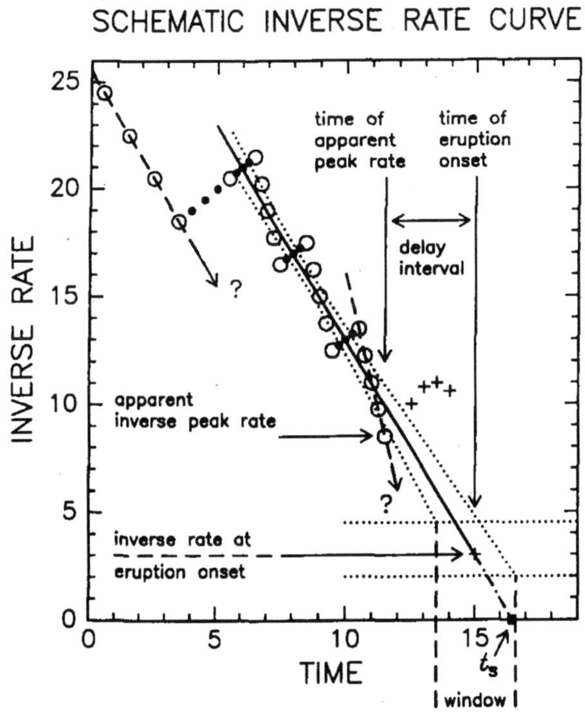


Fig. 14.11. Schematic inverse rate curve. Open circles define inverse downward (accelerating) rate trends, solid dots decelerating jogs. Decelerating jogs may cyclically alternate with accelerating jogs (dashes and arrows), which may be steeper than the overall trend (heavy solid line). Data points beyond the apparent peak rate were excluded from analysis (crosses). Inverse rate trends can be fitted numerically and a data envelope (curved lines of small dots) can be constructed at a specified confidence level. The intersection of the data envelope with a range of expected rates at time of eruption (horizontal lines of small dots), as derived from previous eruptions, may be used to define an eruption window (vertically dashed lines). Alternatively, the time of eruption may be defined in terms of an expected apparent peak rate, plus a delay interval. Time of singularity ts is defined by the intersection of the curve fit with the abscissa, implying infinite rate (square). From Cornelius and Voight (1994).

The graphical technique of forecasting is based on a plot of the inverse rate of the variable of state versus time (Fig. 14.11). The cumulative values of RSAM or SSAM as well as cumulative coda duration or seismic amplitudes can be used as a seismic variable. When the cumulative seismic variable begins to increase sharply, the inverse rate of variable goes to zero. The estimation of a time of the onset of an eruption is based upon successive estimations of the apparent failure times from the approximation of the inverse rate curve according to equation (14.4):

$$(\Omega''^{*})^{\beta} = Ct + D \qquad (14.4)$$

where $\beta = 1 - \alpha$; C is a slope of the inverse rate curve; and D is the intercept. The time of singularity, i.e., the occurrence of an infinite rate, is found by extending the inverse rate data toward the time axis. A finite peak rate at the time of failure may be assumed, or estimated from experience from earlier eruptions. This finite rate, plotted as a horizontal line in the inverse rate plot, can be used for the intersection with the extending trend of the inverse rate. The interval between the apparent time of failure and time of eruption is termed delay interval.

Fig. 14.12 (A-D) shows the applications of the method for three types of seismic variables. These seismic data were observed before the October 22 (Julian day 295.5), 1986 eruptive event at the lava dome of Mount St. Helens. The inverse rates for all variables (coda duration (A), seismic wave amplitude (B) and RSAM (D)) were fitted to Eq. 14.4 using the linearized least-squares technique. The $\alpha$-values derived for the inverse rate curves are comparable (1.87 for coda accumulation; 1.67 for amplitude accumulation; and 1.38 for RSAM).

The rates used for the curve fitting of A and B (solid dots in A and B) were calculated over time periods of two days, as daily accumulations. Calculating the rates over shorter time periods will usually increase data scatter but is feasible when rate becomes larger. Examples (circles) in Fig. 14.12 (A and B) are for accumulations per day calculated over periods of one day between October 14 and 19 (Julian days 288 to 293). Distinct downward trends in the inverse plot are thus recognizable 2.5 days before the extrusion. Eruption began around day 295.5. The prediction windows were $294.6 < 294.9 < 296.1$ for A curve and $295.0 < 297.2 < 301.6$ for B curve. The prediction window is based on 97.5% confidence level for linearized fits according to FFM.

For the RSAM system (Fig. 14.12C and D), the data were arbitrary averaged over three-hour periods. The data window chosen for the RSAM analysis ends at the same time as one data window used for the analysis of coda and amplitude. The prediction window for RSAM is $296.3 < 296.8 < 297.6$.

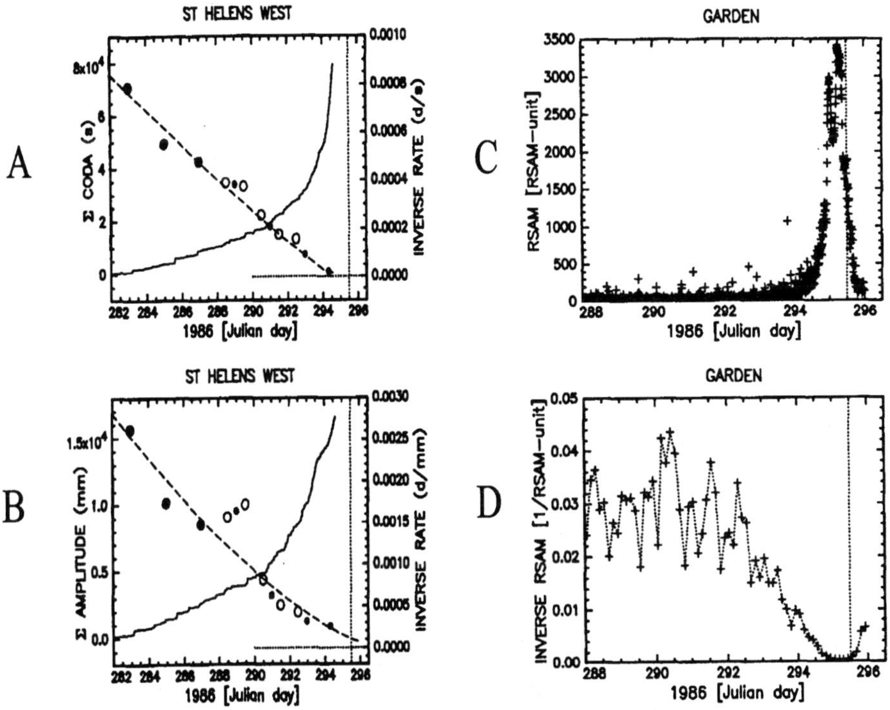


Fig. 14.12. Illustration of the FFM method applied to eruption prediction (The October 22, 1986 Mt. St. Helens volcano lava dome extrusion). Plots of cumulative coda duration (A), cumulative amplitude (B); and cumulative RSAM (C) are shown with their inverse rate curves (A, B and D, respectively). Dashed lines in A and B show the fitted curves. ST HELENS WEST and GARDEN are seismic stations of the Cascade Volcano Observatory. Solid dots are the rates calculated over period of two days; open circles are the mixed data set of rates calculated over periods of two and one days. Dotted vertical lines mark onset of extrusion. From Cornelius and Voight (1995).

These methodologies work if the eruption occurs at the moment of maximum seismic strain release. The cases described in Chapters 3 to 5 show that this type of activity is common for andesitic and dacitic volcanoes but it is rare for basaltic volcanoes when the eruption may begin after the cease of seismic activity.

### 14.2.2. Methods based on a study of the variations in complex behavior of different types of volcanic earthquakes

These methods are usually used for forecasting an increase in eruptive activity during volcanic unrest. It is assumed that the magma is close to the surface; no significant new magmatic uplift can be expected. Now the variations in intensity and duration of volcanic tremor, a change in spectral content of volcano-tectonic earthquakes, or an increase in the number of degassing (explosion) earthquakes may imply a reviving of the subsurface magmatic body and a forthcoming eruptive event. That is more art than science and strongly depends on personal experience.

*1975. Tolbachik fissure eruption, Kamchatka* (See Chapter 5). The appearance of the third cone along the fissure was predicted (Tokarev et al., 1984). On August 15, the number of small volcano-tectonic earthquakes sharply decreased from 110 on August 14 to only 10. At 11:15 a.m. volcanic tremor stopped. In the evening of August 16, it was proposed at the field expedition meeting that in one day a new cone may appear. This forecast was reported at the press conference on August 17 at 10:00 a.m. for journalists, which arrived to the region of the eruption. The eruption began at 21:12 p.m. and the journalists had the possibility to witness the birth of a new volcanic cone.

*1980. Mt. St. Helens, Cascades* (See Chapter 7). Four secondary magmatic eruptions were forecast using information about harmonic tremor and shallow volcano-tectonic earthquakes (Malone et al., 1981). The first of them occurred on June 12. On June 3 an increase in volcanic tremor was observed that was followed by a sudden quiet period; by June 8 the harmonic tremor had died out completely. In mid-afternoon on July 12, harmonic tremor was again observed at low level with a following increase in amplitude. About 19:00 that evening there was a large increase in tremor amplitude that lasted for at least 40 min, followed by a very quiet period. This time (about 20:00) the forecasting of an eruption was carried out by U.S.G.S. The major eruption began at 21:09. The similar tremor precursor was later observed before the August 7 eruption.

The July 22 eruption was preceded by a group of low-frequency earthquakes with epicenters located within the zone that was active before the great explosion of May 18. These earthquakes began on the morning of July 22, with their number obviously increasing by mid-afternoon and sharply increasing by 16:00. At 17:13 the eruption began without any premonitory harmonic tremor. The October 16-18 eruption was also preceded by the appearance of a few hours of low-frequency earthquakes that began on the background of very low seismic activity.

### 14.2.3. Seismic precursors to small explosions

A high-sensitive seismic network installed very close to the active crater allows recording of seismic precursors to small volcanic explosions. These precursors continue for a few minutes to a few hours

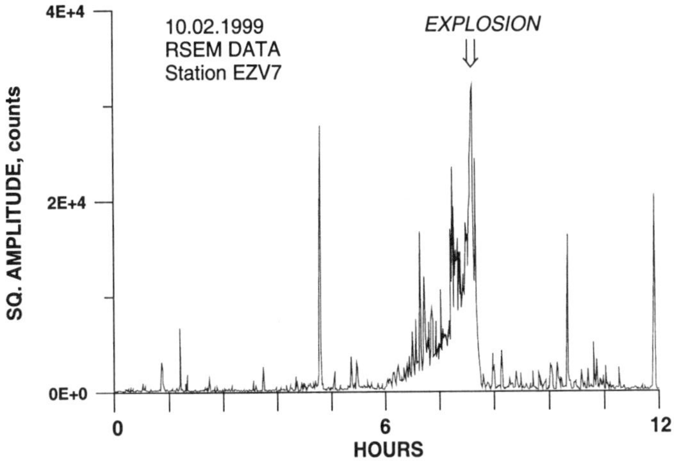


Fig. 14.13. Variations in the square of the amplitudes of seismic signals summarized for each minute (RSEM) before the February 10, 1999 explosion at Volcán de Colima, México. Seismic station EZV7 was situated at a distance of 1.3 km from the crater. From Zobin et al. (2002c).

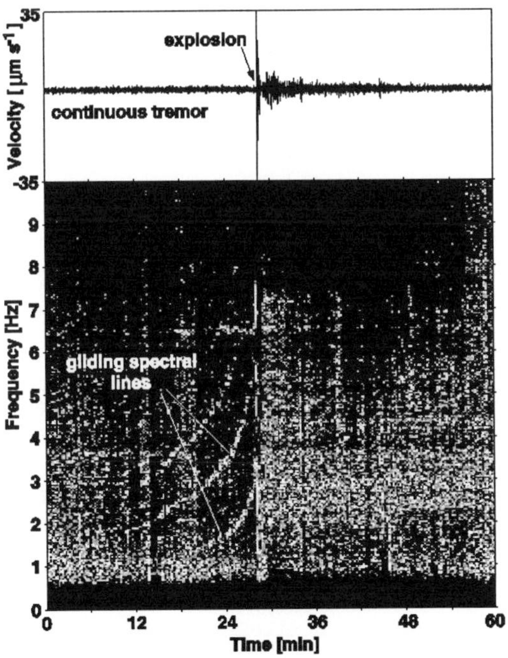


Fig. 14.14. Gliding of the spectral lines on spectrogram before the January 7, 1999 explosion at Soufriére Hills volcano, Montserrat. Seismic station MBRY was situated at a distance of 2 km from the dome. Above, seismogram; Below, spectrogram. Courtesy of J. Neuberg.

but may be used for practical prediction in real time. The first precursor is similar to the usual statistical precursor: there is a sharp increase in the number of microearthquakes just before an explosion. In Fig. 14.13 it is demonstrated for the February 10, 1999 explosion at Volcán de Colima, México (Zobin et al., 2002c). Seismic station EZV7 was situated at a distance of 1.3 km from the crater. The total amplitude of events recorded by this station for each minute began to increase about two hours before the explosion. The explosion occurs at the maximum value of preceding events.

The second type of precursor is a gliding of spectral lines on spectrogram before an explosion. Fig. 14.14 shows this effect that was observed before the January 7, 1999 explosion at Soufriére Hills volcano, Montserrat (Neuberg, 2000). Seismic station MBRY was situated at a distance of 2 km from the dome. The shift in spectral lines at the frequency band from 1 to 5 Hz began about 15 minutes before an explosion.

# Chapter 15
# Some Remarks About Unresolved Problems

This Chapter concludes the book. In fourteen chapters I wrote about things, more or less resolved. Now I want to discuss some points that are not yet resolved, but they are very important for better monitoring of volcanic hazard. There are at least two: (1) the problem of low-frequency volcanic earthquakes and (2) the problem of seismicity at dormant or potential volcanoes.

## *15.1. Variety of low-frequency volcanic earthquakes*

Low-frequency, or long-period volcanic earthquakes are very popular in numerous descriptions of volcanic events. But nobody has yet written clearly, exactly what it is, the low-frequency seismic events at volcanoes. Even the definitions of low-frequency and long-period volcanic earthquakes are different (McNutt, 2000). McNutt (2000) defines long-period events as the "discrete events with very regular, low-frequency (1-5 Hz) waveforms that resonate for many cycles". About low-frequency events, he writes in the same paper that it is the "general term for discrete events with either enriched low frequencies or deficient high frequencies compared with typical earthquakes".

With these definitions, it is impossible to understand the nature of low-frequency earthquakes. In certain papers, low-frequency earthquakes are of tectonic nature; in others, of eruption nature. They may be shallow or deep events. The common opinion is that low-frequency earthquakes are very important for the monitoring of volcanic eruptions, but it is the only common opinion about low-frequency volcanic earthquakes.

Here I wish to discuss this situation at length. I separate low-frequency volcano-tectonic earthquakes and low-frequency eruption earthquakes. Low-frequency volcano-tectonic earthquakes are more or less understandable; low-frequency eruption earthquakes need a better definition of their nature and role in the eruptive process.

### 15.1.1. Low-frequency volcano-tectonic earthquakes

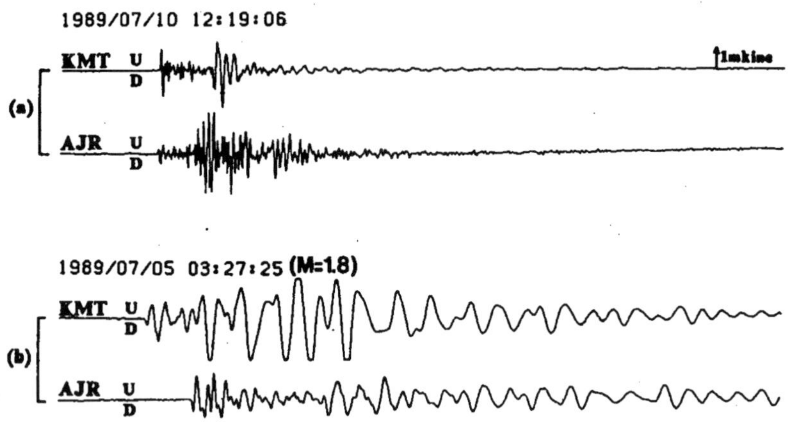


Fig. 15.1. Examples of seismograms of a high-frequency (a) and low-frequency (b) volcano-tectonic earthquakes recorded during the swarm of volcano-tectonic earthquakes preceding the 1989 submarine eruption of Teishi Knoll, Izu islands. From Yamasato et al. (1991).

A change in spectral content of volcano-tectonic earthquakes was noted by Minakami (1960): he defined A-type and B-type volcanic earthquakes as high-frequency and low-frequency events, respectively (See Chapter 2). Minakami showed that the relation in number of A- and B- type events may change during the eruptive process. It was observed at Usu volcano, Hokkaido during the 1943-1945 eruptions with the birth of the new Syowa-Sinzan lava dome. All the volcanic earthquakes that occurred in 1943 before the beginning of the upheaval at the site of the new lava dome were of the A-type. With the beginning of upheaval, B-type events appeared and began to dominate. When the lava dome began to form, the B-type (or low-frequency) events completely took the place of A-type (or high-frequency) earthquakes.

Fig. 15.1 gives examples of short-period records of high- and low-frequency earthquakes preceding the July 13 1989 submarine Teishi Knoll eruption, Japan. The seismic swarm began on June 30 and consisted of high- and low-frequency volcano-tectonic earthquakes (See Chapter 5). Yamasato et al. (1991) showed that the low-frequency earthquake record is similar to that of the high-frequency earthquake and consists of the initial arrivals of the body waves and the surface waves in low-frequency coda. The majority of low-frequency earthquakes occurred at a depth of 0 km while 90% of the hypocenters of high-frequency events were determined within the depth range of 1 to 6 km. The epicenters of the majority of low-frequency events clustered just at the site of forthcoming eruption while the epicenters of high-frequency earthquakes were widely distributed within a 100-km²-area. The majority of low-frequency earthquakes were observed after July 10, during the last 3 days before the beginning of the eruption.

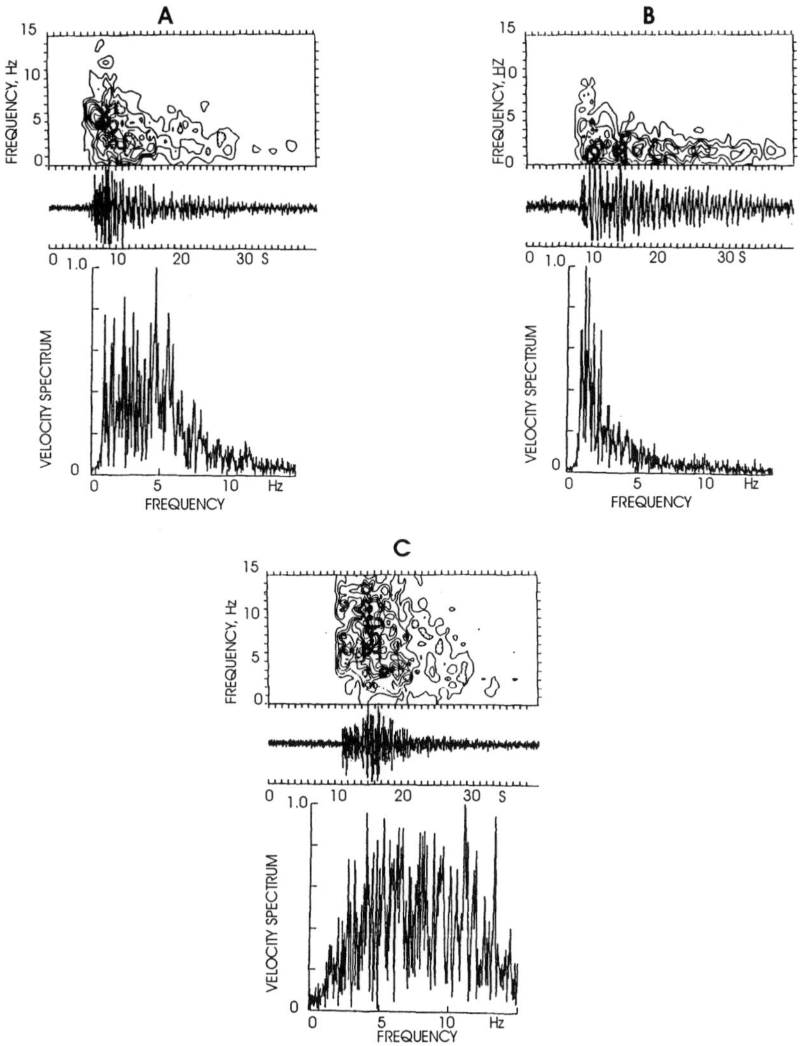


Fig. 15.2. Examples of short-period vertical component seismograms of high-frequency (A), low-frequency (B) and hybrid (C) volcano-tectonic earthquakes that were recorded at Pinatubo volcano, Philippines in August 1992 during dacitic lava dome activity at a distance of about 1 km from the dome. Each seismogram is shown together with a 10 s moving window spectrogram (top) and a whole signal spectrum (bottom). The velocity spectra are normalized. From Ramos et al. (1999).

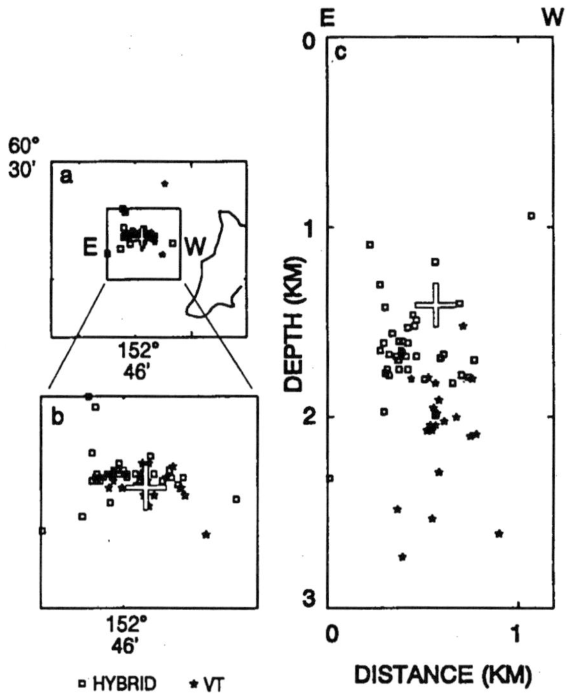


Fig. 15.3. Comparative position of hypocenters of hybrid (squares), high-frequency (stars) and low-frequency (cross) earthquakes with a depth shallower than 3 km and associated with the December 14-15, 1989 eruptions at Redoubt volcano, Alaska. Inner square in (a) indicates region expanded in map (b) and vertical east-west section (c) at the same scale as expanded map. From Lahr et al. (1994).

At Pinatubo volcano, Philippines the low-frequency earthquakes appeared one week prior to the June 1991 cataclysmic eruption, and after more than two months of high-frequency earthquake activity. Later, low-frequency events dominated during the 1992 lava dome growth. They appeared one day before a new dome began to grow and then at least 23 000 low-frequency events accompanied its growth. The low-frequency seismicity ceased with the termination of lava dome growth (Ramos et al., 1999).

Recent studies have revealed the additional, intermediate, type of volcano-tectonic earthquakes: hybrid earthquakes ((Lahr et al., 1994; Ramos et al., 1999; etc). Fig. 15.2 illustrates the differences in the records and spectra of three types of volcano-tectonic earthquakes. The low-frequency earthquake is characterized by the emergent onset of the signal and extended coda in contrast to the distinct P- and S-arrivals and a short coda of the high-frequency earthquake. The high-frequency earthquake spectrum is broad and extends from 2 to 7 Hz; the peaked spectrum of the low-frequency event has a narrower range from 1 to 3 Hz. The hybrid event is more close to the high-frequency earthquake with its largely extended spectrum but has more low-frequency P-waves.

Hybrid earthquakes have been observed at different volcanoes and at different stages of seismo-eruptive activity. They might reflect the processes of magmatic activity at depth, when the high-frequency events exchange in frequency with the low-frequency events. Fig. 15.3 shows the relationship between all three types of volcano-tectonic earthquakes at shallow depths during the 1989-1990 eruptions at Redoubt volcano, Alaska (Lahr et al., 1994). The hybrid events cluster adjacent to the low-frequency events and above the shallowest high-frequency events. The hybrid events are generally confined in occurrence to a brief interval on December 15, 1989, which followed the shallow swarm of high-frequency events associated with the December 15 eruptions at Redoubt. They did not occur in November or December 1-14, when the preliminary seismic activity began.

Therefore, the hybrid earthquakes may serve as an indicator of any change in the spectral content of volcano-tectonic earthquakes: from high-frequency to low-frequency events, and their identification is important in volcanic monitoring.

### 15.1.2. Low-frequency eruption earthquakes

Low-frequency eruption earthquakes are more problematic than volcano-tectonic. Very often it is impossible to say what type of event we have: volcanic tremor or low-frequency event, explosion earthquake or low-frequency event.

McNutt (1992) wrote in his review of volcanic tremor: "volcanic tremor or L(ow) F(requency) events". Lesage et al. (2002) wrote: "L(ong) P(eriod) events and tremor seem to share common source processes and location". "Tremor is closely related to LP events. Distinct types of tremor ... were observed at Galeras volcano with temporal and spectral characteristics similar to those of LP events" (Gil Cruz and Chouet, 1997). Isolated tremors that were observed before the 1989 Teishi Knoll submarine eruption, Izu islands (see Chapter 11) could also be called low-frequency events. A short tremor episode was observed a few minutes before the volcano-tectonic earthquake swarm preceding the 1975 birth of the New Tolbachik volcanoes, Kamchatka (see Chapter 11): what was it – tremor or a low-frequency event?

Lava lake Strombolian explosions at Erebus volcano, Antarctica produced seismic signals. Recorded by a short-period seismograph they are referred to as explosion earthquakes; recorded by a broadband seismograph, they are very long period signals (or low-frequency events) (Rowe et al., 2000). But maybe they are the same, two aspects of the same process?

About 600 deep long-period earthquakes occurred before the 1991 eruption of Pinatubo volcano, Philippine within the depth range of 28 to 35 km beneath the summit. The dominant frequency for most of the events was 2.0 Hz. White (1996) wrote that these seismic signals exhibited clear P- and S-phases. He considers them of the same nature as the signals that were named by Aki and Koyanagi (1981) as deep volcanic tremor (see Chapter 11). So, deep volcanic tremor and deep long-period earthquakes are the same events?

### *15.2. Seismic activity at dormant volcanic structures: A problem of possible eruption*

A volcano can be dormant for thousands of years and then become active again; a new volcano may form. We have a seismic sequence leding up to the eruption. But once in a while we have a seismic sequence in a volcanic zone without any eruptive activity. The problem is: how to recognize if the seismic swarm announces the forthcoming eruption or it is only a seismic crisis without volcanic manifestations?

I give here some well-known cases to illustrate the situation.

#### 15.2.1. Matsushiro swarms

The Matsushiro earthquake swarms (Central Honshu, Japan) began on August 3, 1965 with the earthquakes clustering within a 5-km radius zone around Mt. Minakami that was not among the known active volcanoes of Japan. The swarm activity strongly increased in 1966 (Fig. 15.3) reaching about 7000 small earthquakes per day. During February-June 1966, strong deformations accompanied by ground fissures and spring water were observed. During July-December 1966, the spring water and the expansion of ground fissures caused numerous landslides.

The area of the epicentral zone spread significantly during the first two years of activity (Fig. 15.4). After June 1967, activity quickly dissipated and completely stopped by the end of 1970. More than 700 000 events were recorded during this swarm; more than 60 000 of them were felt. The largest of these earthquakes was of magnitude M 5.4; the total energy of the swarm events was equivalent to the magnitude M 6.4 earthquake. No eruption was observed (Seismic Activity in ..., 1998).

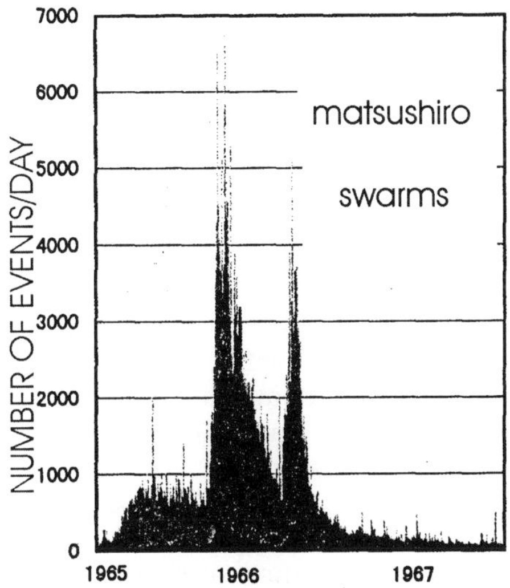


Fig. 15.3. Daily frequency of earthquakes during the Matsushiro earthquake swarm. Data from Nagano Local Meteorological Observatory (Seismic Activity in Japan, 1998).

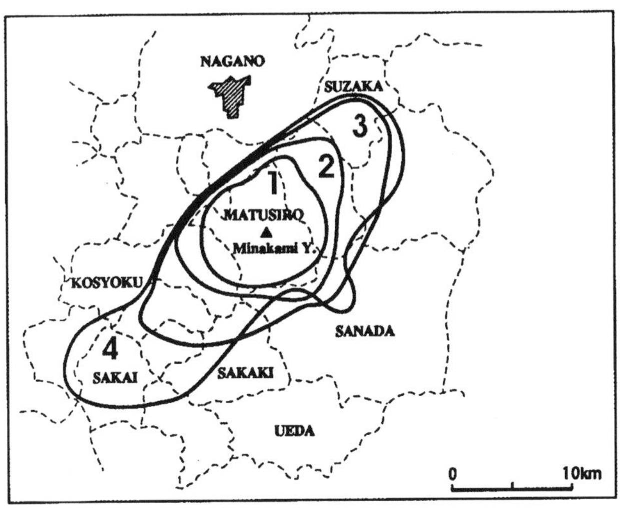


Fig. 15.4. Spreading of the source region during the Matsushiro swarm (1) up to February, 1966; (2) up to July 1966; (3) up to December 1966; and (4) up to May 1997. From Seismic Activity in ... (1998).

15.2.2. Long Valley swarms

The large 17 x 32 km Long Valley caldera (California, USA) formed as a result of the voluminous Bishops Tuff eruption about 730,000 years ago. Resurgent doming in the central part of the caldera occurred shortly afterwards, followed by rhyolitic eruptions from the caldera moat and the eruption of rhyodacite from outer ring fracture vents, ending about 50,000 years ago. The caldera remains thermally active, with many hot springs and fumaroles (Smithsonian Institution, WEB page, 2002).

Seismic swarms began in October 1978 and were accompanied by a broad uplift of the caldera floor and local increase in fumarolic activity. Seismic activity culminated in a sequence of four magnitude M 6 earthquakes that occurred beneath and to the south of the caldera from25 to 27 May 1980. Since a swarm in July 1984, most seismicity has been outside the caldera (Newhall and Dzurisin, 1988).

The majority of seismic events are typical tectonic earthquakes but also the episodes of spasmodic tremor and long-period events were observed (Cramer and McNutt, 1997). Fig. 15.5 shows the temporal characteristics of the 1989 swarm.

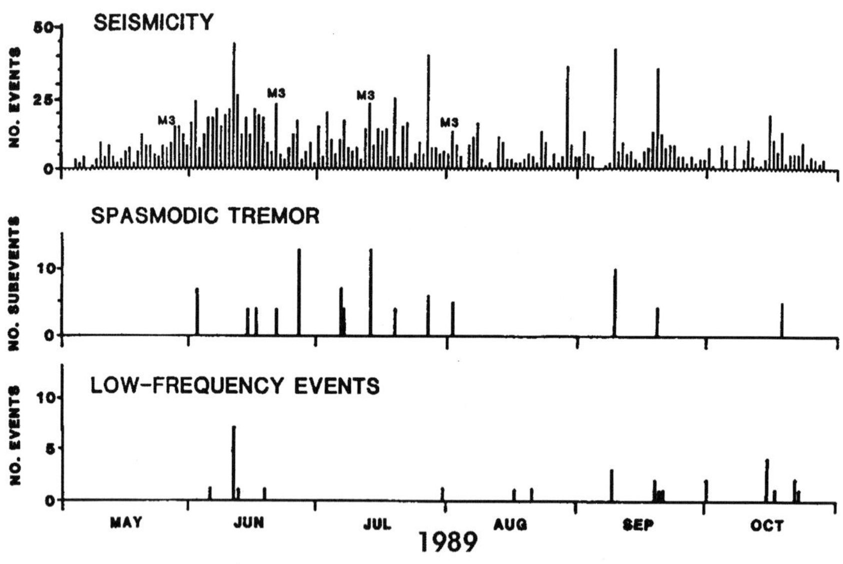


Fig. 15.5. Seismic events recorded during the 1989 swarm beneath Mammoth Mountain near Long Valley caldera, USA. The occurrences of magnitude 3 events are shown. From Cramer and McNutt (1997).

This swarm occurred beneath Mammoth Mountain that is situated at the SW rim of the caldera. It continued for about 6 months; more than 1 500 events of magnitude 3 and less were recorded within the depth range of 2 to 11 km. Only 30 of them were identified as low-frequency events. Fourteen strong bursts of spasmatic tremor were also recorded. No regularities in the occurrence of low-frequency events and tremors were observed, and no volcanic eruption occurred during this period of caldera unrest.

### 15.2.3. Campi Flegrei swarms

Campi Flegrei is a volcanic complex located near Naples, Italy. Its numerous monogenic explosive vents are located within a 13-km-wide caldera formed by multiple collapses. The last eruption was in 1538; it was preceded by uplift that lasted over 30 years. After this eruption, the caldera has been dominantly sinking.

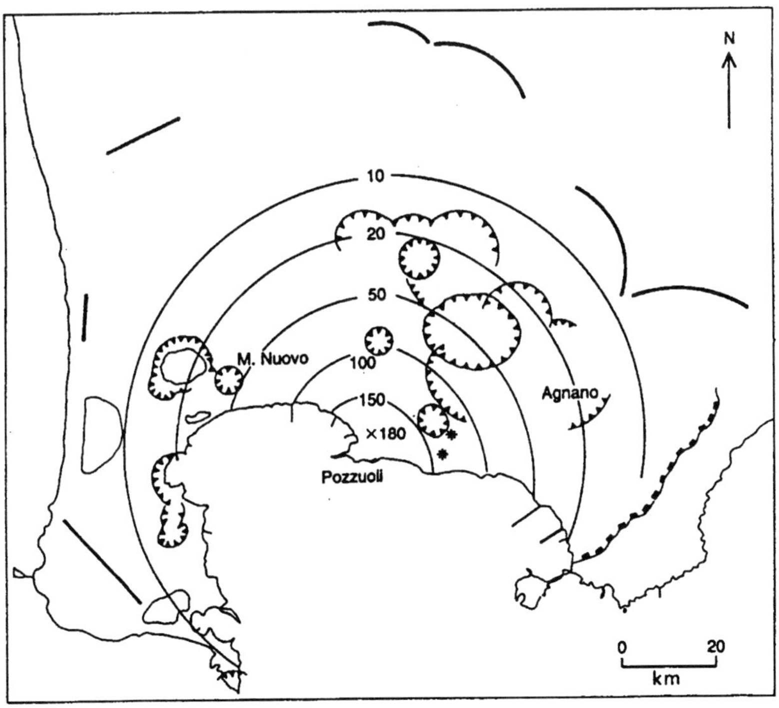


Fig. 15.6. Uplift of Campi Flegrei, Italy during the 1982–1984 unrest. The curves are isolines of vertical uplift in cm. The cross near 180 cm indicates the point of maximum uplift. From Barberi and Carapezza (1996).

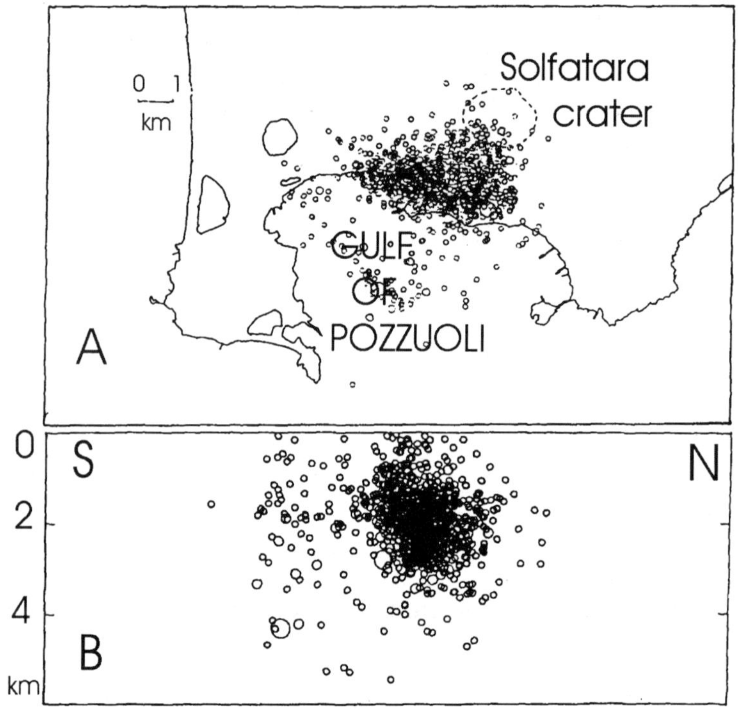


Fig. 15.7. The epicenters (A) and hypocenters (B) of earthquakes located during the 1982-1984 unrest at Campi Fregrei, Italy. From Barberi and Carapezza (1996).

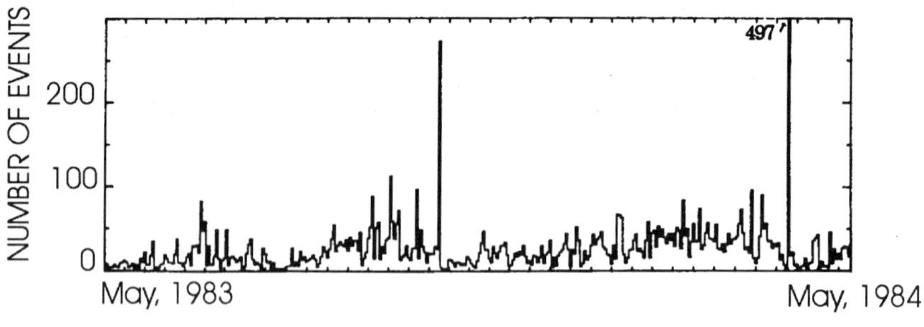


Fig. 15.8. Temporal variation in daily number of earthquakes recorded during the period May 1983 – May 1984 at Campi Fregrei, Italy. From Del Pezzo et al. (1987).

In late 1969, a new period of uplift began near Pozzuoli. It continued for four years and reached 1.5 m. After a short gap, the uplift continued in 1982-1984 with the maximum value of 1.8 m (Fig. 15.6). The uplift was accompanied by numerous small earthquakes (Barberi and Carapezza, 1996). Figs. 15.7 and 15.8 show the temporal and spatial characteristics of the 1982-1984 seismicity. During the period January 1983 to September 1984 more than 10 000 events were recorded, with a magnitude ranging between 0.1 and 4.2. The earthquakes clustered within the depth range of 0 km to 4 km; their epicenters were distributed within the uplift zone. Since January 1985, the area has undergone a phase of general subsidence interrupted by short-duration uplifts. The uplift phases were accompanied by seismic activity while the subsidence phases occur aseismically (Del Pezzo et al., 1987). No volcanic eruption has occurred.

***

As we can see, there is no difference in spatial-temporal distribution between these three cases and the preceding swarms of volcano-tectonic earthquakes described in Chapters 5 to 7. No difference was observed in the shape of the seismic signals; at least, in the short-period field. In the future, we have to search for any interior signs; the statistics of seismic events cannot help us, we need to perform a detailed study of the structure of the seismic signals and their nature. How? I do not know.

The only option available is to install networks of broadband digital seismic stations around those potentially active volcanic zones to discover any very special component of the seismic signal. The same can be done to resolve the problem of low-frequency earthquakes. And, of course, experimental and theoretical research remains very important.

The XX-th century for *Volcanic Seismology* was a century of short-period instruments and the statistics of seismic signals; the new century has to be a century of a broadband seismology and the detailed study of the physics of volcanic earthquakes. For today, it is the only way to mitigate volcanic hazard.

## REFERENCES

Abe, K., 1979, Magnitudes of major volcanic earthquakes of Japan 1901 to 1925, J. Fac. Sci., Hokkaido Univ. Ser. VII, 6, 201-212.

Abe, K., 1992, Seismicity of the caldera-making eruption of Mount Katmai, Alaska in 1912, Bull. Seismol. Soc. Am. 82, 175-191.

Adams, R.D. and R.R. Dibble, 1967, Seismological studies of the Raoul island eruption, 1964, N.Z. J. Geol. Geophys. 19, 1948-1961.

Aki, K., M. Fehler and S. Das, 1977, Source mechanism of volcanic tremor: fluid-driven crack models and their application to the 1963 Kilauea eruption, J. Volcanol. Geotherm. Res. 2, 259-287.

Aki, K. and R. Koyanagi, 1981, Deep volcanic tremor and magma ascent mechanism under Kilauea, Hawaii, J. Geophys. Res. 86, 7095-7109.

Aramaki, S., Ed., 1998, The 1986-1987 Eruption of Izu-Oshima Volcano. Earthquake Research Institute, Tokyo University, Tokyo. 62 p.

Aspinall, W.P., A.D. Miller, L.L. Lynch, J.L. Latchman, R.C. Steward, R.A. White, and J.A. Power, 1998, Soufriére Hills eruption, Montserrat, 1995-1997: volcanic earthquake locations and fault solutions, Geophys. Res. Lett. 25, 3397-3400.

Azarro, R., 1999, Earthquake surface faulting at Mount Etna volcano (Sicily) and implications for active tectonics, J. Geodyn. 28, 193-213.

Barberi, F. and M.L. Carapezza, 1996, The problem of volcanic unrest: The Campi Flegrei case history, Monitoring and Mitigation of Volcano Hazards, Eds. R. Scarpa and R. Tilling, Springer-Verlag, Berlin New York. P. 771-786.

Bardintzeff, J-M and A.R. McBirney, 1999, Volcanology. Jones and Bartlett Publ., Boston. 268 p.

Barrientos, S.E. and P.S. Acevedo-Aranguiz, 1992, Seismological aspects of the 1988-1989 Lonquimay (Chile) volcanic eruption, J. Volcanol. Geotherm. Res. 53, 73-87.

Basham, P.W., D.H. Weichert, F.M. Anglin and M.J. Berry, 1982, New Probabilistic Strong Motion Maps of Canada: A Compilation of Earthquake Source Zones, Methods and Results. Earth Physics Branch Open File No 82-33, Ottawa. 205 p.

Bautista, B.C., M.L.P. Bautista, R.S. Stein, E.S. Barcelona, R.S. Punongbayan, E.P. Laguerta, A.R. Rasdas, G. Ambubuyog and E.Q. Amin, 1996, Relationship of regional and local structures to Mount Pinatubo activity, Fire and Mud. Eruptions and lahars of Mount Pinatubo, Philippines, Eds. C.G. Newhall and R.S. Punongbayan. University of Washington Press, Seattle. Pp. 351-370.

Benoit, J.P., 1998, Pattern and Process in Volcano Seismology, Ph. D. Thesis, University of Alaska, Fairbanks.

Benoit, J.P. and S.R. McNutt, 1996, Global volcanic earthquake swarm database and preliminary analysis of volcanic earthquake swarm duration, Ann. Geofisica 39, 221-229.

Benoit, J.P. and S.R. McNutt, 1997, New constraints on source processes of volcanic tremor at Arenal volcano, Costa Rica, using broadband seismic data, Geophys. Res. Lett. 24, 449-452.

BGVN, Bulletin of Global Volcano Network. Smithsonian Institution, 1990-1996.

Blatt, H. and R.J. Tracy, 1996, Petrology. Freeman and Co., New York. 529 p.

Blong, R.J., 1984, Volcanic Hazards. Academic Press, Sydney. 424 p.

Bogoyavlenskaya, G.E., O.A. Braitzeva, I.V. Melekestsev, A.P. Maximov and B.V. Ivanov, 1991, Bezymianny Volcano, Active Volcanoes of Kamchatka, vol. I, Eds. S.A. Fedotov and Yu.P. Masurenkov, Nauka, Moscow. Pp. 195-209.

Bogoyavlenskaya, G.E. and I.T. Kirsanov, 1981, 25 years of volcanic activity of Bezymianny volcano, Volc. Seismol. 2, 3-13 (in Russian).

Bormann, P. and E. Bergman, 2000, The new IASPEI Manual of seismological observatory practice, Seismol. Res. Lett. 71, 510-518.

Bottari, A., E. Lo Giudice, G. Patane, R. Romano and C. Sturiale, 1975, L'eruzione etnea del gennaro-marzo 1974, Riv. Min. Sic. 154/155, 175-198.

Brandsdóttir, B. and P. Einarsson, 1979, Seismic activity associated with the September 1977 deflation of the Krafla central volcano in northeastern Iceland, J. Volcanol. Geotherm. Res. 6, 197-212.

Brandsdóttir, B. and P. Einarsson, 1992, Volcanic tremor and low-frequency earthquakes in Iceland, Volcanic Seismology, Eds. P. Gasparini, R. Scarpa and K. Aki, Springer, Berlin. Pp. 212-222.

Bretón González, M., J.J. Ramírez and C. Navarro, 2002, Summary of the historical eruptive activity of Volcán de Colima, México 1519-2000, J. Volcanol. Geotherm. Res. 117, 21-46.

Brune, J.N., 1970, Tectonic stress and the spectra of seismic shear waves from earthquakes, J. Geophys. Res. 75, 4997-5009.

Bullard, F.M., 1963, Volcanoes in History, in Theory, in Eruption. Univ. Texas Press, Austin. 441 p.

Calder, E.S., R. Luckett, R.S.J. Sparks and B. Voight, 2002, Mechanisms of lava dome instability and generation of rockfalls and pyroclastic flows at Soufrére Hills volcano, Montserrat, The eruption of Soufriére Hills volcano, Montserrat, from 1995 to 1999, Eds. T.H. Druitt and B.P. Kokelaar, Geol. Soc. London, Memoirs 21, 173-190.

Cashman, K.V., B. Sturtevant, P. Papale and O. Navon, 2000, Magmatic fragmentation, Encyclopedia of Volcanoes, Ed. H. Sigurdsson, Academic Press, San Diego. Pp. 421-430.

Castellanos, G. and Z. Jiménez, 1995, Sismología en Colima. SEP, México. 107 p.

Chouet, B., 1986, Dynamics of fluid-driven crack in three-dimensions by the finite difference method, J. Geophys. Res. 91, 13967-13992.

Chouet, B., 1996a, New methods and future trends in seismological volcano monitoring, Monitoring and Mitigation of Volcano Hazards, Eds. R. Scarpa and R. Tilling, Springer-Verlag, Berlin, New York. Pp. 23-98.

Chouet, B, 1996b, Long-period volcano seismicity: its source and use in eruption forecasting, Nature 380, 309-316.

Chouet, B., P. Dowson, T. Ohminato, M. Martini, G. Saccorotti, F. Giudicepietro, G. De Luca, G. Milana and R. Scarpa, 2002, Source mechanisms of explosions at Stromboli volcano, Italy, determined from moment-tensor inversions of very-long-period data (in press).

Chouet, B., R. Page, R. Stevens, J. Lahr and J. Power, 1994, Precursory swarms of long-period events at Redoubt volcano (1989-1990), Alaska: their origin and use as a forecasting tool, J. Volcanol. Geotherm. Res. 62, 95-135.

Christiansen, , R.L. and D.W. Peterson, 1981, Chronology of the 1980 eruptive activity, The 1980 Eruptions of Mount St. Helens, Eds. P.W. Lipman and D.R. Mullineaux, Washington, Geol. Surv. Prof. Paper 1250. Pp. 17-30.

Chronology of the eruption of Soufriere Hills volcano from January 1992 to early July 1997 (2002). WEB page: http://www.geo.mtu/volcanoes/west.indies/soufriere/govt/miscdocs/chronology.html. 07.06.2002.

Cornelius, R.R. and B. Voight, 1994, Seismological aspects of the 1989-1990 eruption at Redoubt volcano, Alaska: the material failure forecast method (FFM) with RSAM and SSAM data, J. Volcanol. Geotherm. Res. 62, 469-498.

Cornelius, R.R. and B. Voight, 1995, Graphical and PC-software analysis of volcano eruption precursors according to the material failure forecast method (FFM), J. Volcanol. Geotherm. Res. 64, 295-320.

Cosentino, M., S. Gresta, G. Lombardo, G. Patané, M. Riuscetti, R. Schick and A. Viglianisi, 1984, Features of volcanic tremors at Mt. Etna,(Sicily) during the Match-August 1983 eruption, Bull. Volcanol. 47, 929-939.

Cramer, C.H. and S.R. McNutt, 1997, Spectral analysis of earthquakes in the 1989 Mammoth Mountain swarm near Long Valley, California, Bull. Seismol. Soc. Am. 87, 1454-1462.

Cristofolini, R., S. Gresta, S. Imposa and G. Patané, 1988, Feeding mechanism of eruptive activity at Mt. Etna based on seismological and petrological data, (Eds) Modeling of Volcanic Processes, Eds. C.-Yu King and R. Scarpa, F. Vieweg and Sohn, Braunschweig. Pp. 73-93.

Daines, M.J., 2000, Migration of melt, Encyclopedia of Volcanoes, Ed. H. Sigurdsson, Academic Press, San Diego. Pp.69-88.

Decker, R. and B. Decker, 1999, Volcanoes. Freeman and Co., New York. 321 p.

Del Pezzo, E., G. De Natale, M. Martín and A. Zollo, 1987, Source parameters of microearthquakes at Phlegraean Fields (Southern Italy) volcanic area, Phys. Earth Planet. Inter. 47, 25-42.

Domínguez, T., V.M. Zobin and G.A. Reyes-Dávila, 2001, The fracturing in volcanic edifice before an eruption: the June-July 1998 high-frequency earthquake swarm at Volcán de Colima, México, J. Volcanol. Geotherm. Res. 105, 65-75.

Dvigalo, V.N., A.M. Chirkov and S.A. Fedotov, 1984, New summit caldera of Plosky Tolbachik volcano, Large Tolbachik Fissure Eruption, Ed. S.A. Fedotov, Nauka, Moscow. Pp. 537-575 (in Russian).

Eaton, J.P. and K.J. Murata, 1960, How volcanoes grow, Science 132, 925.

Ekström, G., 1994, Anomalous earthquakes on volcano ring-fault structures, Earth Planet. Sci. Lett. 128, 707-712.

Elsworth, D. and B. Voight, 1995, Dike intrusion as a trigger for large earthquakes and the failure of volcano flanks, J. Geophys. Res. 100, 6005-6024.

Endo, E.T., S.D. Malone, L.L. Noson and C.S. Weaver, 1981, Locations, magnitudes, and statistics of the March 20 – May 18 earthquake sequence, The 1980 Eruptions of Mount St. Helens, Eds. P.W. Lipman and D.R. Mullineaux, Washington, Geol. Surv. Prof. Paper 1250. Pp. 93-108.

Endo, E.T. and T. Murray, 1991, Real-time seismic amplitude measurement (RSAM): a volcano monitoring and prediction tool, Bull. Volvanol. 53, 533-545.

Falsaperla, S., H. Langer and S. Spampinato, 1998, Statistical análisis and characteristics of volcanic tremor on Stromboli volcano (Italy), Bull. Volcanol. 60, 75-88.

Fedotov, S.A., Ed., 1984a, Large Tolbachik Fissure Eruption. Nauka, Moscow. 637 p. (in Russian).

Fedotov, S.A., 1984b, Large fissure eruptions, their theory. Mechanism of the Tolbachik eruption, Large Tolbachik Fissure Eruption, Ed. S.A. Fedotov, Nauka, Moscow. Pp. 537-575 (in Russian).

Fedotov, S.A., 1998, Study and mechanism of the simultaneous 1996 karymsky volcano and Akademii Nauk caldera eruptions in Kamchatka, Volc. Seis. 19, 525-566.

Ferrucci, F., 1995, Seismic monitoring at active volcanoes. Monitoring Active Volcanoes, Eds. B. McGuire et al., UCL Press, London. Pp. 60-90.

Ferrucci, F. and D. Patane, 1993, Seismic activity accompanying the outbreak of the 1991-1993 eruption of Mt. Etna (Italy), J. Volcanol. Geotherm. Res. 57, 125-135.

Filson, J., T. Simkin and L.-K. Leu, 1973, Seismicity of a caldera collapse: Galapagos Islands 1968, J. Geophys. Res. 78, 8591-8622.

Fisher, R.V., G. Heiken and J.B. Hulen, 1998, Volcanoes. Princeton Univ. Press, Princeton. 317 p.

Francis, T.J.G., 1974, New interpretation of the 1968 Fernandina caldera collapse and its implications for the mid-oceanic ridges, Geophys. J. R.A.S. 39, 301-318.

Francis, P., 1998, Volcanoes. A Planetary Perspective. Clarendon Press, Oxford. 443 p.

Garduño-Monroy, A., R. Saucedo-Girón, Z. Jiménez, J.C. Gavilanes-Ruiz, A. Cortes-Cortes and R.M. Uribe-Cifuentes, 1998, La falla Tamazula, limite suroriental del bloque de Jalisco y sus relaciones con el complejo volcánico de Colima, Revista Mexicana de Cien. Geol. 15, 132-144.

Gasperini, P., S. Gresta and F. Mulargia, 1990, Statistical analysis of seismic and eruptive activities at Mt. Etna during 1978-1987, J. Volcanol. Geotherm. Res. 40, 317-325.

Gibowicz, S.J., J.H. Latter and G.K. Sutton, 1974, Earthquake swarm associated with volcanic eruption, Curacoa Reef area, Northern Tonga, July 1973, Ann. Geofis. 27, 443-475.

Gil Cruz, F. and B.A. Chouet, 1997, Long-period events, the most characteristic seismicity accompanying the emplacement and extrusion of a lava dome in Galeras volcano, Colombia, in 1991, J. Volcanol. Geotherm. Res. 77, 121-158.

Glot, J.P, S. Gresta, G. Patané and G. Poupinet, 1984, Earthquake activity during the 1983 Etna eruption, Bull. Volcanol. 47, 953-963.

Goldstein P. and B. Chouet, 1994, Array measurements and modeling of sources of shallow volcanic tremor at Kilauea volcano, Hawaii, J. Geophys. Res. 99, 2637-2652.

Gonzalez-Ferran, O., 1995, Volcanes de Chile. Instituto Geográfico Militar, Santiago. 640 p.

Gordeev, E.I. and M. Ripepe, 1998, Sources of volcanic tremor at basaltic volcanoes, Volc. Seis. 19, 353-369.

Gorelchik, V.I., 1989, Seismological study of the Kliuchevskoy flank eruption of 1983 (Kamchatka), J. Volcanol. Geotherm. Res. 38, 269-280.

Gorelchik, V.I., V.T. Garbuzova, D.V. Droznin, V.I. Levina, P.P. Firstov, O.S. Chubarova and V.A. Shirokov, 1995, Shiveluch volcano: deep structure and forecasting eruptions using the 1962-1994 seismic data, Volc. Seismol. 4-5, 54-74 (in Russian).

Gorelchik, V.I., V.A. Shirokov, P.P. Firstov and O.S. Chubarova, 1997, Shiveluch volcano: seismicity, deep structure and forecasting eruptions (Kamchatka), J. Volcanol. Geotherm. Res. 78, 121-132.

Gorelchik, V.I. and V.M. Zobin, 1971, Earthquake swarm development in the region of the active volcanism in Kamchatka. Volcanism and the Earth's Interior, Nauka, Moscow. Pp. 118-120 (in Russian).

Gorelchik, V.I., V.M. Zobin and P.I. Tokarev, 1990, Volcanic earthquakes of Kamchatka: classification, nature of source and spatio-temporal distribution, Tectonophys. 180, 255-271.

Gorshkov, G.S., 1956, About the depth of magmatic chamber of Klyuchevskoy volcano, Dokl. Acad. Sci. USSR 106 (in Russian).

Gorshkov, G.S., 1959, Gigantic eruption of the volcano Bezymianny, Bull. Volcanol. 20, 77-109.

Gorshkov, G.S. and G.E. Bogoyavlenskaya, 1965, Bezymianny Volcano and the Peculiarities of its Last Eruption in 1955-1963. Nauka, Moscow. 171 p. (in Russian).

Gorshkov, G.S. and Y.M. Dubik, 1970, Gigantic directed blast at Shiveluch volcano (Kamchatka), Bull. Volcanol. 34, 261-288.

Goto, K., S. Mori and K. Fukui, 1991, Volcanic tremor accompanying the 1989 submarine eruption off the east coast of Izu peninsula, Japan, J. Phys. Earth 39, 47-63.

Gresta, S., V. Longo and A. Viavattene, 1990, Geodynamic behavior of eastern and western sides of Mount Etna, Tectonoph. 179, 81-92.

Gresta, S., A. Montalto and G. Patané, 1991, Volcanic tremor at Mount Etna (January 1984 – March 1985): its relationship to the eruptive activity and modeling of the summit feeding system, Bull. Volcanol. 53, 309-320.

Gresta, S. and G. Patané, 1983, Changes in b values before the Etnean eruption of March-August 1983, Pure Appl. Geophys. 121, 903-912.

Gresta, S. and G. Patané, 1987, Review of seismological studies at Mount Etna, Pure Appl. Geophys. 125, 951-970.

Gudmundsson, M., 1996, Ice-Volcano interaction at the subglacial Grimsvotn volcano, Iceland. Glaciers, Ice Sheets and Volcanoes: A Tribute to Mark Meier, Ed. S.C. Colbeck, CRREL Special Rep. 96-27. Pp. 34-40.

Gutenberg, B. and C. Richter, 1956, Seismicity of the Earth and associated Phenomena, Princeton University Press, Princeton, New Jersey.

Hamaguchi, H., T. Nishimura and N. Zana, 1992, Process of the Nyiragongo eruption inferred from the analysis of long-period earthquakes and volcanic tremors, Tectonoph. 209, 241-254.

Hashimoto, T., K. Maeda, T. Odaka and M. Seino, 1991, Waveform analysis of volcanic tremors associated with the July 1989 volcanic activity off the east coast of Izu peninsula, J. Phys. Earth 39, 65-78.

Harlow, D.H., J.A. Power, E.P. Laguerta, G. Ambubuyog, R.A. White and R.P. Hoblitt, 1996, Precursory seismicity and forecasting of the June 15, 1991, eruption of Mount Pinatubo, Fire and Mud. Eruptions and lahars of Mount Pinatubo, Philippines, Eds. C.G. Newhall and R.S. Punongbayan, University of Washington Press, Seattle. Pp. 285-305.

Havskov, J. and K. Atakan, 1991, Seismicity and volcanism of Jan Mayen Island, Terra Nova 3, 517-526.

Havskov, J., S. De la Cruz-Reyna, S.K. Singh, F. Medina and C. Gutiérrez, 1983, Seismic activity related to the March-April, 1982 eruption of El Chichón volcano, Chiapas, México, Geophys. Res. Let. 10, 293-296.

Harvard University CMT Data Base, 1977-1999.

Hill, D.P., 1977, A model for earthquake swarms, J. Geophys. Res. 82, 1347-1352.

Icelandic Meteorological Office, Seismological Bulletin, 1996.

Iguchi, M., 1994, A vertical expansion source model for the mechanisms of earthquakes originated in the magma conduit of an andesitic volcano: Sakurajima, Japan, Bull. Volcanol. Soc. Jap. 39, 49-67.

ISC, Bulletin of International Seismological Centre, 1964-1999.

Ishihara, K., 1990, Pressure sources and induced ground deformation associated with explosive eruptions at an andesitic volcano: Sakurajima volcano, Japan, Magma Transport and Storage, Ed. M.P. Ryan, J. Wiley and Sons Ltd. Pp. 335-356.

Jiménez, Z., V.H. Espíndola and J.M. Espíndola, 1999, Evolution of the seismic activity from the 1982 eruption of El Chichón volcano, Chiapas, México, Bull. Volcanol. 61, 411-422.

Jimenez, Z., G. Reyes and J.M. Espíndola, 1995, The July 1994 episode of seismic activity at Colima volcano, Mexico, J. Vocanol. Geotherm. Res. 64, 321-326.

JMA, Magnitudes of Japan Meteorological Agency, published in ISC.

Johnson, J.B. and J.M. Lees, 2000, Plugs and chugs – seismic and acoustic observations of degassing explosions at Karymsky, Russia and Sangay, Ecuador, J. Volcanol. Geotherm. Res. 101, 67-82.

Johnson, J.B., J.M. Lees and E.I. Gordeev, 1998, Degassing explosions at Karymsky volcano, Kamchatka, Geophys. Res. Lett. 25, 3999-4002.

Jolly, A.D., G. Thompson and G.E. Norton, 2002, Locating pyroclastic flows on Soufriére Hills volcano, Montserrat, West Indies, using amplitude signals from high dynamic range instruments, J. Volcanol. Geotherm. Res. 118, 299-317.

Julian, B., A. Miller and G.R. Foulger, 1998, Non-double-couple earthquakes. 1. Theory, Rev. Geophys. 36, 525-549.

Kagan, Y.Y., 1997, Seismic moment-frequency relation for shallow earthquakes: Regional comparison, J. Geophys. Res. 102, 2835-2852.

Kanamori, H. and J.W. Given, 1982, Analysis of long-period seismic waves excited by the May 18, 1980, eruption of Mount St. Helens – A terrestrial monopole? J. Geophys. Res. 87, 5422-5432.

Kanamori, H. and J.W. Given, 1983, Lamb pulse observed in nature, Geophys. Res. Lett. 10, 373-376.

Kanamori, H., J.W. Given and T. Lay, 1984, Analysis of seismic body waves excited by the Mount St. Helens eruption of May 18, 1980, J. Geophys. Res. 89, 1856-1866.

Kaufman, K. and L.J. Burdick, 1980, The reproducing earthquakes of the Galapagos Islands, Bull. Seismol. Soc. 70, 1759-1770.

Kawakatsu, H., S. Kaneshima, H. Matsubayashi, T. Ohminato, Y. Sudo, T. Tsutsui, K. Uhira, H. Yamasato, H. Ito and D. Legrand, 2000, Aso94: Aso seismic observation with broadband instruments, J. Volcanol. Geotherm. Res. 101, 129-154.

Kawakatsu, H., T. Ohminato, H. Ito, Y. Kuwahara, T. Kato, K. Tsuruga, S. Honda and K. Yomogida, 1992, Broadband seismic observations at the Sakurajima volcano, Japan, Geoph. Res. Lett. 19, 1959-1962.

Khrenov, A.P., V.N. Dvigalo, I.T. Kirsanov, S.A. Fedotov, V.I. Gorelchik and N.A. Zharinov, 1991, Klyuchevskoy volcano, Active Volcanoes of Kamchatka, vol. I, Eds. S.A. Fedotov and Yu.P. Masurenkov, Nauka, Moscow. Pp. 146-153.

Khrenov, A.P., Yu.M. Dubik, B.V. Ivanov, A.A. Ovsyannikov, V.P. Pilipenko, Yu. A. Taran, P.P. Firstov and A.M. Chirkov, 1982, Eruptive activity of Karymsky volcano in 1970-1980, Volc. Seismol. 4, 29-48 (in Russian).

Kizawa, T.,1958, A study of earthquakes in relation to volcanic activity (II), Paper Met. Geophys. 9, 204-239.

Klein, F.W., 1984, Eruption forecasting at Kilauea volcano, Hawaii, J. Geophys. Res. 89, 3059-3073.

Klein, F.W. and R.Y Koyanagi, 1989, The seismicity and tectonics of Hawaii, The Eastern Pacific Ocean and Hawaii, The Geology of North America, vol. N, Eds. E.L. Winteler, D.M. Hussong and R.W. Decker, Geol. Soc. Am., Boulder. Pp. 238-252.

Klein, F.W., R.Y. Koyanagi, J.S. Nakata and W.R. Tanigawa, 1987, The seismicity of Kilauea's magma system, Volcanism in Hawaii, US Geol. Surv. Prof. Paper 1350, 1019-1185.

Konstantinou, K.I., G. Nolet, W.J. Morgan; R.M. Allen and M.J. Pritchard, 2000, Seismic phenomena associated with the 1996 Vanajökull eruption, central Iceland, J. Volcanol. Geotherm. Res. 102, 169-187.

Koyanagi, R., B. Chouet and K. Aki, 1987, Origin of volcanic tremor in Hawaii. Part I, Volcanism in Hawaii, US Geol. Surv. Prof. Paper 1350, 1221-1257.

Kubotera, A., 1988, Mechanism of the earthquakes with M equal to or greater than 6 that occurred in volcanoes (including calderas) and their relation to volcanic activity, Proceedings of Kagoshima International Conference on Volcanoes 1988, Kagoshima. Pp. 160-163.

La Delfa, S., G. Patané, R. Clocchiatti, J.-L. Joron and J.-C. Tanguy, 2001, Activity of Mount Etna preceding the February 1999 fissure eruption: inferred mechanism from seismological and geochemical data, J. Volcanol. Geotherm. Res. 105, 121-139.

Lahr, J.C, B.A. Chouet, C.D. Stephens, J.A. Power and R.A. Page, 1994, Earthquake classification, location, and error analysis in a volcanic environment: implications for the magmatic system of the 1989-1990 eruptions at Redoubt volcano, Alaska, J. Volcanol. Geotherm. Res. 62, 137-152.

Lamb, H., 1904, On the propagation of tremors over the surface of an elastic solid, Phil. Trans. Royal Soc. (London), A 203, 1-42.

Latter, J.H., 1979, Volcanological observations at Tongariro national Park. Types and classification of volcanic earthquakes, 1976-1978. Rep. 150, Geoph. Division. Wellington. 60 p.

Lay, T. and T.C. Wallace, 1995, Modern Global Seismology. Academic Press, San Diego. 521 p.

Lees, J.M., 1992, The magma system of Mount St. Helens: non-linear high-resolution P-wave tomography, J. Volcanol. Geotherm. Res. 53, 103-116.

Legrand, D., S. Kaneshima and H. Kawakatsu, 2000, Moment tensor analysis of near-field broadband waveforms observed at aso volcano, Japan, J. Volcanol. Geotherm. Res. 101, 155-169.

Lesage, P., F. Glangeaud and J. Mars, 2002, Applications of autoregressive models and time-frequency analysis to the study of volcanic tremor and long-period events, J. Volcanol. Geotherm. Res. 114, 391-417.

Lipman, P.W. and D.R. Mullineaux, Eds., 1982, The 1980 Eruptions of Mount St. Helens, U.S. Geol. Surv. Prof. Pap. 1250.

Lombardo, G., G. Coco, M. Corrao and S. Gresta., 1996, Features of seismic events and volcanic tremor during the preliminary stages of the 1991-1993 eruption of Mt. Etna, Ann. Geofisica 39, 403-410.

Luckett, R., B. Baptie and J. Neuberg, 2002, The relationship between degassing and rockfall signals at Soufriére Hills volcvano, Montserrat, The eruption of Soufriére Hills volcano, Montserrat, from 1995 to 1999, Eds. T.H. Druitt and B.P. Kokelaar, Geol. Soc. London, Memoirs 21, 595-603.

Macdonald, G.A., 1972, Volcanoes, Prentice-Hall, Inc., Englewood Cliffs, New Jersey, 510 p.

Machado, F., W.H. Parsons, A.F. Richards and J.W. Mulford, 1962, Capelinhos eruption of Fayal volcano, Azores, 1957-1958, J. Geophys. Res. 67, 3519-3529.

Malone, S.D., 1983, Volcanic earthquakes: examples from Mount St. Helens, Earthquakes: Observations, Theory, and Interpretation, Eds. H. Kanamori and E. Boschi, North-Holland Publ. Co., Amsterdam, Pp. 436-455.

Malone, S.D., E.T. Endo, C.S. Weaver and J.W. Ramey, 1981, Seismic monitoring for eruption prediction, The 1980 Eruptions of Mount St. Helens, Washington, Eds. P.W. Lipman and D.R. Mullineaux, Geol. Surv. Prof. Paper 1250. Pp. 803-814.

Matsumura, S., T. Ohkubo and M. Imoto, 1991, Seismic swarm activity in and around the Izu Peninsula preceding the volcanic eruption of July 13, 1989, J. Phys. Earth 39, 93-106.

McNutt, S.R., 1986, Observations and analysis of B-type earthquakes, explosions, and volcanic tremor at Pavlov volcano, Alaska, Bull. Seismol. Soc. Am. 76, 153-175.

McNutt, S.R., 1992, Volcanic tremor, Encyclopedia of Earth System Science, vol. 4, Ed. W.A. Nierenberg, Academic Press, San Diego. Pp. 417-425.

McNutt, S.R., 1994, Volcanic tremor amplitude correlated with eruption explosivity and its potential use in determing ash hazards to aviation, Volcanic ash and aviation safety, U.S. Geol. Surv. Bull. 2047, 377-385.

McNutt, S.R., 1996, Seismic monitoring and eruption forecasting of volcanoes: A review of the state-of-the-art and case histories. Monitoring and Mitigation of Volcano Hazards, Eds. R. Scarpa and R. Tilling, Springer-Verlag, Berlin New York. Pp. 99-146.

McNutt, S.R., 2000a, Volcanic seismicity, Encyclopedia of Volcanoes, Ed. H. Sigurdsson, Academic Press, San Diego. Pp. 1015-1034.

McNutt, S.R., 2000b, Seismic monitoring. Encyclopedia of Volcanoes, Ed. H. Sigurdsson, Academic Press, San Diego. Pp. 1095-1120.

McNutt, S.R., T.P. Miller and J.J. Taber, 1991, Geological and seismological evidence of increased explosivity during the 1986 eruptions of Pavlov volcano, Alaska, Bull. Volcanol. 53, 86-98.

McNutt, S.R., H. Rymer and J. Stix, 2000, Synthesis of volcano monitoring, Encyclopedia of Volcanoes, Ed. H. Sigurdsson, Academic Press, San Diego. Pp. 1165-1184.

Melekestsev, I.V., O.N. Volynets, V.A. Yermakov, T.P. Kirsanova and Yu. P. Masurenkov, 1991, Sheveluch volcano, Active Volcanoes of Kamchatka, vol. I, Eds. S.A. Fedotov and Yu.P. Masurenkov, Nauka, Moscow. Pp.98-103.

Miller, A.D., G.R. Foulger and B.R. Julian, 1998, Non-double-couple earthquakes. 2. Observations, Rev. Geophys. 36, 551-568.

Miller, A.D., R.C. Steward, R.A. White, R. Luckett, B.J. Baptie, W.P. Aspinall, J.L. Latchman, L.L. Lynch and B. Voight, 1998, Seismicity associated with dome grows and collapse at Soufriére Hills volcano, Montserrat, Geophys. Res. Lett. 25, 3401-3404.

Mills, H.H., 1991, Temporal variations of mass-wasting activity in Mount St. Helens crater, Washington, U.S.A., indicated by seismic activity, Arctic and Alpine Res. 23, 417-423.

Mimatsu, M., 1995, Showa-Shinzan Diary. Suda Seihan, Sapporo. 179 p.

Minakami, T., 1960, Fundamental research for predicting volcanic eruptions. Part I, Bull. Earthq. Res. Inst., Tokyo Univ. 38, p. 4, 497-544.

Minakami, T., 1964, The 1962 eruption of Miyake-sima, one of the seven Izu islands, Japan, Bull. Volcanol. 27, 225-234.

Minakami, T., 1974a, Seismology of volcanoes in Japan, Physical Volcanology, Eds. L. Civetta, P. Gasparini, G. Luongo and A. Rapolla, Elsevier, Amsterdam. Pp.1-27.

Minakami, T., 1974b. Prediction of volcanic eruptions. Physical Volcanology, Eds. L. Civetta, P. Gasparini, G. Luongo and A. Rapolla,Elsevier, Amsterdam. Pp. 313-333.

Minakami, T., S. Utibori and S. Hiraga, 1969, The 1968 eruption of volcano Arenal, Costa Rica, Bull. Earthq. Res. Inst., Tokyo Univ. 47, 783-802.

Mogi, 1963, Some discussions on aftershocks, foreshocks and earthquake swarms – the fracture of a semi-infinite body caused by an inner stress origin and its relation to the earthquake phenomena (third paper), Bull. Earthq. Res. Inst., Tokyo Univ. 41, 615-658.

Mooney, W.D., G. Laske and T.G. Masters, 1998, CRUST 5.1: A global crustal model at 5°x5°, J. Geophys. Res. 103, 727-747.

Moore, J.G., P.W. Lipman, D.A. Swanson and Tau Rho Alfa, 1981, Growth of lava domes in the crater, June 1980-January 1981, The 1980 Eruptions of Mount St. Helens, Washington, Eds. P.W. Lipman and D.R. Mullineaux, Geol. Surv. Prof. Paper 1250. Pp. 541-548.

Mori, J., D. Eberhart-Phillips, and D.H. Harlow, 1996a, Three-dimensional velocity structure at Mount Pinatubo: Resolving magma bodies and earthquake hypocenters, Fire and Mud. Eruptions and lahars of Mount Pinatubo, Philippines, Eds. C.G. Newhall and R.S. Punongbayan, University of Washington Press, Seattle. Pp. 371-382.

Mori, J., R.A. White, D.H. Harlow, P. Okubo, J.A. Power, R.P. Hoblitt, E.P. Laguerta, A. Lanuza and B.C. Bautista, 1996b, Volcanic earthquakes following the 1991 climactic eruption of Mount Pinatubo: strong seismicity during a waning period, Fire and Mud. Eruptions and lahars of Mount Pinatubo, Philippines, Eds. C.G. Newhall and R.S. Punongbayan, University of Washington Press, Seattle. p. 339-350.

Morrissey, M., B. Zimanowski, K. Wohletz and R. Buettner, 2000, Encyclopedia of Volcanoes, Ed. H. Sigurdsson, Academic Press, San Diego. Pp. 431-445.

MOS, Seismological Bulletin of Institute of Physics of the Earth, Moscow, 1991.

Mullineaux, D.R. and D.R. Crandell, 1981, The eruptive history of Mount St. Helens, The 1980 Eruptions of Mount St. Helens, Washington, Eds. P.W. Lipman and D.R. Mullineaux, Geol. Surv. Prof. Paper 1250. Pp. 3-16.

Nakada, S., H. Shimizu and K. Ohta, 1999, Overview of the 1990-1995 eruption at Unzen volcano, J. Volcanol. Geotherm. Res. 89, 1-22.

Navarro-Ochoa, C., J.C. Gavilanes-Ruíz, and A. Cortés-Cortés, 2002, Movement and emplacement of lava flows at Volcán de Colima, Mexico: November 1998 - February 1999, J. Volcanol. Geotherm. Res. 117, 155-167.

NEIC, Monthly Listings, 1970-1991.

Nettles, M. and G. Ekström, 1998, Faulting mechanism of anomalous earthquakes near Bárdabunga volcano, Iceland, J. Geophys. Res. 103, 17973-17983.

Neuberg, J., 2000, Characteristics and causes of shallow seismicity in andesitic volcanoes, Phil. Trans. R. Soc. Lond. A 358, 1533-1546.

Newhall, C.G. and D. Dzurisin, 1988, Historical Unrest at Large Calderas of the World. US Geol. Surv. Bull. 1855, Washington. 1108 p.

Newhall, C.G. and S. Self, 1982, The volcanic explosivity index (VEI): an estimate of explosive magnitude for historical volcanism, J. Geophys. Res. (Oceans and Atmosperes) 87, 1231-1238.

Nishimura, T., 1995, Source parameters of the volcanic eruption earthquakes at Mount Tokachi, Hokkaido, Japan, and magma ascending model, J. Geophys. Res. 100, 12456-12473.

Nishimura, T., 1998, Source mechanisms of volcanic explosion earthquakes: single force and implosive sources, J. Volcanol. Geotherm. Res. 86, 97-106.

Nishimura, T. and H. Hamaguchi, 1993, Scaling law of volcanic explosion earthquakes, Geophys. Res. Lett. 20, 2479-2482.

Norris, R.D., 1994, Seismicity of rockfalls and avalanches at three Cascade Range volcanoes: Implications for seismic detection of hazardous mass movements, Bull. Seismol. Soc. Am. 84, 1925-1939.

Okada, H., 1982, Earthquake swarm activity of Usu volcano in 1910, Geophys. Bull. Hokkaido Univ. 41, 53-63 (in Japanese).

Okada, H., 1983, Comparative study of earthquake swarms associated with major volcanic activities, Arc Volcanism: Physics and Tectonics, Eds. D. Shimozuru and I. Yokoyama, Terra Publ. Co., Tokyo. Pp. 43-61.

Okada, H., Y. Nishimura, H. Miyamachi, H. Mori and K. Ishihara, 1990, Geophysical significance of the 1988-1989 explosive eruptions of Mt. Tokachi, Hokkaido, Japan, Bull. Volcanol. Soc. Japan 35, 175-203.

Okada, H., H. Watanabe, H. Yamashita and I. Yokoyama, 1981, Seismological significance of the 1977-1978 eruptions and the magma intrusión prtocess of Usu volcano, Hokkaido, J. Volcanol. Geotherm. Res. 9, 311-334.

Omori, F., 1911-1913, The Usu-san eruption and earthquake and elevation phenomena, Bull. Imper. Earthq. Invest. Com. 5, 1-38.

Omori, F., 1912, The eruptions and earthquakes of the Asama-yama, Bull. Imper. Earthq. Invest. Com. 6, 1-147.

Omori, F., 1914, The Sakura-jima eruptions and earthquakes, I, Bull. Imper. Earthq. Invest. Com. 8, 5-34.

Omori, F., 1922, The Sakura-jima eruptions and earthquakes, VI, Bull. Imper. Earthq. Invest. Com. 8, 467-524.

Oshima, S., M. Tsuchide, S. Kato, S. Okubo, K. Watanabe, K. Kudo and J. Ossaka, 1991, Birth of a submarine volcano "Teishi Knoll", J. Phys. Earth 39, 1-19.

Pacheco, J., S.K. Singh, J. Domínguez, A. Hurtado, L. Quintanar, Z. Jiménez, J. Yamamoto, C. Gutierrez, M. Santoyo, W. Bandy, M. Guzman and V. Kostoglodov, 1997, The October 9, 1995 Colima-Jalisco, Mexico earthquake (Mw 8): An aftershock study and a comparison of this earthquake with those of 1932, Geophys. Res. Lett. 24, 2223-2226.

Panza, G.F. and A. Saraó, 2000, Monitoring volcanic and geothermal areas by full seismic moment tensor inversion: are non-double-couple components always artifacts of modeling? Geophys. J. Int. 143, 353-364.

Parfitt, E.A. and D.C.P. Peacock, 2001, Faulting in the south flank of Kilauea volcano, Hawaii, J. Volcanol. Geotherm. Res. 106, 265-284.

Patané, G., G. Coco, M. Corrao, S. Imposa and A. Montalto, 1995, Source parameters of seismic events at Mount Etna volcano, Italy, during the outburst of the 1991-1993 eruption, Phys. Earth Planet. Int. 89, 149-162.

Patané, G., S. Gresta and S. Imposa, 1984, Seismic activity preceding the 1983 eruption of Mt. Etna, Bull. Volcanol. 47, 941-952.

Perfit, M.R. and J.P. Davidson, 2000, Plate tectonics and volcanism, Encyclopedia of Volcanoes, Ed. H. Sigurdsson, Academic Press, San Diego. Pp. 89-114.

Piip, B.I., 1956, Klyuchevskaya Sopka and its eruptions in 1944-1945 and in the past, Proc. Lab. Volcanol. 13, Pp. 99-119 (in Russian).

Purbawinata, M.A., A. Ratdomopurbo, I.K. Sinulingga, S. Sumarti, and Suharno, Eds., 1997, Merapi Volcano. A guide book. The Volcanological Survey of Indonesia, Bandung. 64 p.

Ramos, E.G., M.W. Hamburger, G.L. Pavlis and E.P. Laguerta, 1999, The low-frequency earthquake swarms at Mount Pinatubo, Philippines: implications for magma dynamics, J. Volcanol. Geotherm. Res. 92, 295-320.

Ratdomopurbo, A. and G. Poupinet, 2000, An overview of the seismicity of Merapi volcano (Java, Indonesia), J. Volcanol. Geotherm. Res. 100, 193-214.

Reyes-Dávila, G.A. and S. De la Cruz-Reyna, 2002, Experience in the short-term eruption forecasting at Volcán de Colima, Mexico, and public response to forecasts, J. Volcanol. Getherm. Res. 117, 121-128.

Ripepe, M. and E. Gordeev, 1999, Gas bubble dynamics model for shallow volcanic tremor at Stromboli, J. Geophys. Res. 104, 10639-10654.

Ripepe, M., P. Poggi, T. Braun and E. Gordeev, 1996, Infrasonic waves and volcanic tremor at Stromboli, Geoph. Res. Lett. 23, 181-184.

Riznichenko, Yu.V., 1977, Computation of the deformation rate during seismic flow of rocks, Izvestia USSR Ac Sci Phys Solid Earth 10, 34-47 (in Russian).

Roger, N. and C. Hawkesworth, 2000, Composition of magmas, Encyclopedia of Volcanoes, Ed. H. Sigurdsson, Academic Press, San Diego. Pp. 133-148.

Rowe, C.A., R.C. Aster, P.R. Kyle, R.R. Dibble and J.W. Schlue, 2000, Seismic and acustic observations at Mount Erebus volcano, Ross Island, Antarctica, 1994-1998, J. Volcanol. Geotherm. Res. 101, 105-128.

Rubin, A.M. and D. Gillard, 1998, Dike-induced earthquakes: theoretical considerations, J. Geophys. Res. 103, 10017-10030.

Rubin, A.M., D. Gillard and J.-L. Got, 1998, A reinterpretation of seismicity associated with the January 1983 dike intrusion at Kilauea volcano, Hawaii, J. Geophys. Res. 103, 10003-10015.

Ruff, L.J., 1983, Fault asperities inferred from seismic body waves, Earthquakes: Observations, Theory, and Interpretation, Eds. H. Kanamori and E. Boschi, North-Holland Publ. Co., Amsterdam, Pp. 251-276.

Rutherford, M.J. and J.E. Gardner, 2000, Rates of magma ascent, Encyclopedia of Volcanoes, Ed. H. Sigurdsson, Academic Press, San Diego. Pp. 207-217.

Saraó, A., G.F. Panza, E. Privitera and O. Cocina, 2001, Non-double-couple mechanisms in the seismicity preceding the 1991-1992 Etna volcano eruption, Geophys. J. Int. 145, 319-335.

Saucedo, R., J.L Macías, M. Bursik, M. Mora, J.C. Gavilanes, A. Cortés, 2002, Emplacement of pyroclastic flows during the 1998-1999 eruption of Volcán de Colima, Mexico, J. Volcanol. Geotherm. Res. 117, 129-153.

Schick, R., 1981, Source mechanism of volcanic tremor, Bull. Volcanol. 44, 491-497.

Schick, R., G. Lombardo and G. Patané, 1982, Volcanic tremors and shocks associated with eruptions at Etna (Sicily), September 1980, J. Volcanol. Geotherm. Res. 14, 261-279.

Scholz, C.H., 1990, The Mechanics of Earthquakes and Faultings. Cambridge Univ. Press, Cambridge. 439 p.

SEAN, 1975-1989. Scientific Event Alert Network Bulletin, Smithsonian Institution.

Seidl, D, R. Schick and M. Riuscetti, 1981, Volcanic tremors at Etna: a model for hydraulic origin, Bull. Volcanol. 44, 43-56.

Seismic Activity in Japan, 1998. Earthquake Research Committee, Tokyo. 222 p.

Shebalin, N.V., 1968, Methods of application of engineering seismology data for seismic zonation, Seismic Zonation of the USSR

Territory. Eds. S.V. Medvedev et al., Nauka, Moscow, Pp. 95-11 (in Russian).

Shimozuru, D., 1971, A seismological approach to the prediction of volcanic eruptions, The surveillance and prediction of volcanic activity, UNESCO, Paris. Pp. 19-45.

Siggerud, T., 1972, The volcanic eruption on Jan Mayen 1970, Norsk Polarinstitutt – Ărbok 1970, Oslo, 5-18.

Simkin, T. and L. Siebert, 1994, Volcanoes of the World (second edition). Geoscience Press, Tucson. 349 p.

Spera, F.J., 2000. Physical properties of magma. Encyclopedia of Volcanoes, Ed. H. Sigurdsson, Academic Press, San Diego. Pp. 171-190.

Staples, R.K., White, R.S., Brandsdottir, B., W. Menke, P.K.H. Maguire and J.H. McBride, 1997, Färoe-Iceland Ridge experiment. 1. Crustal structure of northeastern Iceland, J. Geophys. Res. 102, 7849-7866.

Steinberg, G. and A. Steinberg, 1975, On possible causes of volcanic tremor, J. Geophys. Res. 80, 1600-1604.

Suzuki, S. and M. Kasahara, 1979, Seismic activity immediately before and in the early stage of the 1977 eruption of Usu volcano, Hokkaido, Japan, J. Fac. Sci., Hokk. Univ., ser. VII 6, 239-254.

Tada, T. and K. Nakamura, 1988, Inflation-deflation related with the 1983 eruption of Miyakejima volcano, Japan, Bull. Volcanol. Soc. Japan 33, 173-184.

Takahashi, T. and H. Tsujimoto, 2000, A mechanical model for Merapi-type pyroclastic flow, J. Volcanol. Geotherm. Res. 98, 91-115.

Takeo, M., 1992, The rupture process of the 1989 offshore Ito earthquakes preceding a submarine volcanic eruption, J. Geophys. Res. 97, 6613-6627.

Tanaka, K., M. Kasahara, S. Hori, Z. Suzuki and A. Takagi, 1972, Research on Akita-Komaga-take (I). Summary of its eruptions in 1970-1971, Sci. Rep. Tohoku Univ., Ser. 5 21, 61-75.

Tiedemann, H., 1992, Earthquakes and Volcanic Eruptions: A Handbook on Risk Assessment. Swiss Re., Zurich. 951 p.

Tokarev, P.I., 1963, On a possibility of forecasting of Bezymianny volcano eruptions according to seismic data, Bull. Volcanol. 26, 379-386.

Tokarev, P.I., 1966, Eruptions and Seismic Regime of Volcanoes of Klyuchevskaya Group. Nauka, Moscow (in Russian). 118 p.

Tokarev, P.I., 1967, Gigantic eruption of Sheveluch volcano and its precursors, Izv. USSR Ac. Sci., Fizika Zemli 9, 11-22 (in Russian).

Tokarev, P.I., 1971, Forecasting volcanic eruptions from seismic data, Bull. Volcanol. 35, 243-250.

Tokarev, P.I., 1978, Prediction and characteristics of the 1975 eruption of Tolbachik volcano, Kamchatka, Bull. Volcanol. 41, 1-8.

Tokarev, P.I., 1981, Volcanic Earthquakes of Kamchatka. Nauka, Moscow (in Russian). 164 p.

Tokarev, P.I., 1983, Experience in predicting volcanic eruptions in the USSR, Forecasting Volcanic Events, Eds. H. Tazieff and J.-C. Sabroux, Elsevier, Amsterdam. Pp. 257-268.

Tokarev, P.I., 1985, The prediction of large explosions of andesitic volcanoes, J. Geodynamics 3, 219-244.

Tokarev, P.I., 1989, Eruptions and seismicity of Karymsky volcano in 1965-1986, Volcanol. Seismol. 2, 3-13. (in Russian).

Tokarev, P.I., S.A. Fedotov and V.V. Stepanov, 1984a, Prediction of the beginning and development of the eruption, Large Tolbachik Fissure Eruption, Ed. S.A. Fedotov, Nauka, Moscow. Pp. 373-388 (in Russian).

Tokarev, P.I. and P.P. Firstov, 1967, Seismological study of Karymsky volcano, Bull. Volcanol. St. 43, 9-22 (in Russian).

Tokarev, P.I., E.I. Gordeev and A.I. Farberov, 1984b, Volcanic tremor, Large Tolbachik Fissure Eruption, Ed. S.A. Fedotov, Nauka, Moscow. Pp. 448-471 (in Russian).

Tokarev, P.I., V.A. Shirokov and V.M Zobin, 1968, Seismic events associated with the Piip lateral eruption in October-December, 1966, Bull. Volcanol. St. 44, 30-41 (in Russian).

Uhira, K., H. Yamasato and M. Takeo, 1994, Source mechanism of seismic waves excited by pyroclastic flows observed at Unzen volcano, Japan, J. Geophys. Res. 99, 17757-17773.

Ukawa, M., 1993, Excitation mechanism of large-amplitude volcanic tremor associated with the 1989 Ito-oki submarine eruption, central Japan, J. Volcanol. Geotherm. Res. 55, 33-50.

Umakoshi, K., H. Shimizu and N. Matsuwo, 2001, Volcano-tectonic seismicity at Unzen volcano, Japan, 1985-1999, J. Volcanol. Geotherm. Res. 112, 117-131.

Voight, B., 1988, A method for prediction of volcanic eruptions, Nature 332, 125-130.

Voight, B. and R.R. Cornelius, 1991, Prospects for eruption prediction in near real-time, Nature 350, 695-698.

Walker, G.P.L., 2000, Basaltic volcanoes and volcanic systems. Encyclopedia of Volcanoes, Ed. H. Sigurdsson, Academic Press, San Diego. Pp. 283-289.

Wano, K. and H. Okada, 1980, Peculiar occurrence of Usu earthquake swarm associated with the recent doming activity, Kazan 33, 215-226 (in Japanese).

Watanabe, H., 1988, Seismic activity, The 1986-1987 Eruption of Izu-Oshima Volcano, Ed. S. Aramaki, Earthquake Research Institute, Tokyo University, Tokyo. Pp. 12-21.

Weaver, C.S., W.C. Grant, S.D. Malone and E.T. Endo, 1981, Post-May 18 seismicity: volcanic and tectonic implications, The 1980 Eruptions of Mount St. Helens, Washington, Eds. P.W. Lipman and D.R. Mullineaux, Geol. Surv. Prof. Paper 1250. Pp. 109-121.

White, R.A., 1996, Precursory deep long-period earthquakes at Mount Pinatubo: spatio-temporal link to a basalt trigger, Fire and Mud. Eruptions and lahars of Mount Pinatubo, Philippines, Eds. C.G. Newhall and R.S. Punongbayan (eds) University of Washington Press, Seattle. Pp. 307-327.

Wolfe, E.W. and R.P. Hoblitt, 1996, Overview of the eruptions. Fire and Mud. Eruptions and lahars of Mount Pinatubo, Philippines, Eds. C.G. Newhall and R.S. Punongbayan (eds) University of Washington Press, Seattle. Pp. 3-20.

Wood, H.O., 1913, Hawaiian volcano observatory, Bull. Seismol. Soc. 3, 14-19.

Wood, H.O., 1915, The seismic prelude to the 1914 eruption of Mauna Loa, Bull. Seismol. Soc. Am. 5, 39-51.

Yamaoka, K., J. Oikawa and Y. Ida, 1991, An isotropic source of volcanic tremor – observation with a dense seismic network at Izu-Oshima volcano, Japan, J. Volcanol. Geotherm. Res. 47, 329-336.

Yamaoka, K., H. Watanabe and S. Sakashita, 1988, Seismicity during the 1986 eruption of Izu-Oshima volcano, Bull. Volcanol. Soc. Japan 33, S91-S101.

Yamasato, H., 1997, Quantitative analysis of pyroclastic flows using infrasonic and seismic data at Unzen volcano, Japan, J. Phys. Earth 45, 397-416.

Yamasato, H., K. Fukui, K. Uhira, T. Hashimoto and H. Mori, 1993, Analysis of seismic and acoustic signals excited by pyroclastic flows at Unzen volcano, Bull. Volcanol. Soc. Jap. 38, 79-90 (in Japanese with English abstract).

Yamasato, H., T. Yokota and S. Kashiwabara, 1991, Earthquake swarm and volcanic tremors off Eastern Izu Peninsula in 1989. Spectral investigation and characteristics of waveforms, J. Phys. Earth 39, 79-92.

Yokoyama, I., 1997, An interpretation of the 1914 eruption of Sakurajima volcano, Proc. Jap. Acad. 73, Ser.B, 53-58.

Yokoyama, I., 2001, The largest magnitudes of earthquakes associated with some historical volcanic eruptions and their volcanological significance, Ann. Geof. 44, 1021-1029.

Yokoyama, I. and S. De la Cruz-Reyna, 1990, Precursory earthquakes of the 1943 eruption of Paricutin volcano, Michoacan, Mexico, J. Volcan. Geotherm. Res. 44, 265-281.

Yokoyama, I., S. De la Cruz-Reyna and J.M. Espindola, 1992, Energy partition in the 1982 eruption of El Chichón volcano, Chiapas, México, J. Volcanol. Geotherm. Res. 51, 1-21.

Yokoyama, I. and M. Seino, 2000, Geophysical comparison of the three eruptions in the 20th century of Usu volcano, Japan, Earth Planet Space 52, 73-89.

Yokoyama, I., H. Yamashita, H. Watanabe and Hm. Okada, 1981, Geophysical characteristics of dacite volcanism - The 1977-1978 eruption of Usu volcano, J. Volcanol. Geotherm. Res. 9, 335-358.

Zobin, V.M., 1970, Focal mechanism of volcanic earthquakes related to the Sheveluch volcano eruption in November 1964, Izvestia, Solid Earth Physics, USSR Acad. Sci., 3, 31-36 (in Russian). A.G.U. Transl., p. 161-164.

Zobin, V.M., 1972, Focal mechanism of volcanic earthquakes, Bull. Volcanol. 36, 561-571.

Zobin, V.M., 1976, Seismotectonic conditions of the Tyatya Volcano, Kunashir Island, eruption in July 1973, Bull. Volcan. Stations 52: 24-29 (in Russian).

Zobin, V.M., 1979a, Source Dynamics of Volcanic Earthquakes. Nauka, Moscow. 92 p. (in Russian).

Zobin, V.M., 1979b, Variations of volcanic earthquake source parameters before volcanic eruptions, J. Volcanol. Geotherm. Res. 6, 279-293.

Zobin, V.M., 1988, Discriminations between volcanic earthquakes associated with central and lateral eruptions, J. Volcanol. Geotherm. Res. 34, 307-315.

Zobin, V.M., 1990, Seismotectonic crustal deformations during the Large Tolbachik fissure eruption, J. Volcanol. Geotherm. Res. 43, 271-278.

Zobin, V.M., 1992. Volcanic earthquakes and their role in volcano-tectonic process, Volcanol. Seismol. 5-6, 87-100 (in Russian).

Zobin, V.M., 1996, The epicentral areas of volcanic earthquakes and their volcanotectonic significance, Geofis. Internacion. 35, 95-103.

Zobin, V.M., 1999, The fault nature of the Ms 5.4 volcanic earthquake preceding the 1996 subglacial eruption of Grimsvøtn volcano, Iceland, J. Volcanol. Geotherm. Res. 92, 349-358.

Zobin, V.M., 2001, Seismic hazard of volcanic activity, J. Volcanol. Geotherm. Res. 112, 1-14.

Zobin, V.M. and V.N. Chirkova, 1984, Catalog of source parameters of earthquakes based on P-wave spectra (Klyuchi station records), Large Tolbachik Fissure Eruption, Ed. S.A. Fedotov, Nauka, Moscow. Pp. 607-609 (in Russian).

Zobin, V.M., M. González-Amezcua, and G.A. Reyes-Dávila, 2002a, Seismotectonic deformations of volcanic edifice prior to the 1998 eruption of Volcán de Colima, México, Bull. Volcanol. 64, 349-355.

Zobin, V.M., M. González Amezcua, G.A. Reyes-Dávila, T. Domínguez, J.C. Cerda Chacón and J.M. Chávez Álvarez, 2002b, The comparative characteristics of the 1997-1998 seismic swarms preceding the November 1998 eruption of Volcán de Colima, Mexico, J. Volcanol. Geotherm. Res. 117, 47-60.

Zobin, V.M. and V.I. Gorelchik, 1982, Seismicity and source parameters of earthquakes in the region of the Large Tolbachik fissure eruption, Bull. Volcanol. 45, 99-113.

Zobin, V.M. and V.I. Levina, 1998, Rupture history of the 1996 January 1 Ms 6.6 volcanic earthquake preceding the simultaneous eruption of Karymsky and Akademia Nauk volcanoes in Kamchatka, Russia, J. Geophys. Res. 103, 18315-18324.

Zobin, V., J.F. Luhr, Y.A. Taran, M. Bretón, A. Cortés, S. De la Cruz-Reyna, T. Domínguez, I. Galindo, J.C. Gavilanes, J.J. Muñiz, C. Navarro, J.J. Ramírez, G.A. Reyes, M. Ursúa, J. Velasco, E. Alatorre and H. Santiago, 2002c, Overview of the 1997-2000 activity of Volcán de Colima, México, J. Volcanol. Geotherm. Res. 117, 1-19.

# Index of Volcanic Events

**N**



**O**



**P**



**R**

**S**



**T**



**U**

①忘 694864929 18500571853
②佩 452221361 18810356404
③润 1937531789 13521532859
④丹 694020248 15733210917
⑤弹 1017466004
⑥Max 578287803 18500239229
⑦魏 270640967 18531127698
⑧小陈 339784237 18618361499
⑨瑞 84726867 18632765502